G000139752

Preface

Physicists pretend not only to know everything, but also to know everything better. This applies in particular to computational statistical physicists like US. Thus many of our colleagues have applied their computer simulation techniques to fields outside of physics, and have published sometimes in biological, economic or sociological journals, and publication flow in the opposite direction has also started.

If one sets plates, knifes, and forks onto a dinner table, one has to put in human organisation to order the pieces properly. The magnetic atoms in iron, on the other hand, order their magnetic orientation parallel to each other (over small distances) by themselves, and similarly water molecules in vapour cluster all by themselves into small drops when it rains. Such effects are called "self-organisation" (or "emergence") and are typical for "complex systems" of many simple elements, often different from each other, which altogether generate effects which cannot be seen from the properties of a single element. The whole is not the simple superposition of its many parts. 65 years after van der Waals wrote his thesis with what may be regarded as the first theory to explain self-organisation in complex systems, computers became available and simulations on them triggered the systematic research activity on this field. These studies flourished during the last few decades.

Many other natural phenomena outside physics are related to the terms "complexity", "emergence", etc., in particular, evolutionary dynamic systems, where a population of agents evolves in time, the behaviour of each influencing the behaviour of others. Biological evolution through natural selection is the master example, but the same general concepts apply also to distant subjects, such as the occurrence of earthquakes. Other examples can be seen in diverse social behaviour and human activities such as the distribution and evolution of languages, elections, the diffusion of opinion, terrorism, etc.

The present book reviews selected applications to evolutionary biology (Chapters 3 and 4), social sciences (5 and 6) and geosciences (7), while Chapter 2 explains the general concepts of evolutionary dynamical systems, and why computer simulations of agent-based-models are the basic tool for these studies. The book as a whole is intended for graduate students and researchers not only in physics. No deep knowledge concerning the many different subjects or computer programming is required to follow the book, which can therefore be useful (we hope) to a wide and general audience. The parts we marked with asterisks contain mostly additional information, not fundamental for the comprehension of the book as a whole.

stauffer@thp.uni-koeln.de
suzana@if.uff.br
pmco@if.uff.br
jssm@if.uff.br

Instituto de Física, Universidade Federal Fluminense; Av. Litorânea s/n, Boa Viagem, Niterói 24210-340, RJ, Brazil; September 2005. DS permanently at Theoretical Physics, Cologne University, D-50923 Köln, Euroland.

Contents

Chapter 1

Introduction

Computational Physics is now a multidisciplinary line of research. A long time ago, it was not like that, computers were used by physicists only in order to solve classical problems resistant to the analytical approach: One is able to model the problem at hands through, say, a set of coupled differential equations, but the analytical solution for these equations is not available. The solace is the numerical solution, and the old branch of Computational Physics consists in providing fast and precise numerical methods to be applied to these cases.

Some systems, however, resist even to this numerical strategy, for instance the life cycle of a bacterium with 10 thousand proteins, a rather "simple" biological organism. The concentration of each protein varies in time according to the current concentrations of the others, and also to external stimuli. Suppose one models how each concentration depends on all others by writing down a set of 10 thousand coupled differential equations, which depend also on the possible external stimuli. Furthermore, suppose one is able to solve this mathematical problem on a computer, within a reasonable time: then, one can run the program for a given set of initial concentrations and external stimuli. In order to study the behaviour of this "simple" system, one would need to run the program again and again, for different conditions, and try to extract some useful information. Some particular stimulus may result in some particular behaviour, if it occurs when the concentration of some particular protein is high. Has it the same effect when this concentration is low? Does it depend on the concentrations of other proteins? Is the effect of two superimposed stimuli obtained simply by adding the individual effects of each one? How long does one need to wait, after the onset of some stimulus, in order for the system to have reached a state in which a renewed triggering of the stimulus would generate the same effect anew? How does this waiting time depend on the concentrations? This is an endless approach, which in some very lucky cases may be circumvented by a reductionist reasoning: to consider only the effect of some dozen proteins and stimuli, forgetting all the rest. Beyond a simple bacterium, the reader can imagine the mess one reaches in the study of a bacterial colony. Also, even worse than a 10 thousand protein life cycle is a system for which there are no fundamental equations relating the various important quantities. Biological evolution, where there is no Darwin equation, is an example. Social behaviour within a human population is another, as well as the dynamics of economics. Normally, one cannot even model such a system by a set of coupled differential equations.

An alternative is population dynamics: one keeps on the computer memory the current features of each individual, and simulates the whole dynamic evolution by programming the interaction rules governing the influences of these individuals on each other, as well as external stimuli. A crucial ingredient is randomness, which is included through the use of some pseudo-random number generator. The long-term evolution of the same system is repeated many times, for different randomness and initial conditions, and averages are taken at the end. Besides the numerical solution of equations, this simulational approach is the second, modern branch of Computational Physics, introduced by the pioneering work of Metropolis, Rosenbluth, Rosenbluth, Teller and Teller (1953) half a century ago. First applied to equilibrium models of Statistical Physics, simulations are now applied to many different problems outside physics and out of equilibrium. The reason for this success is a subtle concept known as universality, discovered within the study of critical phenomena where the simulational approach has a fundamental position.

Equilibrium critical phenomena occur in macroscopic systems which present long-range correlations: The behaviour at some position X depends on the current state of another far position Y. If one slightly "shakes", or perturbs, the system at Y, an observer at X feels the effect of the shake in spite of the long distance X-Y. Indeed, as the distance increases, the intensity of the effects of the perturbation decays, but not according to the normal exponential decay, for which there is a characteristic correlation length beyond which correlations can be neglected. Within critical systems, instead, the decay normally follows a power-law, lacking any characteristic length: no matter how distant X is from Y, the effects of the perturbation cannot be neglected. All length scales are equally important. The reductionist approach of taking only a small, localised piece of the system clearly does not work: critical systems must be studied as a whole. Outside equilibrium, in most cases, the power-law decays responsible for the long-range spatial correlations also appear defining the time dependence of the various quantities of interest: they produce long-memory effects, small contingencies occurred a long time ago can be determinant for the present situation of the critical system. All time scales are equally important. These features turn it very difficult to model a critical system through simple space-time differential equations.

On the other hand, the length and time scale-free behaviour of critical phenomena provides an interesting feature: Both the microscopic details and the short-term dynamics are not crucial for the long-range and long-term evolution of the system under study. Only some general characteristics as the spatial dimension and symmetries matter. Systems sharing the same spatial dimension and symmetries fall into the same universality class, in spite of the big differences in what concerns the microscopic and short-term behaviour of each one. For critical systems in equilibrium, this universality concept was already well understood through the Wilson's renormalisation group. An equivalent general theory for systems out of equilibrium is still lacking, but the evidences of universal time dependent behaviour joining together completely different systems are ubiquitous. Universality allows us to model complicated real systems by toy models belonging to the same universality class, simplifying a lot the study of these systems. Even so, due to the long-range and long-memory features, one cannot hope to solve the toy model by the reductionist paradigm, by following only a small piece of the system during a small interval of time. The population dynamics simulational approach appears instead as the most important instrument for these studies.

This book shows some examples of critical dynamic systems studied through computer simulations. They belong to different fields, not just Physics, and are connected by two very general features. First, they are critical, presenting long-range correlations and long-term memories. Second, they are modelled by simple rules one can easily program on a computer, turning it possible to follow in seconds what corresponds to centuries of the real system under study.

We wanted to avoid, also because of the way references had to be put in this edition, to present a book which would mainly be a list of references surrounded by little text. Thus, not only have we selected a few fields of interdisciplinary computer simulations with which we are more familiar, but also chose to reference within these fields papers which we feel are the most important, as seen both from today's perspective and from our restricted interests and knowledge. We are aware of the fact that important papers will for sure be missing from our list of references, and we apologize to the reader and to the authors for that mostly unwanted omission.

We start in Chapter 2 with general principles of evolution, and then apply them to biology in Chapters 3 (ageing) and 4 (speciation). Then comes the presently fashionable field of languages (Chapter 5) and the related one of sociophysics (Chapter 6). Finally, Chapter 7 gives applications to earthquakes. Our summary in Chapter 8 tries to point out the similarity in the methods used in the previous chapters. An appendix, Chapter 9, lists and explains selected complete computer programs, written in Fortran – for the desperation of half the authors and as an early example of Galam conservatism model explained in Section 6.3.1.

Chapter 2

Evolution

The word "evolution" is directly linked with time. Something which evolves is not static, it varies as time goes by. A variable quantity x describing such a dynamic system is a function of time, $x(t)$. The speed of its variation is measured by the first derivative $\mathrm{d}x/\mathrm{d}t$ of this function, the acceleration by the second derivative $\mathrm{d}^2x/\mathrm{d}t^2$, and so on. The canonical way to study this kind of problems is through the so-called differential equations, i.e., mathematical relations linking x with $\mathrm{d}x/\mathrm{d}t$, $\mathrm{d}^2x/\mathrm{d}t^2$, etc.

A famous example is Newton's law

$$m\frac{\mathrm{d}^2x}{\mathrm{d}t^2} = F(x)$$

which describes the movement of a particle with mass m along the X axis. $F(x)$ is the external force which drives the movement.

Another famous example is the Schrödinger equation

$$i\hbar\frac{\mathrm{d}|\psi\rangle}{\mathrm{d}t} = H|\psi\rangle$$

which tells us how the state $|\psi\rangle$ of a quantum system evolves in time. H is the Hamiltonian operator for this system, essentially its energy. $\hbar$ is the Planck constant, and $i = \sqrt{-1}$ is the imaginary unit for complex numbers (nothing to do with complexity).

Diffusion also obeys a differential equation

$$\frac{\partial\rho}{\partial t} = D\nabla^2\rho$$

where $\rho(\vec{r}, t)$ represents the local density of the diffusing material at position $\vec{r} = (x, y, z)$ and time t. The Laplacian operator ∇^2 sums up the second derivatives with respect to x, y and z, and D is the diffusion coefficient. This problem was studied by Einstein, in one of his five famous papers published in 1905, "Einstein's Miraculous Year" (Einstein, 1998). Now, exactly one century later, UN and UNESCO commemorate the "World Year of Physics, WYP2005", with a lot of events all over the world. Diffusion describes, for instance, how an ink drop diffuses in a glass of water: as time goes by, the ink rapidly occupies the whole glass, resulting in a homogeneous mixture at the final equilibrium situation.

Unfortunately, we are not able to write down a fourth example, because Darwin legacy did not include a "Darwin equation". However, would this equation exist, it would certainly be a differential one, involving the time. Instead, his famous book (Darwin, 1859) describes a series of concepts and rules for biological evolution.

Fortunately, dynamic systems can also be studied by tools other than differential equations. One important such a tool is computer programming, where the computer is instructed to follow some dynamic rules imposed by the researcher/programmer, for instance some of the rules one can read in Darwin's book. At the two last sections of this chapter, we will treat some very simple evolutionary models under this point of view.

First, we discuss why evolution is a subject which resists analytical treatments through differential equations. We emphasise the very special dynamics followed by evolutionary systems, during which the space of possibilities is not completely covered. Unlike the ink drop's fast diffusion through the water glass, evolutionary paths slowly grow like a tree. They do not spread over the whole space of possibilities. Only a tiny fraction, a fractal, is actually covered by the evolutionary dynamics. In between the tree branches, the great majority of the space remains unvisited forever. As a consequence, no final equilibrium exists, the evolving tips of the growing tree continue their slow walk inside this space, forever.

2.1. Linearity

This book belongs to the series "Nonlinear Science and Complexity". Why these two concepts, nonlinearity and complex behaviour, are put together? In order to answer this question, we need first to treat linearity, the basic, simplest possible behaviour for a system which evolves in time.

A linear dynamic system is one described by the simplest possible differential equation of the form

$$x + \tau \frac{\mathrm{d}x}{\mathrm{d}t} + \gamma \frac{\mathrm{d}^2 x}{\mathrm{d}t^2} + \cdots = K$$

where neither the dynamic variable $x(t)$ itself nor its derivatives appear under complicated forms like squares, square roots, etc. All these variables are proportional to each other, through the multiplicative constants τ, γ, etc. In short, their dependence is linear. In what concerns the constants dimensions, τ is a time, γ is a time squared, etc. K is another constant sharing the same dimension of $x(t)$, maybe a distance, a number counting, or whatsoever.

Among the simplest, let's take the simplest case

$$x + \tau \frac{\mathrm{d}x}{\mathrm{d}t} = 0 \tag{2.1}$$

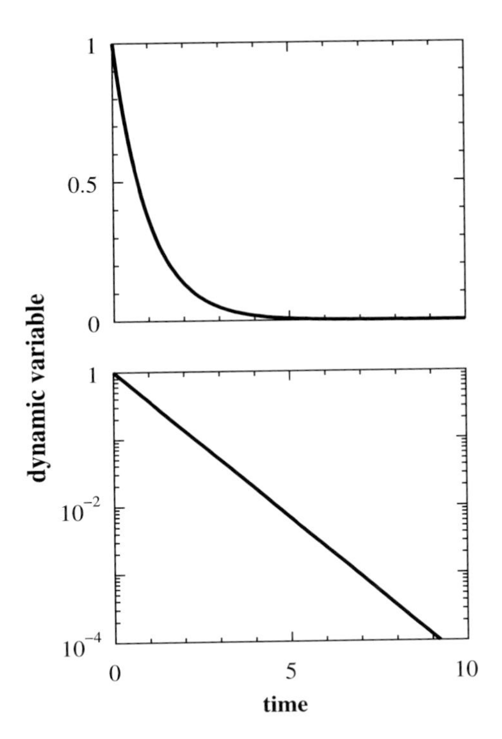

Figure 2.1. Two different plots for equation (2.2) with $x_0 = 1$. The normal plot, with linear scales in both axis, appears on the top. For this rapid decaying function, it does not allow a good visualisation of the final tail. Below, with logarithmic scale along the vertical axis, one can better appreciate this tail: the upper curve becomes a straight line. In both cases, the time t is measured in units of τ.

for which the solution is

$$x(t) = x_0 \, \mathrm{e}^{-t/\tau} \tag{2.2}$$

where the new constant x_0 is the initial value of x, at time $t = 0$. In what concerns the time flow, x_0 is unimportant, as we shall see in the next paragraph. In our conceptual analysis, the only important constant is τ, which defines the system's characteristic time scale. It is the natural unit to measure the time $t = \tau, 2\tau, 3\tau$ etc. Figure 2.1 shows the plot of this solution, in two different representations.

A concrete example is radioactivity. Nuclide tables show lifetimes of $\tau = 2$ min for ^{82}Rb, or $\tau = 43$ years for ^{137}Cs. The radioactivity of these materials virtually ceases after a time of, say, 10τ, as shown in Figure 2.1 (the factor 10 is only an estimate, maybe also 5, 8, 15, etc.). In this case, x_0 represents the sample's initial radioactivity, determined by the starting number of not-yet-decayed nuclei. Being proportional to its mass, x_0 is a measure for the sample's size. Curiously,

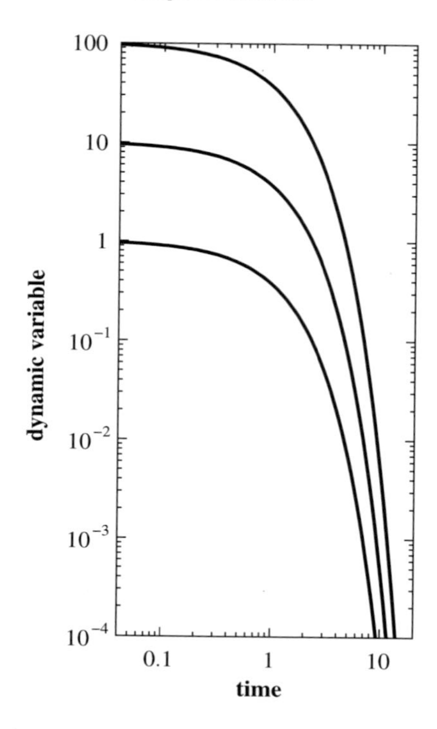

Figure 2.2. Plot of equation (2.2) again, now with $x_0 = 100$ (top curve); $x_0 = 10$ (middle); and $x_0 = 1$ (bottom). Logarithmic scales were adopted in both axis.

these nuclid tables do not mention how much material they refer to. Maybe a sample of ^{82}Rb with mass of 1 g, another sample of 10 g, or a third one of 100 g: after $10\tau = 20$ min, the radioactivity has already nearly vanished for any of these different sized samples. Figure 2.2 explains why.

The bottom curve shows the same data on Figure 2.1, now with logarithmic scales in both axis. By performing a scaling transformation on the vertical axis, i.e., by multiplying all its values with the same factor 10 or 100, for instance, this curve is risen as a whole, generating the two upper curves in Figure 2.2. The effect on the horizontal axis, however, is much smaller: only an additive, not multiplicative increment is observed on the waiting time ($\approx 10\,\tau$) needed to reach the same final level of radioactivity (10^{-4}).

We have used the name "lifetime" for the parameter τ, based on a naive analysis of Figures 2.1 and 2.2. It refers to the macroscopic sample's radioactivity as a whole. However, one can also formally define the average lifetime of a single nucleus, how much time one needs to wait for its decay, averaged over all nuclei.

$x(t)$ represents the number of not-yet-decayed nuclei at time t. Thus,

$$-\mathrm{d}x = x(t) - x(t + \mathrm{d}t) = \frac{x_0\,\mathrm{e}^{-t/\tau}\,\mathrm{d}t}{\tau}$$

is the number of nuclei which decay in between t and $t + \mathrm{d}t$, those contributing with lifetime t to the average over all nuclei. So, the average lifetime is given by the integral

$$\frac{1}{x_0}\int_0^\infty \frac{x_0\,\mathrm{e}^{-t/\tau}\,\mathrm{d}t}{\tau}\,t$$

the result of which coincides with τ, justifying its name.

The lifetime τ of a linear dynamical system does not depend on its size. Due to linearity, there is no relation between time and size scales of the same system. The scaling properties of the variable x have nothing to do with the scaling properties of the time t. In order to better understand this important concept, let's take again the radioactivity example. Each not-yet-decayed nucleus will decay in some unknown future time. One can know only its probability of decaying in the next $\mathrm{d}t$ seconds (or whichever unit of time one uses). This probability is $\mathrm{d}t/\tau$, and depends only on the internal features of the nuclei itself, nothing to do with other neighbouring nuclei. In other words, there is no correlation at all between different nuclei located at different positions of the sample. Without spatial correlations, one can conclude that the number $-\mathrm{d}x$ of nuclei decaying together within the same time interval $\mathrm{d}t$ is proportional only to the current number $x(t)$ of not-yet-decayed nuclei so far. This is precisely what the linear equation (2.1) states.

Let's compare the decaying population of radioactive nuclei with a population of living individuals going towards extinction. Could we follow the same reasoning above? To consider each individual as an isolated entity whose destiny is completely independent from other individuals? Certainly not! First, unlike radioactive nuclei, living individuals reproduce, creating new individuals of the same species. Second, they compete against each other for many different reasons. Third, groups of living individuals collaborate among themselves, sometimes against other groups, sometimes in favour. These intricate connections between different individuals of the same population represent some kind of spatial correlation. As we shall see later, long-range correlations drive this kind of problems out of linearity.

Another, independent remark concerns the finite lifetime τ of the linear systems. In Darwinian sense, they cannot represent an evolutionary system, for which the eternal search for new forms, better than the current one, is imperative. Evolutionary systems should obey another kind of dynamical rule, certainly not linear, in order to allow such "infinite lifetimes", as required.

2.2. Chaos

In the example of equations (2.1) and (2.2), the dynamic variable $x(t)$ vanishes at the end. Other systems would converge to other stationary states. For instance, by keeping the constant K in the right-hand side of equation (2.1), instead of zero, the solution would change to $x(t) = K + x_0\,\mathrm{e}^{-t/\tau}$. The fast, exponential rate of convergence, however, is the same. The final state K is called the attractor. Other linear or nonlinear systems could converge to more complicated attractors, for instance a cycle where the dynamic variable $x(t)$ becomes a periodic function at the end. Yet more complicated are the strange attractors, final situations which are not periodic, and occupy a fractal portion of the whole space of (in principle available) possibilities.

Some of these systems are called chaotic, a denomination which refers to the speed they reach their final attractor, also exponentially fast, not to the kind of attractor itself. Let's consider the same system evolving from two slightly different initial conditions, $x_0^{(1)}$ and $x_0^{(2)}$ distant $\Delta_0 = x_0^{(1)} - x_0^{(2)}$ from each other, at time $t = 0$.

As time goes by, the distance $\Delta(t) = x^{(1)}(t) - x^{(2)}(t)$ also evolves. The simplest possible dynamics is

$$\frac{\mathrm{d}\Delta}{\mathrm{d}t} = \lambda\Delta \tag{2.3}$$

which is formally the same equation (2.1) if one replaces the letters λ by $-1/\tau$ and Δ by x. Of course, the solution is also the same

$$\Delta(t) = \Delta_0\,\mathrm{e}^{\lambda t} \tag{2.4}$$

where λ is the so-called Lyapunov exponent which can be positive or negative.

Regular systems are those with a negative Lyapunov exponent, for which $\Delta(t)$ fast vanishes within a lifetime $\tau = -1/\lambda$. An example is a clock pendulum following its characteristic go-and-back movement. Let's consider a very precise mechanism which keeps the pendulum reaching its rightmost position every integer second (and consequently every integer minute, hour, etc.). At some time $t = 0$, when it is precisely passing through that position $x_0^{(1)}$, somebody incidentally gives it an additional impulse, suddenly changing the position to $x_0^{(2)}$. For a while, the subsequent movement also changes, but exponentially fast the former trajectory is restored. After a time of $10\,\tau$, nobody can notice there was some incident in the past, it was forgotten. This system presents a short-term memory, or, in other words, a finite lifetime before reaching the final equilibrium.

Chaotic systems are those with a positive Lyapunov exponent. The system follows its normal trajectory $x^{(1)}(t)$. At $t = 0$, some external agent promotes a very small instantaneous perturbation, slightly changing its position from $x_0^{(1)}$ to $x_0^{(2)}$.

At the very beginning, the new trajectory $x^{(2)}(t)$ nearly follows the old one, both are still correlated to each other. However, their distance $\Delta(t)$ increases exponentially fast, while the quoted correlation decreases at the same rate. The lifetime for this decay is $\tau = 1/\lambda$. After $t = 10\,\tau$, the correlation is already negligible, the new trajectory does not keep any memory of the old one. Although very different from regular systems in what concerns the final destiny, chaotic systems also present short-term memory.

Consider a gas confined in a box. A classical example of chaotic system is the zig-zag movement of its molecules. They frenetically collide with each other and against the box walls. Imagine it would be possible to take a movie of them, a sequence of instantaneous pictures. Stretching a little bit the imagination, suppose one can restart the same movement with all molecules at their original positions and velocities, but one particular molecule A which starts from a slightly different initial position and/or velocity. Take a second movie. At the very beginning, both movies seem to be the same, they are correlated. Only molecule A presents slightly different trajectories, comparing one movie to the other. This holds up to the first collision between molecule A with B. From now on, A and B will present slightly different trajectories in the second movie, compared to the first, up to the next collision between molecule A or B with C. And so on. After the finite lifetime τ (or $10\,\tau$ to be sure), the movies are no longer correlated.

Although the final microscopic situation is not static, after a time of $10\,\tau$ the gas (or the chaotic system, in general) is considered in "equilibrium". More precisely, the system is in thermodynamic equilibrium. This concept concerns the macroscopic behaviour of the system, not the microscopic detailed movement of each molecule. All macroscopic quantities such as the internal energy, pressure, density, temperature, mean molecular speed, entropy and so on no longer evolve in time, after the equilibrium is reached. Although the frenetic movement of the molecules goes on, in a continuous change from one microscopic state to another, all these micro-states correspond to the same macroscopic situation, the same averaged quantities, independent of the initial condition which was completely forgotten.

Moreover, all possible microscopic states compatible with the external constraints (volume, temperature, etc.) are likely to be visited by the chaotic system, an important property denominated ergodicity. Consider, for instance, all gas molecules initially located at one half of the box, the other half completely empty at $t = 0$. Surely, this is not an equilibrium situation, after $10\,\tau$ all the box volume will be occupied. Ergodicity, in simple words, is the property a chaotic system has to visit all the available points in the final space of possibilities, covering all its regions.

The word *final* in the last phrase has an important meaning. Consider the gas not only confined inside the box, but also isolated from the rest of the world. The box is a Dewar vessel avoiding any energy exchange through the walls. Thus, the internal energy is rigorously fixed, the system is closed. The initial situation could be any distribution of molecular positions and speeds compatible with the fixed energy. This is what we call the initial space of possibilities. Independent of which particular initial situation one chooses, maybe all molecules into the left half of the Dewar vessel, or any other, after the final equilibrium is reached all possible micro-states compatible with the fixed energy are likely to be visited. Thus, for closed chaotic systems like this example, the final space of possibilities is exactly the same as the one at beginning.

Obviously, biological evolution cannot be classified as a closed chaotic system. First, one cannot study the evolution of all potentially living beings of the universe, past, present and future. One necessarily needs to restrict the study to a particular population, a single species, a group of species, or something like that. Then, this restricted set cannot be considered closed, one needs to include the environment. Second, within a closed chaotic system the final equilibrium is fast reached, and evolution would be stopped from this moment on.

However, chaotic systems are not closed in general, some interactions with the environment could be allowed. Of course, the final space of possibilities cannot be larger than the initial one, but it can be shorter. The environment influence can shrink this space as time goes by, gradually forbidding the system to return back to some micro-states which were allowed in the past. The final space of possibilities is called the attractor, an already quoted notation. Let's call these open systems dissipative (we don't mean dissipative in energy, but in entropy, the quantity which measures the number of available micro-states compatible with the observed macro-state). Normally, these dissipative chaotic systems present a lower-dimensional attractor, a sub-space of the whole initial space of possibilities. A lower dimension sub-space s of a larger space S means the following: (1) take a random point of S; (2) the probability to find this point inside s vanishes. A simple example is a straight line (s, with dimension 1) inside a plane (S, with dimension 2). Would the dimension of s be a non-integer number, it is called a strange attractor.

Again, biological evolution does not fit into the class of chaotic dissipative systems. Being rapidly trapped into the tiny attractor, such a system loses forever the chance to visit other parts of the whole space of possibilities. This fast behaviour could be useful in optimisation processes, where the interest is to extract the best options among the whole set of possibilities, nothing to do with evolution which requires diversity. The attractor depends on the environment, and one cannot suppose the environment is fixed. It certainly varies. Within a chaotic dissipative dynamics, the current tiny attractor which corresponds to the current "best" options is no longer the best as soon as the environment changes a little bit. The current population trapped into the former tiny attractor is no longer adapted to the new environment, and could not survive enough to re-adapt. That is why diversity is a key ingredient for evolution. Somehow, the dynamics should preserve other forms different from the current supposed "optimum". Definitely, evolution and eugenics are not the same concept.

In short, biological evolution certainly does not follow any chaotic dynamics ($\lambda > 0$). On the other hand, neither a regular dynamics ($\lambda < 0$) can describe biological evolution: in this case the final situation would be a completely uniform population.

2.3. Nonlinearity

Beyond linearity, the simplest nonlinear form is a square. Thus,

$$x^2 + x_1 \tau \frac{dx}{dt} = 0 \tag{2.5}$$

is the simplest possible nonlinear differential equation. However, it is not so simple. Compared to equation (2.1), now the multiplicative constant in front of the derivative can no longer have the dimension of time. The form $x_1\tau$ is chosen for this constant in order to explicitly show the coupling of two different scales, namely the already used time unit τ and the new constant x_1 which is the scale for the variable x (distance, number counting, or whatsoever). The solution for this equation is

$$x(t) = x_1 \left(\frac{t}{\tau}\right)^{-1}. \tag{2.6}$$

The mathematical form *variable* raised to *constant exponent* on the right-hand side of equation (2.6) is called a power-law (in the present particular case, the exponent is -1, in general, any other constant). It is the reverse of the exponential form *constant* raised to *variable exponent* obtained as solutions for linear differential equations in general, Section 2.1.

Power-laws are ubiquitous is Nature, not only to describe time dependences as equation (2.6), but also relations between other quantities. As an example, Figure 2.3 shows the number N of earthquakes in Southern California, classified according to the energy E they released. For each value of E, the counter N includes all earthquakes which released more energy than E, during the period 2000–2004 (data from www.scec.org). The straight line behaviour is the signature of a power-law

$$N \propto E^{-b}$$

where the symbol $\propto$ means proportionality, generally used in order to omit multiplicative constants. The slope of the straight line measures the power-law exponent b, in this case $b \approx 1$. The so-called Richter scale defines the magnitude M for earthquakes as the logarithm of the released energy measured in a proper unit. Thus, along the horizontal axis, the exponents displaying the powers of 10

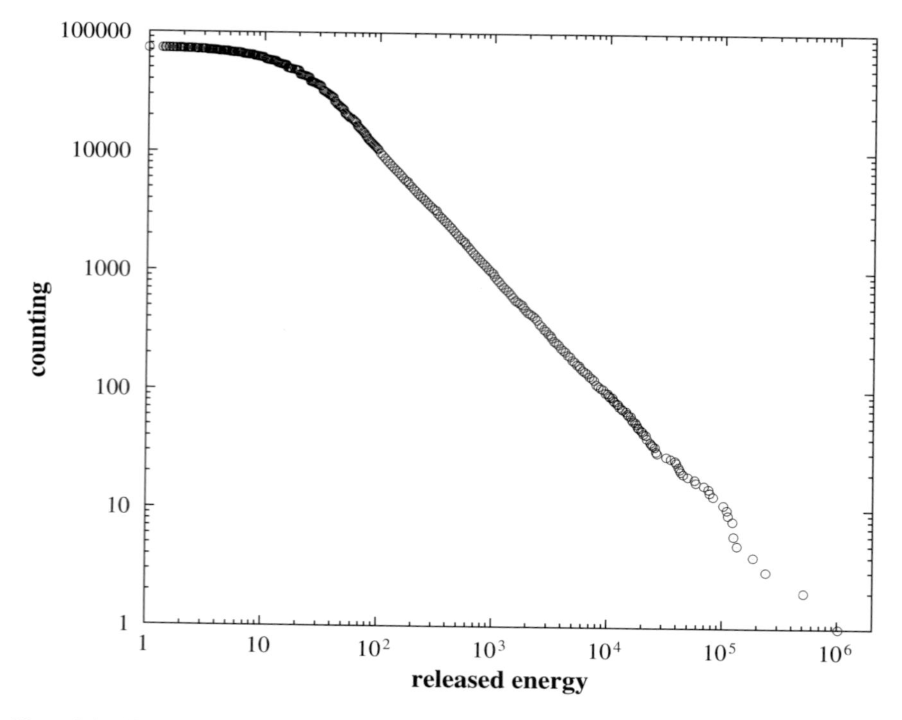

Figure 2.3. Earthquakes occurred in Southern California in the period 2000–2004, classified according to their magnitudes.

correspond to the Richter magnitudes from 0 to 6. Earthquakes are discussed at Chapter 7.

The saturation effect seen at the leftmost part of Figure 2.3 is supposedly due to the loss of sensibility of seismographs to the smallest earthquakes. On the other hand, the dangerous events are displayed by the rightmost points, the last one with magnitude $M \approx 6$. Although the statistics is not so good at this region, one cannot see any trend of deviation from the straight line.

Let's return back to the nonlinear time evolution theoretical example, equations (2.5) and (2.6). Note that x_1 is the natural unit for $x(t)$ in the same way as τ is the natural unit for t. However, contrary to x_0 which appears only in the solution (2.2), not in the linear differential equation (2.1) itself, now x_1 appears already in the nonlinear differential equation (2.5) as well as in its solution (2.6). As we shall see now, for nonlinear evolving systems, the scales for t and $x(t)$ are unavoidably linked to each other. Figure 2.4 shows the plot of equation (2.6) in the same two different representations of Figure 2.1, repeated now in dashed lines, by choos-

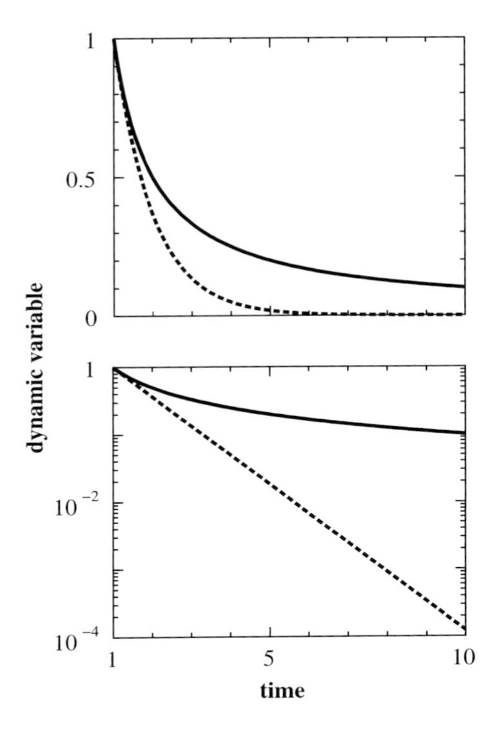

Figure 2.4. Two different plots for equation (2.6) with $x_1 = 1$, in solid lines. For comparison, dashed lines correspond again to equation (2.2) with $x_0 = e$. The time t is measured in units of τ.

ing $x_0 = \mathrm{e}x_1$. With this choice, both the exponential form, equation (2.2), and the power-law, equation (2.6), start from the same position with the same slope at $t = \tau$, allowing thus a fair comparison between both decaying rates from this moment on—note that equation (2.6) forbids to start at $t = 0$, thus we choose the arbitrary initial time as $t = \tau$. As we can see, the exponential decay, dashed lines, is much faster than the power-law one, solid lines. Times up to $t = 10\,\tau$ are not enough to appreciate the much longer power-law tail. The question is: how much faster is the exponential decay, compared with the power-law?

In order to answer this question, let's imagine a radioactive sample decaying according to equation (2.6) instead of equation (2.2). Following the same reasoning of Section 2.1, the number of nuclei decaying between t and $t + \mathrm{d}t$ would be

$$-\mathrm{d}x = x(t) - x(t + \mathrm{d}t) = \frac{x_1\,\tau\,\mathrm{d}t}{t^2}$$

and the lifetime average over all nuclei would be given by the integral

$$\frac{1}{x_1} \int_{\tau}^{\infty} \frac{x_1 \, \tau \, \mathrm{d}t}{t^2} \, (t - \tau)$$

the result of which diverges to infinity! Fortunately for all living beings on the planet, mother Nature did not follow our crazy imagination. All radioactive samples actually decay within some finite lifetime, according to equation (2.2), maybe as large as $\tau = 43$ years for ^{137}Cs, but finite. Instead, equation (2.6) describes an endless dynamics. Thus, the correct answer to the question posed at the end of last paragraph is: the comparison is not possible, these decays are qualitatively distinct, and cannot be quantitatively compared.

In practice, however, a dynamics with "infinite" lifetime is nonsense. Where is the puzzle? Figure 2.5 guides the answer.

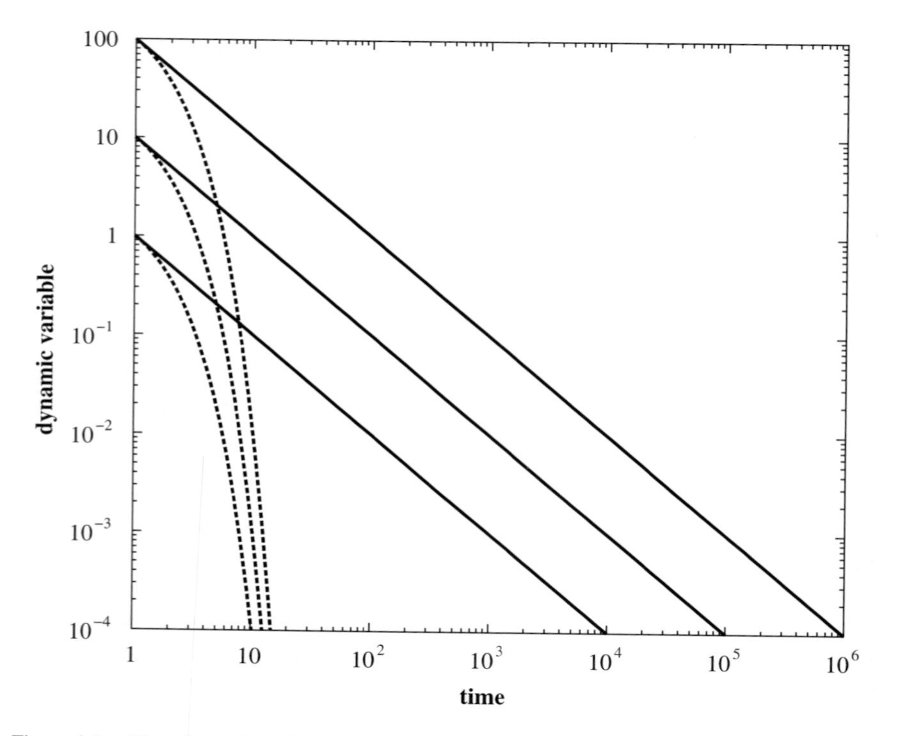

Figure 2.5. Plot of equation (2.6), in solid lines, with $x_1 = 100$ (top line); $x_1 = 10$ (middle); and $x_1 = 1$ (bottom). Logarithmic scales were adopted in both axis. Curved dashed lines display exponential decays, equation (2.2) with $x_0 = \mathrm{e}\, x_1$, for comparison.

The lowest solid and dashed lines show the same data as Figure 2.4, now with a much wider time scale, far beyond $10\,\tau$, allowing one to appreciate the power-law tail along the straight line. The other lines correspond to different size scalings performed on the vertical axis, variable x, in proportion $1:10:100$. It is exactly the same scaling transformation already done in Figure 2.2 for the exponential decay, now shown in dashed lines for easy comparison. Different from the exponential decay, where the multiplicative scaling performed on the system's size x (vertical axis) generates only a small, unimportant additive increment in the lifetime (horizontal axis), the power-law decay shows the size scaling fully reflected in the time scaling. No surprise, we have already noticed that size and time scales are entangled with each other through the product $x_1\tau$, since the original nonlinear differential equation (2.5) was written down.

The size of any physical (biological, social or whatsoever) nonlinear system is certainly finite. Then, its lifetime is also finite. Nonlinearities in size, such as the simple x^2 term exemplified in equation (2.5), appear because the many individual components of the system (its "molecules") are spatially correlated to each other. The behaviour of such a component located at position A directly influences another neighbouring component at position B, which also directly influences another neighbouring component at position C, and so on. Although the direct action of each component concerns only its nearest neighbours, the net result is the emergence of a long-range correlation involving a macroscopic set of components. The information needs time to propagate from one component to all others. The larger the correlation range, the larger the corresponding time. However, even within a would-be infinite correlation range, the system itself is finite: its boundaries impose a cutoff on this range, and a consequent cutoff on its lifetime. A biological species will become extinct in some finite future, because its current population is finite. Would the same species have a smaller (larger) population now, by evolving under the same conditions it would be extinct earlier (later). Only a would-be infinite population could escape from extinction.

Nonlinear differential equations like (2.5) can hold only for "infinite" systems, a crazy but very useful concept which exists on the imagination of physicists and mathematicians, not in reality. Within such a system, an infinite-range correlation would be conceivable, all system's components influencing all others, directly or indirectly. Accordingly, this imaginary system would present an infinite lifetime, and could be represented by nonlinear differential equations of the same kind of (2.5). Although out of reality, infinite size models are useful because they can also represent the corresponding finite real system, provided one does not overflow the maximum allowed size and time scales, namely the size of the finite system itself and its corresponding finite lifetime. Beyond these limits, equations like (2.5) no longer hold. In reality, plots like Figure 2.5 always bend downwards at the rightmost part, deviating from the straight line when the system's finite lifetime is approached. Figure 2.6 is an example. The data used to construct the plot were the

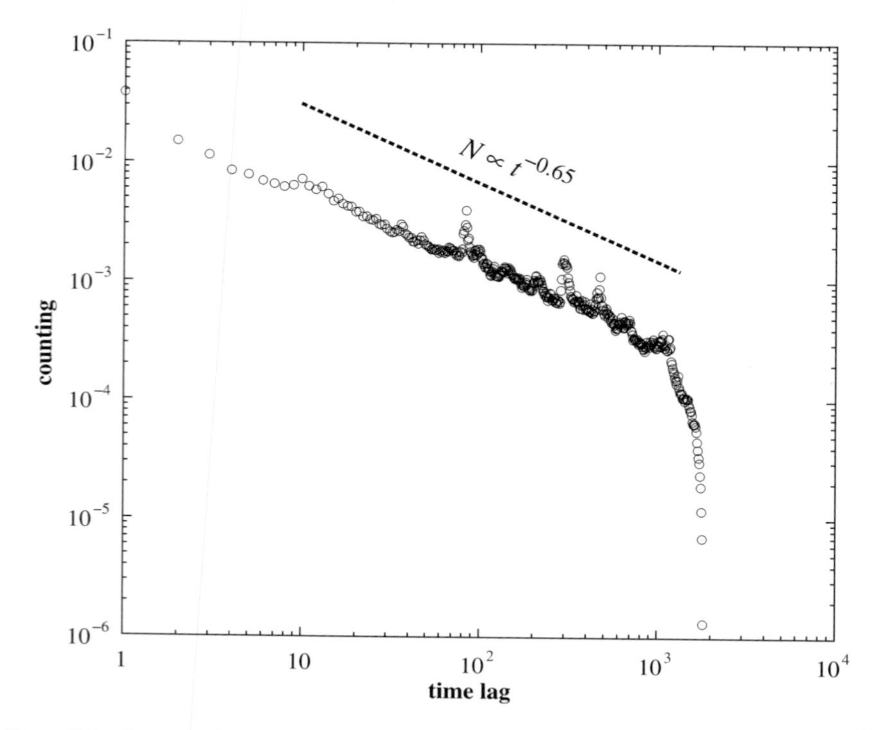

Figure 2.6. Return probability of earthquakes with the same magnitude to the same region, in this case Southern California, measured during the period 2000–2004. The time lag t is the number of days between successive occurrences. The dashed line guides the eyes, and its slope measures the power-law exponent.

same used before in Figure 2.3, www.scec.org, collected during the period 2000–2004. This restricted dataset limits the time lag t to approximately 1800 days, just the end point of the rightmost bending tail of Figure 2.6. No surprise, it is only an example of the cutoff always present in any power-law behaviour, due to the size and time limits of the system itself.

Figure 2.3 does not bend downwards! Would this behaviour remain after improving the statistics? In www.scec.org one can find registers for earthquakes since 1932! Thus, instead of the 5-years period 2000–2004, one could superimpose a new plot to Figure 2.3, with data corresponding to all earthquakes occurred during the 50-years period 1955–2004, a 10-fold time scaling (we will let this task as a homework for the reader). In principle, we could also stretch the time scaling once more, by searching for earthquake registers since 1505, a 100-fold time scaling (fortunately for the exhausted reader, there are no earthquake registers before Columbus). The result of superposing two further plots on Figure 2.3 is easily

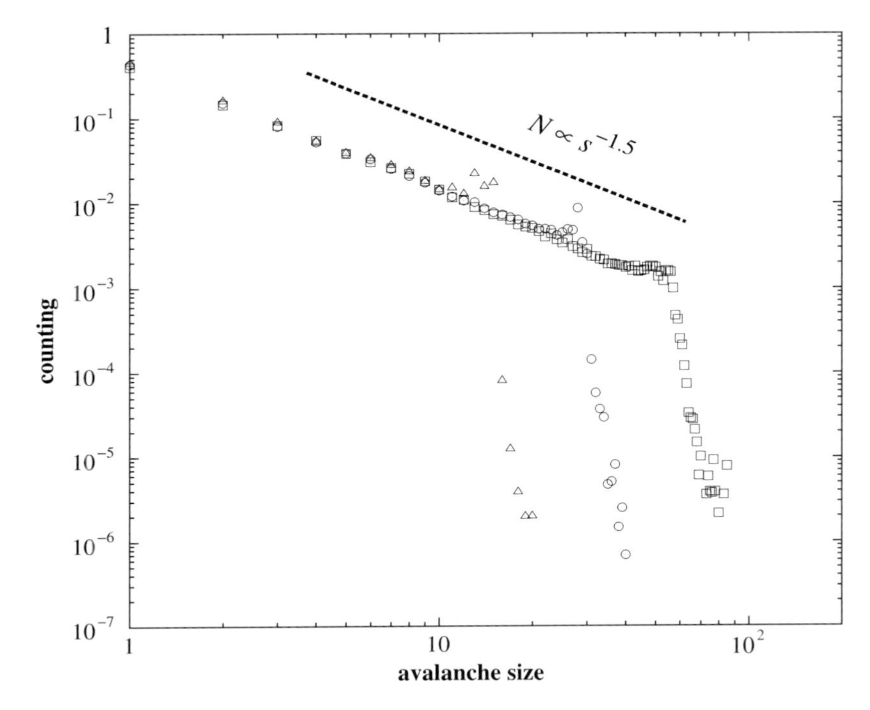

Figure 2.7. Neuronal avalanches classified by size, experimental data kindly provided by professor Dietmar Plenz from the National Institute of Mental Health, Bethesda, Maryland, USA.

predictable: it would present three parallel straight lines, similar to Figure 2.5. The 50-years period plot would display earthquakes up to magnitude $M \approx 7$, and the 500-years data up to $M \approx 8$, a terrifying scenario: simply by scaling up the observation time, the maximum earthquake strength would be scaled up by the same factor! Fortunately, this scaling certainly has a cutoff due to the finite size of the San Andreas fault, forcing the straight line plot to bend downwards for large enough earthquakes. Unfortunately, this upper bound has not yet been reached by the available data registered by seismographs up to now.

A nice example of finite size cutoff can be seen in Figure 2.7. Slices of rat cortical tissue were placed over a square grid of tiny electrodes distant 200 μm from each other. These electrodes measure neural activity which occurs in cascade: the activity of one neuron can trigger activity on another, which could activate a third one, and so on. After some time, the whole system becomes quiet until a second cascade suddenly starts again, and so on. Each electrode is activated as a response to neural activity above it. The size of the avalanche is measured by the number of electrodes activated during each neural cascade. Three different measurements were displayed with different symbols. On the left plot, a grid with 15 electrodes was used, 30 on the middle plot, and 60 on the right one, defining three different size limits for the whole system. In each case, before reaching the corresponding limit, the number of avalanches follows a power-law as a function of the avalanche size, the same straight line for all in Figure 2.7. Beyond the limiting grid size, the finite size cutoff appears. A complete description of the experiment can be found in the original paper (Beggs and Plenz, 2003).

2.4. The edge of chaos

Equation (2.3) is incomplete when the Lyapunov exponent λ vanishes. Indeed, the linear form on its right-hand side is only the first term of a series like $\lambda\Delta + \gamma\Delta^\eta + \cdots$. In Section 2.2, we treated only the cases where this first term dominates, the others were omitted, and the dynamic systems were classified as regular ($\lambda < 0$) or chaotic ($\lambda > 0$). At the edge of chaos ($\lambda = 0$), however, those possible nonlinear terms cannot be omitted anymore.

Systems evolving in time with zero Lyapunov exponent are said to follow a critical dynamics. For them, the distance $\Delta(t) = x^{(1)}(t) - x^{(2)}(t)$ between two initially neighbouring trajectories evolves according to

$$\frac{\mathrm{d}\Delta}{\mathrm{d}t} = \gamma\Delta^\eta = \frac{z\Delta_1^{1/z}}{\tau}\Delta^{1-1/z} \tag{2.7}$$

where $\Delta_1 = x_1^{(1)} - x_1^{(2)}$ is its value at time $t = \tau$, the same time unit already introduced before. On the right-hand side, the constants γ and η were replaced by convenient combinations of Δ_1, τ and z, explicit showing the proper dimension of γ. Of course, the exponents η and z are dimensionless. The solution is the power-law

$$\Delta(t) = \Delta_1\left(\frac{t}{\tau}\right)^z \tag{2.8}$$

which replaces the exponential solution (2.4) for the linear equation (2.3), valid for regular or chaotic dynamics. Now, z is the critical dynamic exponent.

The first comment concerning such a critical dynamics is the intrinsic coupling between the scales for both quantities involved, namely the time t and the dynamic variable x (or Δ), the same characteristic entanglement we have already found at the beginning of last section. The nonlinear character of equation (2.7) is again responsible for that.

The second comment is the endless behaviour of critical dynamics, imposed by the power-law mathematical form. This feature is called long-term memory, in contrast with short-term memory characteristic of the exponential form.

We have already illustrated this point in the last section, in the frustrated attempt to calculate the average lifetime for the dynamic evolution described by equations (2.5) or (2.6): the result is infinite! Before, in Section 2.2, we have obtained a finite memory time $1/|\lambda|$ for both regular or chaotic dynamics. Now, with $\lambda = 0$, there is no characteristic time scale after which the system "loses its memory". On the contrary, the very first initial deviation Δ_1 is "remembered" forever, as we shall see now.

First, let's consider the discrete version of dynamic evolutions in general, by following the time sequence

$$t = 0, 1, 2, 3, \ldots, n, n+1, \ldots$$

in units of τ. Accordingly, a trajectory is described by the sequence

$$x_0, x_1, x_2, x_3, \ldots, x_n, x_{n+1}, \ldots$$

where we have used the short notation x_n for $x(t = n\tau)$. The presence of only the first derivative in the corresponding differential equations such as equation (2.1) or (2.5) characterises the system as Markovian, i.e., for any trajectory, the next entry x_{n+1} depends only on the current x_n, not on the past x_{n-1}, x_{n-2}, etc. Mathematically, one can write

$$x_{n+1} = f(x_n)$$

where f represents some function defining the dynamics, and plays the same role as the differential equation. Figure 2.8 illustrates this time evolution for a chaotic dynamics.

An example of Markovian system is the genetic evolution of a population, if we take the simplified version of non-overlapping generations. The genetic pool x_{n+1} of the next generation is defined exclusively by the genetic pool x_n of their parents. Of course, in this case, the dynamic variable x is not a simple number,

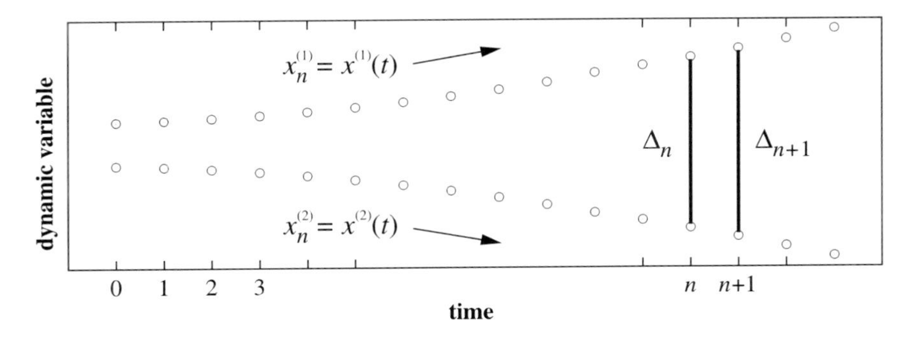

Figure 2.8. Chaotic dynamics. Starting from slightly different initial points, two possible histories $x_n^{(1)}$ and $x_n^{(2)}$ of the same Markovian system $x_{n+1} = f(x_n)$ are shown, for which the next value x_{n+1} depends only on the current one, x_n. The next dispersion Δ_{n+1} also depends only on the current one, Δ_n.

but a distribution of genes among the population. Analogously, instead of a simple difference, Δ_n represents the genetic diversity characterising generation n, where we have used again the short notation Δ_n for $\Delta(t = n\tau)$. Physicists and mathematicians usually refer to dispersion, instead of diversity.

Restricting ourselves to the chaotic or regular cases, we can use equation (2.4) to express Δ_{n+1} as a function of Δ_n, namely

$$\Delta_{n+1} = e^{\lambda\tau}\Delta_n \quad \text{regular or chaotic}$$

Thus, for a Markovian evolution of the dynamic variable x, the conclusion is: if the system is regular or chaotic ($\lambda \neq 0$), the evolution of its diversity or dispersion Δ is also Markovian.

What about critical dynamics? Figure 2.9 shows the picture. In this case, we should use equation (2.8), trying to express Δ_{n+1} as a function of Δ_n. It is not possible! The best one can do is

$$\Delta_{n+1} = \left(\Delta_n^{1/z} + \Delta_1^{1/z}\right)^z \quad \text{critical}$$

By comparing the two last equations, we observe the eternal influence of the very first dispersion Δ_1 on all subsequent future evolution for critical dynamics, a feature not shared neither by regular nor chaotic cases for which the initial dispersion Δ_0 is forgotten. Even being Markovian, the system which follows a critical dynamics presents long-term memory concerning its dispersion, as time goes by. Again, a good example is the genetic evolution of a population. The genes of each individual of generation $n + 1$ were copied only from individuals of generation n, not $n - 1$, nor $n - 2$, etc. The genetic diversity of generation $n + 1$, however, cannot be defined only from the current genetic diversity of generation

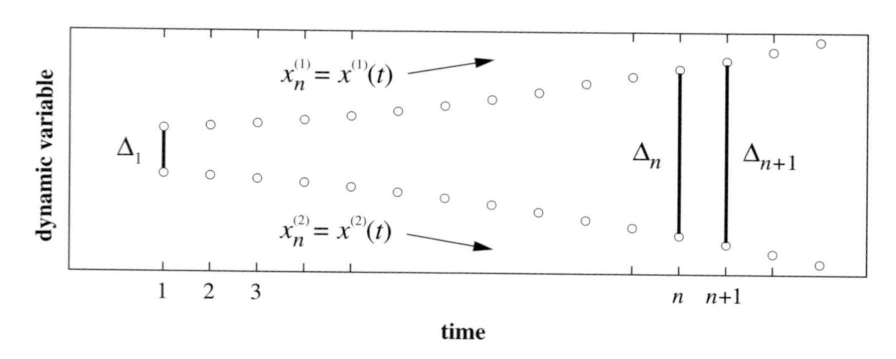

Figure 2.9. Critical dynamics. As in the previous figure, the next value x_{n+1} depends only on the current x_n. The diversity or dispersion Δ_n, however, follows equation (2.8), starting from the initial value Δ_1 at $t = \tau$. Contrary to the previous figure, now the next Δ_{n+1} depends not only on the current one, Δ_n, but also on the very first diversity Δ_1.

n. The evolution of diversity depends on small contingencies occurred on a remote past, an effect geneticists normally call the "founders effect".

There is a legend according to which Ghengis-Khan had a lot of children in many places, much more than any other human being of his time. Thus, his own genes were widespread over the population during the following generations. Let's suppose this is true, and raise the possibility of an alternative history according to which Ghengis-Khan was dead at an age of 10 years, with no children at all. According to this alternative scenario, the current human genetic pool would be different than it actually is, due to a single minor contingency which occurred a long time ago! Critical dynamics present this important feature: dependence on minor contingencies occurred in a remote past. The system evolves in trees, does not occupy the whole space of possibilities, and always leaves most part of the space unvisited for further explorations. Mathematically, the exponential, explosive growth of diversity characteristic of chaotic systems, equation (2.4), is responsible for the fast occupation of the whole space of final possibilities (the attractor), like a drop of ink inside a glass of water. For critical dynamics, however, this relation is replaced by the much slower power-law, equation (2.8), which allows only a tree-like growth where branches bifurcate from each other, keeping the most part of the possibilities unvisited. Some branches can also die. The actual branches occupied during the evolution of a particular history are not necessarily the same for another alternative history.

The same story of Ghengis-Khan is usually told concerning the first Brazilian emperor, Dom Pedro I, who ruled the country from 1822 until 1831. In this much more recent case, the huge number of children widespread over the country is easily verifiable, and really true. Undoubtedly, the genes inherited from D. Pedro I are strongly present in the Brazilian population nowadays. He was a 10 years old child when he arrived in Brazil in 1808, inside a ship coming from Portugal, when the royal family was transferred in order to escape from Napoleon. Had this ship be sunk during this trip ... (the reader already knows the tale end). But let's tell another tale. In principle, D. Pedro I could be a descendant from Ghengis-Khan, who knows? Within this hypothesis, the genetic pool of the Brazilian population could have a strong influence from Ghengis-Khan!

Jealous because of the Brazilian tale, the German author wants to tell his own, not related to genetics, but to historical evolution: if Adolf Hitler had died at the age of 10 years ...

Of course, in what concerns the past, a single history matters, the true one which really occurred. However, in what concerns the possibility to foresee the future, the many-histories scenario should be taken into account. For chaotic dynamics, reaching equilibrium after a finite time, one can make averages over all the potentially possible current situations, at present, in order to predict the probabilities of the various possible futures. For critical dynamics, this average over potentially possible presents is not useful to predict anything: one needs to follow the real history since a remote past. It is not the *average* genetic pool of the Brazilian population at the beginning of the XIX century which was strongly passed on to the future generations. They were the genes of D. Pedro I himself, a single individual!

2.5. Complexity and criticality

The title above is the same as a recently published book (Christensen and Moloney, 2005), and also the same as the entire issue of Physica A dedicated to Per Bak (Bak, 2004). In both, the reader can find examples of complex, critical systems covering a wide set of subjects: evolution, speciation, genetic regulation, epidemics, neuroscience, earthquakes, forest fires, astrophysics, cosmology, turbulent flow, plasma physics, magnetism, traffic, surface physics, economic market, networks, adaptive learning, and also (of course) computer modelling. The pioneering book by Per Bak himself (1997) also shows a lot of examples. The title links two distinct concepts which are nevertheless entangled to each other in the very same way as time and size scales do in nonlinear systems, Section 2.3.

Complexity (de Oliveira, 2005) is not an easy concept, the precise definition is not yet settled by the scientific community. Let's take a simple definition, as follows. A complex system has a large number of components and evolves in time. Each component exerts direct influence on some neighbours, and its behaviour also depends on direct influences exerted by others. The intricate network of influences, direct or indirect, is spatially long-ranged. Some different influences acting on the same component can generate conflicts. These are the basic properties a system should have in order to be classified as complex. A much richer analysis concerning the meaning of complexity can be found in the excellent paper by Giorgio Parisi, entitled "Complex Systems: a Physicist's Viewpoint" (Parisi, 1999).

Criticality is a much older concept, well studied by physicists since the nineteenth century, see Stanley (1971). Boiling water at a temperature of 100 degrees Celsius (absolute temperature $T = 373$ K) and pressure of 1 atmosphere (I do not use Pascal, though it is not as bad as Fortran) is a combination of liquid and vapour, two phases with different densities sharing the same closed vessel. Let's denote the density difference, liquid minus vapour, by m. By warming the whole system to a higher temperature, say $T = 393$ K, the liquid-vapour coexistence remains, provided one tunes the proper pressure higher than 1 atm. Hotter than before, the almost incompressible liquid suffers a small dilatation, its density decreases, while the compressible vapour becomes denser. The density difference $m(T)$ itself decreases, it is a decreasing function of the temperature. Warming more and more, the liquid-vapour coexistence is maintained by controlling the proper pressure, up to the critical value $T_c = 647$ K, where the liquid

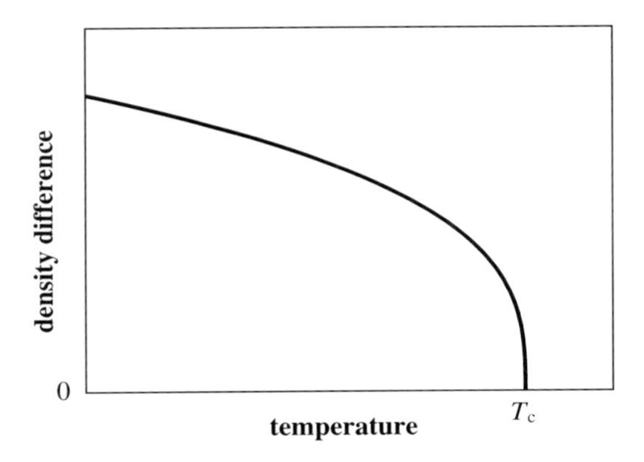

Figure 2.10. Density difference m between liquid and vapour water.

and vapour collapse into a single density $\rho_c = 322\,\text{kg/m}^3$, under the pressure $p_c = 217.7$ atm. Above this point, water is found in only one homogeneous phase we will hereafter call *gas* in order to distinguish from the vapour which can coexist with liquid below T_c. The critical temperature defines a phase transition, not to be confounded with the ordinary transformation of liquid water in vapour. We refer to the coexistence of two distinguishable phases, only possible below T_c, versus the single homogeneous gas above. Figure 2.10 shows the plot for $m(T)$ near T_c.

Phase transitions can also be interpreted as bifurcations. Instead of the density difference, we can plot the fluid density itself, Figure 2.11, a single curve for

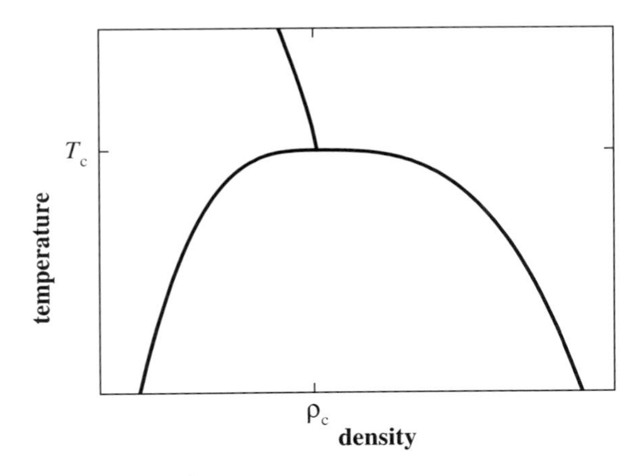

Figure 2.11. Densities for coexisting vapour (left) and liquid (right) water, below the critical temperature. Above, the homogeneous gas density under constant pressure.

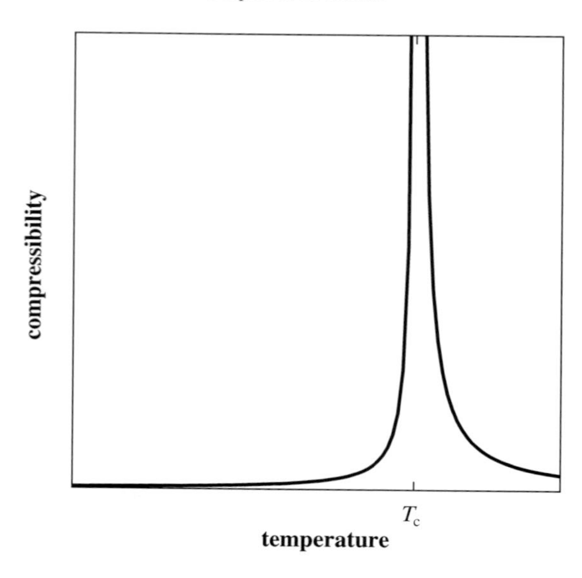

Figure 2.12. Plot of water compressibility κ.

the homogeneous gas above the critical temperature, and two separate curves for liquid and vapour, below it.

At room temperature and pressure, only small vapour bubbles appear. By pressing the fluid beyond 1 atm, but keeping the room temperature, the small bubbles shrink and disappear, the liquid phase remains alone. The compressibility is not high, because the liquid is almost incompressible and its volume remains nearly the same. Relaxing back the pressure to 1 atm, the small bubbles reappear.

By warming the vessel to higher temperatures, and keeping the proper liquid-vapour coexistence pressures, vapour bubbles inside the liquid increase in size. They grow more and more as the temperature increases towards the critical value. Thus, the compressibility also increases, as shown in Figure 2.12.

Near the critical temperature, the compressibility κ becomes enormous. The fluid becomes completely soft, responds with a large volume decrement (increment) to any small compression (decompression). For the gas, above T_c, κ decreases again, but remains larger than the corresponding values for the liquid-vapour coexisting phases below T_c.

In order to estimate the typical diameter of the bubbles, more precisely the Coniglio–Klein droplets (Coniglio and Klein, 1980), one can resort to the so-called correlation length $\xi(T)$, obtained by simultaneously measuring the density fluctuations at different positions inside the vessel, as a function of the distance. Figure 2.13 shows the plot of this typical length. Above the critical temperature, of course, there are no longer vapour bubbles surrounded by liquid, only a single

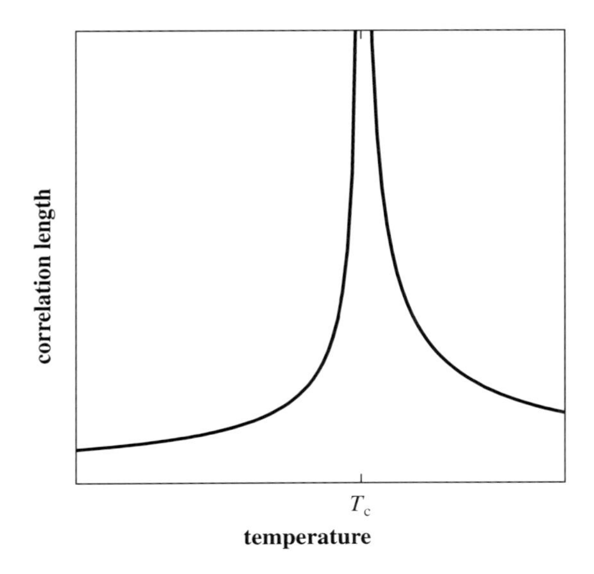

Figure 2.13. Plot of the correlation length ξ. Below the critical temperature, it is an indirect measure for the mean diameter of vapour bubbles observed inside liquid water.

gas big bubble exists, the size of which is limited by the vessel itself. However, the density–density fluctuations do exist along the whole range of temperatures, so its characteristic length $\xi(T)$ continues to exist beyond T_c. For a would-be infinite system, the correlation length really explodes to infinity at T_c. In practice, only the finite size of the vessel itself limits the otherwise endless growth of $\xi(T)$ as one approaches criticality. There, the single, critical big bubble is an entity as a whole, any shake at one side of the vessel is reflected on the far opposite side. Above T_c, the correlation length shrinks back, becomes smaller than the vessel size again, and criticality is lost. Although one continues to have a single gas bubble occupying the whole vessel, the long-range correlation no longer holds, the shake on one side is no longer felt on the opposite one.

From a practical point of view, the first remarkable feature of a fluid like water near its critical point is its already mentioned softness: a tiny increment on the external pressure generates an enormous decrement on the volume, and vice-versa. The reason for that is the very large size of the vapour bubbles, completely susceptible to compression. This property is also shared by many other systems with practical applications, for instance to construct artificial muscles or micro-engines. All modern electronics is based on this same phenomenon, translated into other physical quantities. Electronic devices give a measurable electric current as response to a tiny voltage increment. The plot *current* versus *voltage* is similar to Figure 2.10, read from right to left: the vanishing current for voltages below the critical value suddenly grows up as soon as it is crossed. Magnetic data storage devices (hard disks, for instance) are also based on this. Small magnetic fields are able to produce enormous magnetisations on localised places of the storage media. We could generalise the first phrase of this paragraph: the first remarkable feature of a critical system is the ability to give large responses to very small inputs.

From a conceptual point of view, the second, related feature of all critical systems is the long-range correlation between its component units, i.e., the explosion of the correlation length near the critical point, shown in Figure 2.13. We can define critical systems as the ones for which the correlation length overflows the system size itself. The reader certainly remembers the discussion in Sections 2.3 and 2.4 concerning critical dynamics. Here, we are using the same word *critical* for static systems which (macroscopically) do no evolve in time. The reader also remembers the intrinsic dependence of long-term memory (time) and long-range correlations (size) necessarily present in any nonlinear dynamic system. That is why the word *critical* is the same. Water near its critical point is an example of systems for which one cannot apply the reductionist approach of dividing the system into small isolated pieces. On the contrary, the macroscopic system should be treated as a whole, because any small perturbation performed at a given position propagates through the whole sample. The various microscopic components do not behave independently from each other. In studying the behaviour of a given individual component, one cannot neglect the influence of any other, even those very far from it. Of course, under these circumstances, the system will take a long time to reach the equilibrium situation we assumed in describing the water properties, above. The dynamic evolution of a critical system is also critical.

The third, also related feature of critical systems is the mathematical description through power-laws. The plot on Figure 2.10, for instance, corresponds to

$$m \propto (T_c - T)^\beta \quad \text{with } \beta = 0.326 \pm 0.004 \tag{2.9}$$

near the critical point. Below the critical temperature, Figure 2.11 also follows the same mathematical form, $\rho - \rho_c \propto \pm(T_c - T)^\beta$ with the same exponent β for liquid (+) and vapour (−) phases.

Analogously, near the critical point, Figure 2.12 corresponds to

$$\kappa \propto |T - T_c|^{-\gamma} \quad \text{with } \gamma = 1.239 \pm 0.003 \tag{2.10}$$

and Figure 2.13 to

$$\xi \propto |T - T_c|^{-\nu} \quad \text{with } \nu = 0.627 \pm 0.002 \tag{2.11}$$

where the symbol $|x|$ denotes the absolute value of x.

The numerical values of the so-called critical exponents, β, γ, ν, etc. were obtained through extensive computer work, including both mathematical series expansions and (mainly) Monte Carlo simulations (see, for instance, Ferrenberg and Landau (1991)). That is why the error bars appear in equations (2.9), (2.10) and (2.11).

Critical exponents are universal, i.e., this same set of values is obtained for completely distinct systems sharing only two general properties: (1) three-dimensional geometry, as the volume of water; and (2) one-dimensional order parameter, as the liquid-vapour density difference (a simple scalar number, not a multidimensional vector). That CO_2 shares with water the same set of critical exponents, in spite of a different critical point, $T_c = 304$ K, $p_c = 73.0\,\mathrm{atm}$ and $\rho_c = 468\,\mathrm{kg/m^3}$, may be not a surprise for the reader: both are gases with triatomic molecules. The surprise is that also mono-atomic helium has these exponents: microscopic details as the molecular form do not play any role in defining the critical exponents. Equations (2.9), (2.10) and (2.11), including the exponent's numerical values, are valid also for uniaxial magnetic solid materials as the anti-ferromagnet MnF_2 for which the critical temperature is $T_c = 67.3$ K. In this case, the order parameter m is the spontaneous staggered magnetisation (spontaneous means to keep the system under zero magnetic field, the equivalent of controlling the liquid-vapour coexistence pressure on the fluid). Of course, m vanishes above the critical temperature, meaning that a macroscopic magnet loses its magnetisation when heated too much. These magnetic materials share with water almost nothing but the two very general properties mentioned at the beginning of this paragraph (remember that both the liquid-vapour density difference m and the uniaxial magnetisation m are simple scalar numbers, not vectors).

A very simple model, the so-called Ising model (Ising, 1925) also shares the same general features with fluids and uniaxial magnets. One considers a large three-dimensional lattice, each site i hosting a microscopic magnet which can point either up or down, denoted by $s_i = +1$ or $s_i = -1$, respectively. The magnetic interaction holds only for nearest neighbours (direct influence). Consider a pair of neighbouring sites. If the corresponding magnets point in the same sense, both up or both down, the pair contributes with a negative value $-J$ to the total energy. Otherwise, one magnet pointing up and the other down, the contribution $+J$ of this pair is positive. The magnetisation is simply the thermal average of the sum

$$\frac{1}{N}\sum_i s_i$$

a simple scalar number. Thus, this model also falls into the same universality class of all ordinary fluids and all uniaxial magnets (including anti-ferromagnets like MnF_2, for which J is negative). Indeed, the numerical values presented in equations (2.9), (2.10) and (2.11) correspond to Monte Carlo simulations of this model.

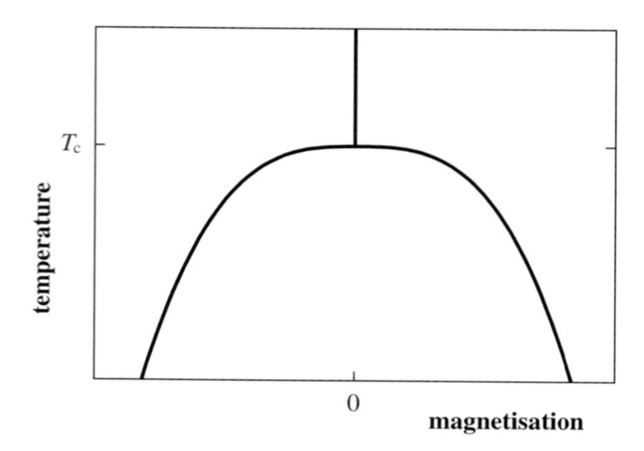

Figure 2.14. Three-dimensional Ising model. Plot of the magnetisation, to be compared with Figure 2.11. The curvature corresponding to the critical exponent β is the same for both.

Of course, it is much easier to study the simple Ising model than, for instance, the complicated quantum behaviour of water molecules, each one subjected to translation, rotation, vibration, etc., besides their mutual interactions and collisions, a real mess. The critical behaviour, however, is the same, including the critical exponents' numerical values. Figure 2.14 shows the Ising model magnetisation as a function of the temperature. Above T_c, the magnetisation vanishes. By cooling the system below T_c, it suddenly becomes positive or negative: magnetic domains (bubbles) pointing either up or down coexist inside the same macroscopic sample.

If site i is interpreted as "occupied" for $s_i = +1$ or "empty" for $s_i = -1$, the same model is called a lattice gas. Perhaps this alternative interpretation helps the reader to accept that a complicated system like water could be described by such a simple model. However, the only relevant features are the scalar character of the order parameter and the three-dimensional geometry, not the particular microscopic interpretation.

Unfortunately, in spite of the extreme simplicity and the huge amount of work performed during the last 80 years, no analytical solution is available for the three-dimensional Ising problem. Only in two dimensions under zero magnetic field we do have analytical solution (Onsager, 1944; Lee and Yang, 1952). Concerning other models, except for some particular cases most of them in two dimensions, analytical solutions are very rare in the whole field of statistical mechanics (Baxter, 1982). The ubiquitous presence of long-range correlations turns things difficult within this discipline. Perhaps this fact could explain why critical behaviour was studied since the nineteenth century, but the universality behind it was understood only at the end of twentieth century (Wilson, 1971, 1979), when computers started to be available. Even so, this understanding concerns only static systems without time evolution, as the phase transitions commented above.

However, criticality and the corresponding universality are also ubiquitous in dynamic evolving systems, and the understanding of this amplified phenomenon is not yet complete. Besides spatial long-range correlations, to take into account also long-term memory is a further difficulty. The relaxation time τ also diverges, similarly to the correlation length, Figure 2.13. Near the critical point, these systems suffer from what physicists call a critical slowing down. In some cases, the dynamical system automatically tunes their internal parameters in order to remain always near the critical point, a phenomenon called self-organised criticality (Bak, 1997), for which biological evolution is a special example.

Nowadays, the prime tool to study complex systems, besides experiments and real observations, is the computer. In particular, for dynamic evolving systems, agent-based models are simulated as follows. One keeps on the machine memory the individual features of N agents. These features are updated as time goes by, according to some dynamic rules describing the problem at hands. The action of each agent depends on other's. Fluctuations can also be introduced during the evolution through random numbers which help to decide the action of each agent. By running the same program many times, possibly starting from different initial conditions, one can appreciate the many possible final situations, and measure the quantities of interest.

These models are usually criticised based on two arguments: (1) they are considered too simple to reproduce the behaviour of so complicated real systems; and (2) they are reductionist. We will try to convince the reader that none of these criticisms are valid. The first argument ignores universality. In order to reproduce the critical behaviour of the complicated real system under study, the simplified model does not need to share all its complicated features, only the very general characteristics defining the universality class to which both belong (for instance, the three-dimensional geometry and one-dimensionality of the order parameter, shared by the various fluids and the Ising model). The researcher's duty is to invent the proper model which respects the general features characterising the universality class to which the real system belongs. This is not easy, particularly for dynamic evolving systems for which one does not completely understand the mechanisms leading to universality. Also the model should be simple enough to be programmed on a computer, where it is supposed to run within an acceptable time. The second argument confounds simplicity with reductionism. The simple model supposed to reproduce the critical behaviour of the real system should present the same long-range correlation properties. Thus, it cannot be solved by breaking the whole into small pieces and summing up the various pieces at the end, which would be just the reductionist approach. It is impossible to exactly solve the Ising model for a large $3000 \times 3000 \times 3000$ cube. As a solace, the exact solution for a tiny $3 \times 3 \times 3$ cube is feasible, one can write down its thermal averages through analytical equations. Unfortunately, the result for the larger $3000 \times 3000 \times 3000$ cube is not the superposition of 10^9 tiny cubes. Size scaling is not trivial. The intrinsic nonlinearity of critical systems can be simply stated as: the whole is not the sum of the parts.

Analogously, a dynamical model supposed to reproduce the complex behaviour of a real system should share with the latter the same long-term memory properties. One can study the evolution of this model within a small time interval, repeat this task for slightly different environments (in order to include fluctuations), and take the average over them after this small interval. From this averaged situation, one can proceed the evolution during a further small time interval, take the average again, and so on. Unfortunately, this strategy does not work, remember Ghengis-Khan and D. Pedro I in Section 2.4: the final result is not the same one would obtain by taking the average only after a very large time interval. Time scaling is not trivial. No reductionist strategies could work in the study of complex systems, in both space and time.

2.6. Mean-field theories

In order to treat a critical system, one needs to consider it as a whole, one cannot break the system into smaller pieces. An attempt to circumvent this difficulty is the so-called mean-field strategy, which unfortunately is quite often unreliable and always gives wrong quantitative predictions for the critical exponents and other important issues. Anyway, it is a very intuitive approach which gives some insights on the problem itself. It is very often used in models of population dynamics. However, mean-field approaches should be used with care, it is not easy to separate among the results what is trustable from what is only an artifact of the approach itself.

Let's take the simplest example, the Ising model mentioned in last section: a cubic lattice with N magnetic atoms, each site surrounded by 6 neighbours. Picking a particular configuration c for the N magnets pointing up or down, the total energy is

$$E_c = -J \sum_{\langle ij \rangle} s_i s_j \tag{2.12}$$

where the sum runs over all pairs $\langle ij \rangle$ of neighbouring sites i and j. Similarly, the magnetisation for this particular configuration is

$$m_c = \frac{1}{N} \sum_i s_i$$

the sum running over all sites.

Without approximations, in order to calculate the thermally averaged magnetisation m, one needs to compute E_c and m_c for each configuration c, and the sums

$$m = \frac{\sum_c m_c \, \mathrm{e}^{-E_c/T}}{\sum_c \mathrm{e}^{-E_c/T}} \tag{2.13}$$

with 2^N terms. The temperature enters into the scene through the Boltzmann factor $\mathrm{e}^{-E_c/T}$ which properly weights the configurations for the thermal average (the Boltzmann constant which simply transforms temperature into energy is omitted, i.e., $k_B = 1$, for simplicity). Even for moderate lattice sizes, this is an impossible task.

In mean-field approximation, one replaces each of the 6 neighbour magnets s_j surrounding site i by the average m, in equation (2.12) which is then transformed into the friendly form

$$2E_c = -6Jm \sum_i s_i$$

and renders feasible the (mean-field) solution for equation (2.13). The result is

$$m = \tanh\left(\frac{6Jm}{T}\right)$$

where the desired quantity m appears in both sides.

The last equation can be numerically solved for m, leading to the plot in Figure 2.15, which looks qualitatively correct, if compared with Figure 2.14. A careful look, however, reveals a different curvature, near the critical point. Indeed, by expanding the hyperbolic tangent up to the term in m^3, one realises that m correctly follows a power-law $m \propto (T_\mathrm{c} - T)^{\overline{\beta}}$ but with the wrong critical exponent $\overline{\beta} = 0.5$ instead of $\beta = 0.326$. Furthermore, the mean-field critical temperature $T_\mathrm{c} = 6J$ is overestimated.

The reason for this drawback is simple to understand: *by replacing the fluctuating sense of the 6 magnets surrounding a given site by the fixed average m, one neglects fluctuations which are just responsible for the long-range correlation leading to critical behaviour.*

Sometimes, mean-field approaches also qualitatively fail, leading to more serious problems. A classical example of this further drawback is the mean-field prediction of a spurious transition in one geometrical dimension (it is enough to replace above the constant 6 by 2, the number of neighbours along a chain). However, at non-zero temperature one cannot observe order in one geometrical dimension, because a single broken link along the chain is enough to destroy the long-range order. Mean-field approaches can wrongly introduce phase transitions or bifurcations where they really do not occur.

Biological speciation, for instance, falls into the general bifurcation description sketched in Figures 2.11 or 2.14, where the vertical axis represents the time

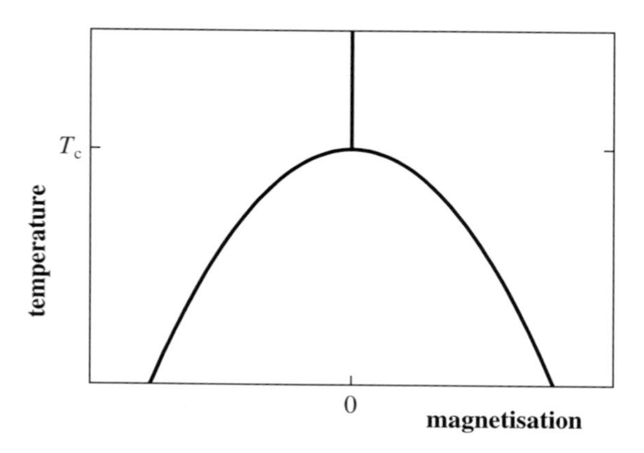

Figure 2.15. Plot of the magnetisation m for the Ising model within the mean-field approximation.

downwards. Mean-field approaches could induce the researcher to predict speciation events where they do not exist, underestimating the required conditions to observe this phenomenon. Moreover, to replace the curvature of Figure 2.14 with the wrong curvature of Figure 2.15 could be a disaster if the purpose is to study the speed of speciation: the researcher would falsely conclude in favour of a much slower process.

Also the dynamics of economic markets usually face bifurcations of the same kind of Figures 2.11 or 2.14 (Arthur, 1990; Anderson, Arrow and Pines, 1988), and mean-field approaches represent the same danger. Fluctuations are also important in population dynamics, thus one cannot completely trust in such models where the influence of each individual on others is replaced by some kind of average. Analytical approaches, even the more sophisticated ones, normally hide such an approximation, sometimes unnoticed.

2.7. Scaling

For systems in thermodynamic equilibrium, universality was understood after the so-called renormalisation group theory, invented by Kenneth Wilson and based on earlier works by Leo Kadanoff, Michael Fisher, Ben Widom and others. The first is awarded with the Nobel prize, and all four won the Boltzmann medal, the most important award within Statistical Physics. Below, there is an intuitive view of the corresponding concepts, the fundamental ingredient being the presence of long-range correlations, $\xi = \infty$, and the basic tool being a scaling transformation successively performed on the system under study. Soon, the reader will realise that similar reasoning could be extended to dynamic evolving systems where long-range correlations and long-term memory are present, in particular biological evolution.

The reductionist approach of dividing a system into smaller parts, studying each part separately, and finally joining the pieces together, is generally adopted because it allows one to reduce the number of variables of the problem, rendering it solvable. However, we have already verified the inadequacy of this approach in the study of (infinite size) critical systems, due to the presence of long-range correlations: the artificial boundaries introduced when a small piece is separated from the larger system also introduce a cutoff on these long-range correlations, destroying the criticality itself. The renormalisation group strategy is to reduce the number of variables (degrees of freedom) without transforming the initially infinite system into a finite one. One applies a length scaling transformation which keeps the renormalised system still infinite. The same procedure is applied again and again, gradually reducing the degrees of freedom.

Let's take a concrete example, the Ising model on an infinite cubic lattice, where each magnetic site interacts with many others in the neighbourhood, according to some set of couplings $J(r)$ which depend on the distance r (not necessarily only the 6 first neighbours). These couplings measure the system's magnetic energy which aligns the individual magnets: for low enough temperatures, the majority of them point into the same sense. For higher temperatures, the thermal energy dominates and breaks the magnetic order, giving rise to the phase transition at the precise critical temperature T_c, which therefore depends on the set of couplings $J(r)$.

In our imagination, we can group these sites into small $3 \times 3 \times 3$ cubic cells, assigning to each cell a single magnet pointing up or down, according to the majority inside the cell. Instead of the original lattice of sites, we have now a cubic lattice of cells, each one collapsed into a single renormalised magnet. As the original number of sites is infinite, the number of cells remains infinite, although 27 times smaller than the former number of sites. Of course, the function $J(r)$ defining the couplings will be transformed into another function $J'(r)$. The ironic reader may argue we have transformed a problem which is unsolvable due to its infinite number of variables into another problem with 27-fold-less variables, which nevertheless remains equally unsolvable. Right! But we are not trying to solve the problem, only to understand why different problems fall into the same universality class.

In order to make the lattice of (collapsed) cells closer yet to the original lattice of sites, we can see it through a 3-times reducing lens, a negative zoom transforming all original distances r into $r' = r/3$. The couplings of the new lattice will be noted by $J'(r')$. The already quoted correlation length $\xi(T)$ is one particularly important distance which will be transformed into

$$\xi(T') = \frac{\xi(T)}{3} \tag{2.14}$$

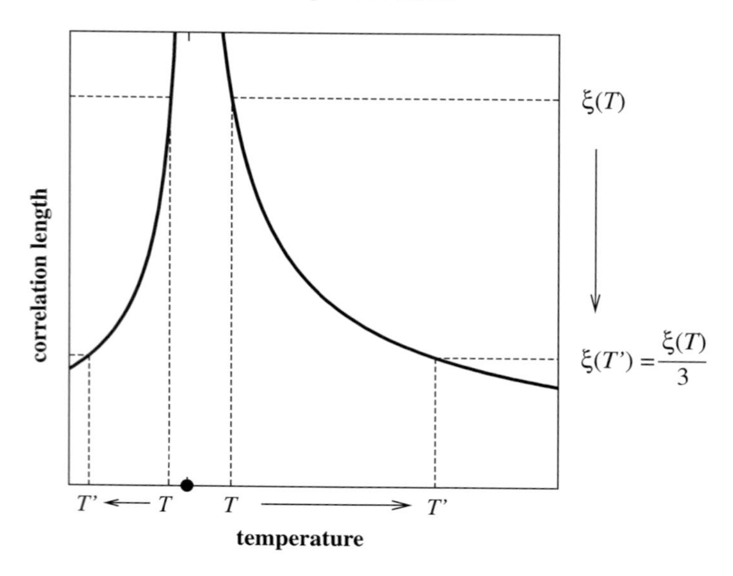

Figure 2.16. Scaling of the correlation length by a factor of, say, $1/3$: the temperature T runs away from the critical point (filled circle).

where a new temperature

$$T' = R(T) \tag{2.15}$$

appears as a consequence of the whole scaling transformation. Figure 2.16 helps to understand what is going on.

For temperatures $T = 0$ and $T = \infty$ (both out of range in Figure 2.16), the correlation length vanishes, thus the system is invariant under scaling transformations at these extreme temperatures. The same invariance also occurs at criticality, $T = T_c$, where the correlation length diverges, Figures 2.13 or 2.16. Because only $\xi = 0$ or $\xi = \infty$ are insensitive to the scaling transformation, these three temperatures, $T = 0$, $T = T_c$ and $T = \infty$ correspond to the only three situations where the system is scaling invariant. For any other temperature, the negative zoom decreases the correlation length. Therefore, according to Figure 2.16, above T_c the temperature increases towards the attractor $T = \infty$, by iteratively repeating the scaling transformation. On the other hand, below T_c the temperature decreases towards the other attractor $T = 0$. In this way, the thermodynamic phases are identified with the renormalisation group basins of attraction. All temperatures corresponding to the ordered phase, where the spontaneous magnetisation appears, are attracted towards $T = 0$, whereas the disordered phase corresponds to temperatures attracted towards $T = \infty$. Figure 2.17 sketches this behaviour.

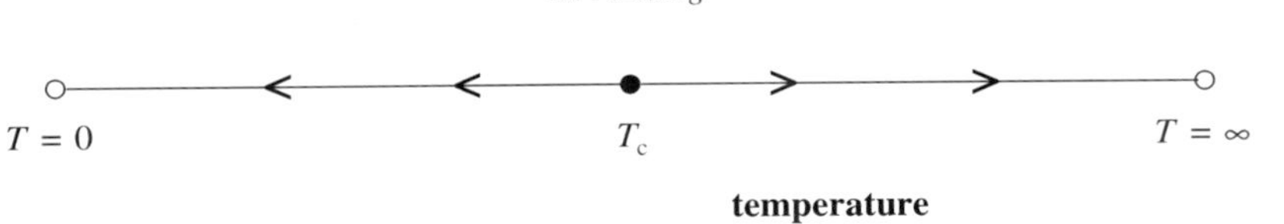

Figure 2.17. Flux diagram of the temperature under repeated scaling transformations. The critical point T_c remains unchanged, while temperatures below (above) it are attracted to $T = 0$ ($T = \infty$).

Consider a temperature below T_c. Then, the symmetry is broken, the total magnetisation is not zero. Let's consider it positive, for instance, the majority of the individual magnets pointing up. The lattice is a big sea of up-magnets with many islands of down-magnets. By looking to this lattice through a negative zoom, those minority islands shrink inside the still infinite big sea of up-magnets. A further negative zoom makes the islands even more unimportant, and so on. After many successive zooms, one sees no more islands, all (renormalised) magnets are aligned up. This is just the characteristic situation at $T = 0$, where correlations were completely washed out.

Let's open a parenthesis for a technical point. The smart reader may argue the final words of the last paragraph are wrong, the situation where all magnets point into the same sense would correspond to the maximum possible correlation, not to the complete absence of correlations as we affirm. The correlation function $C(r)$ between two magnets s_i and s_j distant r from each other is

$$C(r) = \langle s_i s_j \rangle - \langle s_i \rangle \langle s_j \rangle$$

where the symbol $\langle \cdots \rangle$ means thermal average. Why did we not write simply $C(r) = \langle s_i s_j \rangle$? Answer: because this last form includes trivial correlations which do not depend on the distance r! We are interested just on the correlation length ξ which measures how much correlations decay with distance. If one has a majority of magnets pointing in the same sense, the local averages themselves do not vanish, i.e., $\langle s_i \rangle = \langle s_j \rangle \neq 0$, and therefore $\langle s_i s_j \rangle \neq 0$ too, independent of the positions of these two magnets on the lattice. They would be correlated, no matter how far they are from each other, generating an infinite correlation length for any temperature below T_c. Thus, in order to retain only the contributions which really depends on the distance r, the product $\langle s_i \rangle \langle s_j \rangle$ is subtracted in the above definition of the correlation function. The correlation length becomes finite also below T_c. Closed parenthesis.

On the other hand, if the temperature is above T_c, the magnetisation is zero. Within fluctuations determined by the temperature itself, any large enough region of the lattice presents as many up- as down-magnets. The successive application of negative zooms would not change this behaviour, except for the shrinking size of the quoted regions. After many zooms, each (renormalised) magnet points up or down independent of the neighbourhood, the characteristic situation at $T = \infty$. Again, correlations were completely washed out.

In short, the renormalisation group transformation consists in applying a negative zoom to the system, and comparing the new configuration with the former in order to determine the new temperature, equation (2.15). By successively applying this transformation, the corresponding flux of the temperature is monitored. The overall effect is to wash out any trace of correlations the initial system could present. The final destiny is two fold, either $T = 0$ or $T = \infty$. In both cases, the former correlations no longer exist. The only exception occurs if the initial temperature equals the critical value, $T = T_c$. In this case, the correlation length diverges, $\xi = \infty$, being thus insensitive to the zoom.

To obtain the renormalisation group transformation R, equation (2.15), is not an easy task. Some approximations are available, with different accuracy degrees. Suppose we know this transformation. Then, the critical exponent ν can be obtained by combining equations (2.11), (2.14) and (2.15), which yields

$$\frac{dR(T)}{dT} = \frac{T' - T_c}{T - T_c} = 3^{1/\nu}$$

where the derivative of $R(T)$ is taken at T_c. The important point is to realise that ν depends on the rate according to which T runs away from T_c after the scaling transformation, in the neighbourhood of T_c, Figure 2.17.

The one-dimensional Figure 2.17 does not add much more information. One can better appreciate the power of renormalisation group by looking at its counterpart in the multidimensional space of the couplings $J(r)$. Instead of a single quantity T which varies under the scaling transformation, $T \rightarrow T'$, we can follow the behaviour of the whole set of couplings $J(r) \rightarrow J'(r')$, a multidimensional flux. Figure 2.18 is again a sketch of this behaviour. As the paper sheet of this book has only two dimensions, Figure 2.18 considers a two-dimensional space of couplings, but in general this space has much more dimensions.

The derivative of R along the steepest descent direction (indicated by the straight arrows at C, in Figure 2.18) provides the value for the critical exponent ν. Further directions, not shown in this two-dimensional diagram, could be included in order to provide the values for the other critical exponents. Technical details like how to obtain the R transformation or how to derive it in order to calculate the critical exponents do not concern our purposes here. The important information comes from realising that each point sufficiently near the critical line, which represents a generic system near criticality, goes first towards point C. Only after reaching the neighbourhood of C, it gets away from the critical line. Thus, the critical exponent defined by the speed of this run-away is the same for all these different systems. The universality class is set by point C.

An important conceptual point concerns the irrelevance of microscopic details, for instance the fact that each site has just 6 nearest neighbours within the cubic lattice. Also, how exactly each site interacts with its neighbours is irrelevant. When a group of 27 neighbouring individual magnets belonging to the same

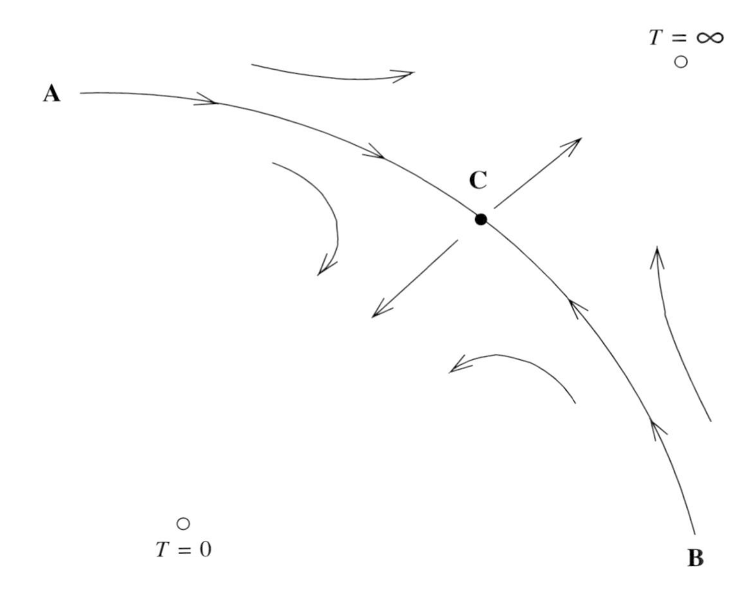

Figure 2.18. Flux diagram in the space of couplings. Two attractors denoted by $T = 0$ and $T = \infty$ define the magnetic and disordered phases, respectively, separated by the frontier line ACB. All points along this line are critical with $\xi = \infty$, each one representing a different critical system. Point C, however, plays the special role of attracting all other critical points: it defines the same set of critical exponents for all these systems.

$3 \times 3 \times 3$ cell were replaced by a single magnet, such microscopic details were washed out. They do not play any role after many successive applications of the scaling transformation R, the successive zooms turn them irrelevant. The complicated interaction mechanism between water molecules plays exactly the same role as the simple $\pm J$ pair energy between Ising magnets.

Physicists call "scale-free" these systems for which $\xi = \infty$. They are invariant when seen through a reducing lens which omits the microscopic details. By superimposing such a lens over another, another yet, and so on, one sees always the same scenario. Geometric fractals are old examples. Scale-free networks such as the Internet are modern, fashionable examples (see Chapter 5 and Section 6.2). One can classify the various Internet nodes widespread over the world according to their "sizes", i.e., how much each one is accessed during a day. Then, one can count how many nodes fall in each class, and construct a plot of these countings versus the "size". We are sure the reader will guess the resulting double logarithmic plot: a straight line, determining a power-law relation.

The kind of correlations observed in such isotropic materials as fluids and simple magnets is the simplest possible: the correlation between two points depends only on the distance they are from each other, not on the direction. The influence exerted by a given point spreads out in all directions, equally, as the wave observed on a liquid surface when a small rock drops from above. Anisotropic crystals may behave differently, correlations spread more effectively along certain directions than others, giving rise to different geometric patterns. Examples are stripes, swirlings, etc. More complex patterns could be observed in the so-called liquid crystals, used in computer screens and other displays, which in reality are not crystals. In principle, one can design the underlying material in order to distribute the correlations (positive, negative, weak, strong, etc.) over each pair of points according to a previously defined, programmed pattern. This task is just what mother Nature dynamically does during embryo development. An initial single cell is successively divided into others, which gradually acquire different forms and properties, specific for each different position inside the embryo. Of course, in these more complex cases, the equivalent to the renormalisation group could not be the simple global scaling transformation obtained simply by dividing all lengths by the same factor. A more sophisticated transformation should be applied. However, the idea seems to be the same, according to the famous biologist John Maynard Smith (1998): "... during development the embryo is successively divided into smaller and smaller regions, whose subsequent growth is to a degree autonomous, although signals do pass between regions, serving to integrate the whole process ... modularity has important consequences for evolution". Anyway, the simplest isotropic critical materials and the much more sophisticated embryos share the same important feature: long range correlations. Does it also imply some kind of universality for embryo development? For evolution?

2.8. Biological evolution

The concept of biological evolution was introduced by the French naturalist Jean-Baptiste de Lamarck more than two centuries ago (Lamarck, 1802). The term *evolution* itself was not yet used at his time, but the word *Biologie* (in French) was invented by Lamarck. According to him, the various currently living species are neither static nor independent entities. They are the result of many small modifications occurred in ancient species, accumulated during very long times. Furthermore, species living today are also under this slow modification process. The whole system of living beings eternally changes. Evolution is a long-term, endless process. A very interesting interpretation of Lamarck's work and ideas can be found in the amusing book by André Langaney (1999).

During his lifetime, Lamarck felt into disgrace because his theory contradicts most religious dogmas. Nowadays, Lamarck also felt into disgrace, because he believed the traits acquired during one individual's life could be passed on to its descendants as a genetic inheritance. In reality, this criticism is not fair. Except for Weissman, all scientists of the nineteenth century believed the same, including Darwin according to his own book (1859). The reason is simple: genetics was not known during this time, only after the pioneering work of Gregor Mendel (1866) which became completely unknown up to the beginning of the twentieth century. Then, chromosomes could be observed on microscopes, and the concept of physical transmission of genetic characteristics to the offspring was settled down. Before this, no clear difference between, for instance, genetic and cultural inheritance could be made.

In order to put Lamarck's idea about what nowadays we call biological evolution close to our language, let's consider the space of all possible living forms. Later in this section, we will make this concept a little bit more concrete. For a while, we ask the reader to imagine a multidimensional mathematical space where each point is a possible living form. Better yet, let's consider a small lamp located at each site of this space, the majority of which are currently off. Only a few lamps are on, emitting light, and correspond to the currently living beings over the Earth surface, belonging to all living species. The lamp corresponding to each individual is switched on at birth, and off at death.

Each offspring of a given individual corresponds to a lamp near the parent's, differing from it and the siblings by small mutations. (For the sake of simplicity, let's consider a single parent for each newborn, because sex does not play any role in our present discussion. Sex will introduce a further source of diversity, besides mutations, but the newborn lamp position will be located near the parent's anyway.) The word *near* introduces a metric in our space, the concept of genetic distance which will also be made concrete later in this section. In order to retain this concept, we will refer hereafter to the *genetic space* of all possible living forms. Each offspring can generate its own offspring, and so on, according to a branching process of lamps switched on and off at the same region, sharing the same root, the same grand-grand- . . . -parent. As time goes by, some dangling-end branches stop to emit light (extinction), while others grow forever. Most lamps are never switched on, but they are there, each one representing a potential living form.

The space would look as a crowded soccer stadium at night, where smokers continuously set light in their cigarettes, producing a succession of small flares here and there all the time. However, different from the crowded soccer stadium where smokers are located everywhere, our space presents no light over large regions. Blinking light comes only from some specific concentrated clouds of lamps which correspond to the currently living species. The space is sparsely populated by such clouds, where some lamps currently blink, one cloud separated from the others by much larger dark regions.

Let's concentrate our attention in one of these clouds, taking a zoom inside it, an instantaneous snapshot of a single species. One sees a sea of lamps, some few of them emitting light. Some time thereafter, a lamp which was off before is suddenly switched on, a newborn. Following the behaviour of this single cloud during the short-term of few generations, one sees some lamps being switched on, while others are switched off, like the blinking lights on the crowded soccer stadium. By tuning a small negative zoom, one can see the blinking cloud as a whole inside the field of vision. It does not seem to move, its "centre of mass" seems to stay at rest. This is exactly the image dominating the religious minds of the nineteen century, static species.

Extending the observation time to some not-so-few generations, however, one can note a slow movement of the blinking cloud as a whole. One does not expect it will collide with another cloud in the future, because the highly inhomogeneous occupation of the genetic space provides an extremely sparse pattern for the clouds. Alternatively, we can imagine the movement back to the past, by reversing the arrow of time. In this case, our cloud certainly collides with another similar one, merging themselves into a single cloud from this moment back, until colliding with another cloud, back again and so on. For any two randomly chosen individuals, one can traceback their ascendents: it is certain to find a common ancestor. In the same way, two species have a common ancestor species. Returning to the normal sense of the time arrow, we conclude that nowadays living species are descendants of ancient species which suffered a cascade of bifurcation processes, the so-called speciation. A currently living species, if not caught by extinction, will also suffer the same process, generating new species in the future, and so on. Evolution follows a branching dynamics not only on the scale of an individual and its offspring, but also on the much larger scale of species.

Lamarck's idea, however, did not inform us how evolution acts on the individual level. This key ingredient was provided by Darwin's concept of natural selection (Darwin, 1859). Individuals more adapted than others to the current environment generate more offspring, on average, spreading more effectively their genetic information over the following generations. The overall effect is to provide the slow movement of clouds (species) inside the genetic space. The population density increases in a region where life is more adapted to the current environment, and the cloud's "centre of mass" is driven towards this region. In cases where two or more such better adapted regions appear at the same time, one could observe speciation.

It is important to note that other species belong to what we call *environment*, in the study of a single species. Therefore, the slow movement of one cloud is not completely independent of the others. However, in order to study this interdependence, we need to extend more our size scale, to take a further negative zoom allowing us to see many clouds, not just one. How many? Where should we stop the zoom-out process? Which is, then, the new zoom scale?

Nature itself answers these questions for us. We are now focused on a single blinking cloud under our field of vision, a single species. This is the first step of a sequence of zoom processes, as follows. By further zooming-out, we find a first neighbouring cloud entering into our field of vision, another similar species.

A little bit further, other clouds successively appear. After a certain zoom degree, no more clouds enter through the borders of our field of vision, an enormous dark space surrounds the already caught group of species. Then, we stop the zoom process at this point, for a while. We reached the second level, an isolated group of neighbouring species, called a genus. One sees a set of small blinking clouds (species), separated from each other by large dark regions. Now, we interpret each small cloud where some lamps are currently on as a single, renormalised lamp, also switched on, a whole-species lamp. They are separated from each other by large dark regions. Déjà-vu! One sees the very same scenario we have already seen at the first zoom level of a single species.

After the second level focusing on a single genus, we can proceed the zoom-out process. Other genera will enter into our field of vision up to the situation where a complete set of neighbouring genera is already caught. This is the third level, and the set of neighbouring genera inside our field of vision is called a family. Again, each genus is a blinking cloud (of clouds) separated from the others by large dark regions. By interpreting each such a cloud (of clouds) as a single (re-renormalised) lamp switched on, a whole-genus lamp, we observe again the same scenario. The zoom-out process proceeds. A set of neighbouring families is called an order. A set of neighbouring orders is a class.

The scaling reasoning presented above is very similar to the renormalisation group treated in last section. However, it has a further ingredient: the time evolution. Now, we are dealing with a more sophisticated phenomenon. For systems in equilibrium, after Kenneth Wilson and others, physicists were able to put the length scaling process under complete control. A complete theory for equilibrium critical phenomena exists, its most important feature being the explanation for the observed universality of critical behaviour. On the other hand, for evolving critical systems, for which one needs to control also the time scaling process, physicists were not able to construct a complete theory up to now. Nevertheless, an enormous progress was achieved during the two last decades, mainly through results obtained from computer simulations. In particular, we have now a lot of examples where universality insists to appear also in these out-of-equilibrium situations.

An important feature concerns the relations between size and time scalings, in the process leading from individuals to species, from species to genera, etc. Imagine the instantaneous picture of a given species, with all its currently alive individuals. One can traceback all parents, grand-parents, grand-grand-parents etc., and pick-up the first common ancestor to all currently alive individuals. (Note that this common ancestor did not live alone. Other individuals belonging to the same species lived at the same time, but their lineages were extinct, only the lineage of the quoted common ancestor survived. See Section 3.5.) The average time one needs to go back in order to find this common ancestor depends on the species size. The corresponding time to find the common species ancestor for a whole current genus is much larger. To trace back the genus ancestor of a given family spends a larger yet time, and so on. Size and time scalings are intrinsically linked to each other.

Let's return to our genetic space. How to define such a mathematical entity? The answer to this question was given by Gregor Mendel (1866), considered the founder of genetics. He worked with sexual reproducing pea plants, and published a short paper entitled "Experiments with Plant Hybrids". Even without knowing chromosomes, he was able to set the main concept: inherited traits do not mix. A single parent's gene (a word out of Mendel's vocabulary) is either inherited by the offspring or not, with no intermediate possibilities. There is not such a thing as half a gene, or any other fraction. John Maynard Smith (1998), again: "The philosophy behind this approach is that the genes carry, in digital form, the instructions for making an organism". The word *digital* should not be interpreted as dealing with numbers expressed in decimal basis. The correct meaning intended by Maynard Smith is discreteness, not the numerical basis. He means the same storage strategy used to record music over *digital* CDs, not the *analog* storage over vinyl. The gene is a discrete entity. Either it is there or not. Genetic information is coded according to a *yes/no* protocol, 1-bits and 0-bits. Perhaps the adjective *binary* would be better than *digital* as applied both to genetic information and CDs.

Therefore, the genetic information an individual carries in its chromosomes is a bit-string. Let's take the simplest interpretation. Consider an ordered sequence of all possible alleles for all possible genes. The genetic information of a given individual is a bit-string with a 1-bit on every position which corresponds to alleles/genes this individual indeed carries along its chromosomes, and 0-bits on every other position. We will call "genome" this bit-string. Now, we can understand what the genetic space is: the set of all possible (very long) bit-strings. The (enormous) dimension of this space is the bit-string length, virtually infinite in our simple theory.

The distance between two different points of this space is counted by bit-to-bit comparison. The simplest option is the Hamming distance which sums the number of positions along the two compared chains where bits differ, divided by the total length, Figure 2.19. We will interpret this quantity as the genetic distance. Two nearest neighbours on the genetic space differ from each other by just one bit. More sophisticated definitions taking different statistical weights to different bits, or different sets of bits, can also be used.

In order to satisfy the taste of some readers, we can also quote an alternative interpretation for the genetic bit-strings, by using Mother Nature's alphabet: each adjacent pair of bits 00, 01, 10 or 11 corresponds to one of the four chemical bases A, T, G or C, in just the same way each three adjacent such bases correspond to some aminoacid, a simple translation code. This interpretation is mandatory for simulations based on real DNA data, because a single genetic code was found on Earth. On the other hand, by no means this is the only possible or even the "best" codification. A lot of other equivalent interpretations could be invented, giving to

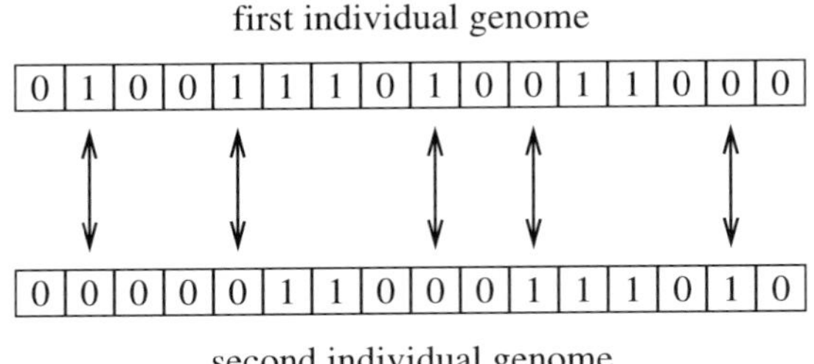

Figure 2.19. Genetic distance between two asexually reproducing individuals, counted by comparing their genomes bit-to-bit. In this example, we count 5 different bits along the genome length of 16, the genetic distance reads $d = 5/16$.

the researcher the possibility of modelling different problems, according to what she/he judges the important features at hands. Independent of any particular interpretation, the important feature is the discrete form in which genetic information is stored and used, as we learned from Mendel, whose consequence is the possibility of coding this information along bit-strings.

For a population of sexually reproducing, diploid individuals, each genome is represented by two homologous, parallel bit-strings A and B. The genetic space is the set of all possible such pairs. In order to define the genetic distance separating two individuals, one can perform the comparison in two ways: (1) chromosomes A × A and B × B; or (2) A × B and B × A, Figure 2.20. Perhaps the better choice is to take the genetic distance as the smallest among the values obtained within these two ways.

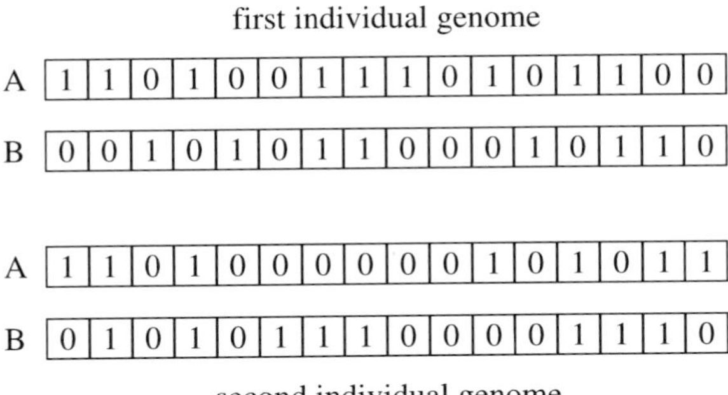

Figure 2.20. Genetic distance between two sexually reproducing individuals. By comparing chromosomes A × A and B × B, one finds 6 and 7 different bits, a total of 13. The alternative comparison would be A × B and B × A, resulting in 5 and 12, a total of 17 different bits. One chooses the smallest result, 13 different bits along a genome of 32, thus $d = 13/32$.

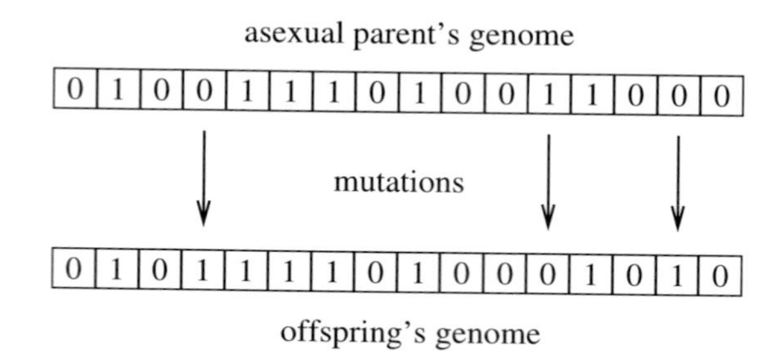

Figure 2.21. Asexual reproduction. Mutations on the offspring genome as compared to the parent's.

Thanks to Mendel's idea, we can now translate biological evolution to computer language, and invent models. Each individual of a population is characterised by a long bit-string, its genome. Sexually reproducing, diploid individuals carry two parallel bit-strings. They are born, live, reproduce and die according to some rules where chance and necessity play crucial roles (Monod, 1973). During reproduction, chance appears as mutations, for instance a bit randomly chosen along the parent's genome which is flipped on the offspring's genome, Figure 2.21.

Also, for sexual reproduction, after crossing and recombination were performed on one parent's genome, chance decides which among the two possible gametes will be passed on to the offspring, Figure 2.22. The same process repeated on the other parent's genome provides the second gamete which completes the offspring diploid genome. During an individual's life, chance can

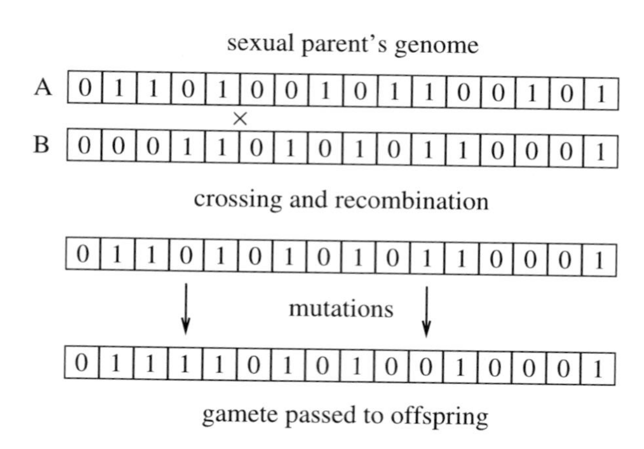

Figure 2.22. Sexual reproduction. In this example, the genome of one parent is crossed at the position marked by $\times$, and the left part of chromosome A is recombined with the right part of B to form one gamete. Finally, mutations are set.

also decide part of its destiny, for instance which contingencies would influence it, when and where this occurs, which other individuals would it meet, etc. Chance is introduced through random number generators, the equivalent to coin-toss.

Necessity corresponds to Darwin's natural selection, the phenotypic features of an individual, at least in part determined by its genome, define how it deals with the contingencies occurred during life, genetic diseases, resistance to other diseases, capacity to escape from predators, etc. One can invent such rules and program them in some computer language (modern Brazilian researchers in C, other old-fashioned German Herr Professors in Fortran). For each individual, the phenotype is defined as a function of the genome. It is used to determine the death probability and fertility of that individual. Also the environment influence should be programmed. Many other features can be included as ageing, sexual selection, maternal care, etc. Further bit-strings could be included in order to represent particular important phenotypic traits, as exemplified in Chapter 4 for speciation. These further bit-strings can also represent non-genetic characteristics which nevertheless play important roles in evolution, an example is the learning of cultural traits introduced by Ticona and de Oliveira (2004), as follows. A "cultural" bit-string is assigned to each newborn, the nth bit corresponding to the future nth birthday of that individual. Initially, all bits are set to zero. At each new year, the individual can learn something which helps its survival, and then the bit corresponding to its current age is set to one. If it misses the opportunity to learn, the corresponding bit remains zero. Then, the distribution of 1-bits accumulated so far is one of the ingredients used in order to determine its death probability within the next year. In the computer program, the manipulation of bitwise fast operations (de Oliveira (1991), see also the first section of Appendix) is important, most times decisive concerning the feasibility of the computer simulation within the available time and memory facilities.

The rest of this book describes some of such computer models, invented not only for the study of biological evolution, but also for other related systems which evolve in time. In general, the same program with the same parameters runs many times, starting from different initial populations and/or following different contingencies occurred along the individuals' lives. From these many runs, one can determine the proper statistical averages and fluctuations, at the end. Some advantages of the simulational approach, in what concerns evolutionary systems, are discussed in de Oliveira (2002). This computer strategy is an important tool for these studies for at least two reasons. First, as already commented before, because we have not a Darwin equation to solve. Second, because this agent-based strategy does not neglect fluctuations at all, therefore avoiding the already quoted problems introduced by mean-field alternative approaches, Section 2.6. These problems are present in virtually any formulation based on differential equations, for which some kind of population-average is always hidden behind the formulas, most of them unnoticed.

Let's give an illustrative example of such a mean-field approximation. Instead of keeping the genetic information of all alive individuals, by storing every genome, it is much cheaper to record only the current frequency with which each possible gene/allele appears among the whole population. Instead of storing one bit-string per alive individual, one records just a single array displaying, for each gene/allele, the number of individuals sharing it, i.e., the current genetic distribution counting how often each gene/allele is spread over the whole population. In order to describe the whole genetic dynamic evolution, one simply updates this array, generation after generation, according to some model rules. This is indeed a very useful approach, in reality very often applied to evolutionary problems (see, for instance, Redfield (1994), Dieckman and Doebeli (1999), Kondrashov and Kondrashov (1999)). However, it implies a crucial assumption: the genetic distribution of the next generation should be determined exclusively from the knowledge of the current genetic distribution. In other words, this approach corresponds to assume a Markovian dynamics for the genetic distribution itself. Is this assumption plausible? Yes, **if** the corresponding dynamics is either regular or chaotic, as discussed in Section 2.4. On the other hand, this assumption **fails** for critical dynamics, as also discussed in Section 2.4. As a matter of fact, the latter is just the kind of dynamics which describes biological evolution. Thus, this approach is at least dangerous as applied to evolutionary problems. In particular, any kind of founders effect could not be described by the genetic frequency single-array approach, which completely rules out Ghengis-Khan and D. Pedro I (Section 2.4).

Why do we classify this approach as a mean-field approximation? We can interpret it as follows. In reality, the genetic information is stored on **all** genomes of **all** currently alive individuals, a bit-string for each. By contracting this information into a single array storing the frequency of each gene/allele among the whole population, one is replacing the set of all individuals by a single "average individual", exactly the same procedure we have exemplified in Section 2.6 for the simple Ising model, neglecting fluctuations. Indeed, the consequences should be the same, for instance the wrong curvature of Figure 2.15 as compared with the correct, non-mean-field Figure 2.14, and many other false conclusions one could reach. As already commented before, due to its simplicity, the mean-field approach can be very useful in order to shed light into some cumbersome issues, sometimes helping a qualitative understanding, but its results are not trustable at all (on the other hand, also simulational computer programs contain errors, particularly those written in Fortran by some German Herr Professors). Special care should be taken with possible mean-field false-positive diagnosis in favour of the existence of a phase transition, a speciation process, a bifurcation and alike.

2.9. A simple evolutionary model

Let's consider the evolution of individuals which die and breed at the same rate, keeping nearly constant the population which fluctuates around $P_0 = 1000$ individuals. First, each of them produces one offspring, with the help of a sexual partner, doubling the population for a while. Then, half of the population is killed, on average, according to a selection rule described later, restoring the original number P_0.

Let's first describe the breeding process. Sexual reproduction is adopted with diploid individuals. Each genome is a pair of bit-strings with $L = 1024$ bits each. In order to breed, both bit-strings of individual M (the mother) are cut at the same random position, crossed and recombined (see Figure 2.22, last section). Two gametes are then formed, two new bit-strings also with $L = 1024$ bits each. One of them, randomly chosen, will be passed on to the offspring, after mutations are set as follows. Just one random bit of this gamete is flipped from 0 to 1 or vice-versa. Besides this gamete, the offspring also inherits the mother's family name, without mutations. Another individual F (the father) is randomly chosen, and the same process of crossing, recombination and mutation is performed in order to produce the second gamete for the quoted offspring, which is now complete. The whole procedure is repeated for a new individual, the new mother, and so on, generating as many offspring as parents. These offspring are then included in the population.

In order to save computer time and memory, we do not divide the population into two genders, neither males nor females separately. Every individual breeds once as mother, but can also be chosen as father, the partner of another mother.

Now comes the death step, where selection acts. Sequentially following the two parallel bit-strings representing individual i, we count the number N_i of homozygous loci where two homologous bits 11 are found. The larger this quantity, the larger the death probability of the individual. Although the particular biological interpretation for phenotypes is not important within such a simple model, one can here interpret 1-bits as representing harmful mutated genes, and suppose the existence of only recessive diseases which reduce the life expectancy. This quantity N_i will be referred to as the genetic load of individual i. Normally, the term "genetic load" is used for the population average of N_i/L, see, for instance, Ridley (2003). We assume a survival probability which exponentially decreases with the genetic load. Let's call x the survival probability for individuals with $N_i = 0$, carrying only 00, 01, or 10 loci along the whole genome. Then, the survival probability for $N_i = 1$ individuals, those with just one 11 pair, is x^2. That for individuals with $N_i = 2$ is x^3, and so on. After breeding, the doubled population P should be reduced back to the former value P_0. First, we need to find the proper value of x by solving the polynomial equation

$$\sum_{i=1}^{P} x^{N_i+1} = P_0$$

where the sum runs over all individuals (including newborns), or alternatively

$$\sum_{N=0}^{L} H(N)x^{N+1} = P_0$$

where $H(N)$ counts the current number of individuals with N homologous 11 bit pairs. Besides the genome of each alive individual, we keep in the computer memory the current histogram $H(N)$.

After the above polynomial equation is numerically solved for x, finding the root just below 1, the death roulette starts once per individual. A random number r is tossed between 0 and 1 and compared with x^{N_i+1}. Individual i survives only if $r < x^{N_i+1}$. After the roulette has already passed through all individuals, the normal population (near) P_0 is restored. One complete simulational step, or generation, is then finished. It is time to record the interesting quantities in accumulating registers, for later averages and statistics. Finally, the evolutionary process restarts towards the next generation.

After many generations, say $T = 10\,000$ (any number a little bit larger than $P_0 = 1000$ suffices), a sort of dynamic equilibrium is reached, the distribution $H(N)$ stabilises with fluctuations. Then, we start to count the time, with $t = 0$. Each individual belonging to this generation founds a new family, and receives a family name. We suppose no family records were stored before. These 1000 different names will be passed on to offspring, every time the individual breeds as mother, following the maternal lineage. Here, the difference concerning genetic inheritance is only the absence of mutations in family names.

But names become extinct. Now and then, some family disappears forever from the population. The number of alive families monotonically decreases. At the end, after many generations, all individuals belong to a single family. A single individual who lived at generation $t = 0$ is ancestor of the whole current population. Let's call it Eve (see also Section 3.5.1). This is not a novelty, it is exactly the predictions of the coalescence theory, see Excoffier (1997) for a friendly and excellent review. It is worth to remark that Eve was not a single female in paradise, just other 999 reproductive "females" lived together at $t = 0$. However, all their 999 lineages were extinct sooner or later during the past history. Would we repeat the history following distinct contingencies, different coin-tosses, Eve would be another individual who also lived at $t = 0$. Even by repeating the result of all coin-tosses but one (Ghengis-Khan, D. Pedro I, etc., Section 2.4) perhaps Eve could be another individual.

The reader could ask whether this behaviour limits the necessary genetic diversity required by evolution. This question will be answered soon. First, we should realise that it is unavoidable, a mathematical constraint based on rigorous theorems. We should recognise also that mathematical constraints are imposed over any God or Nature, who both should obey the rules established by mathematicians as Kolmogorov, in his theorems concerning the coalescence theory. Genetic diversity should be kept in spite of the common ancestor coalescence. How Nature deals with this constraint?

One of the big advantages of a computer program is the possibility to run it again, getting exactly the same result. Even when contingencies play an important role, one can also repeat the same sequence of random numbers. Our program simulating the evolution model described above works like that. At the end of the first run, we only verify which is the common family name of all alive individuals, in order to identify who was Eve among all possible ancestors alive at $t = 0$. Now, for the second run, starting exactly from the same initial population at $t = 0$, we already know a priori who will be Eve.

During this second run, after each new generation t is complete, we measure the genetic distance d_i between each alive individual i and Eve (Figure 2.20, last section), and determine the population average $\langle d \rangle_{\text{all}}$ over all alive individuals. Some of them carry the same family name as Eve, and we also perform a restricted population average $\langle d \rangle_{\text{Eve}}$ only over her descendants. Finally, we define the genetic similarity with Eve

$$g(t) = \frac{\langle d \rangle_{\text{all}} - \langle d \rangle_{\text{Eve}}}{\langle d \rangle_{\text{all}}} \tag{2.16}$$

which starts with $g(0) = 1$, and eventually vanishes after many generations when all alive individuals descend from Eve. We also compute the fraction of alive families

$$f(t) = \frac{\text{number of alive families}}{P_0} \tag{2.17}$$

which also starts with $f(0) = 1$, and decreases towards the final steady value $1/P_0 \ll 1$. Figure 2.23 shows the plots of these two quantities as functions of time, averaged over $A = 1000$ samples, each of them corresponding to a different initial population and to another sequence of random numbers. Eve's genetic trace remains among the population only during a few initial generations, compared with the much larger time required to reach a single family. Indeed, after 64 generations we count 57.8 alive families, on average, still far from the final situation.

Which is the mathematical functional form of $g(t)$? In order to answer this question, Figure 2.24 shows the same plots again, now with vertical logarithmic scale. The straight line at the beginning denotes an exponential decay,

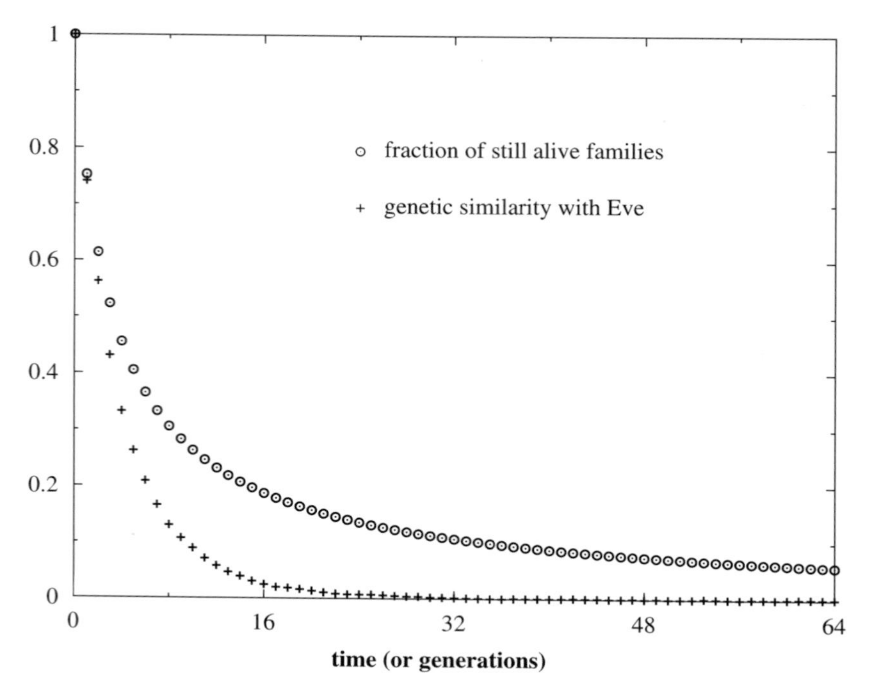

Figure 2.23. Fraction of still alive families, equation (2.17), and genetic similarity with Eve, equation (2.16). Both decay as time goes by, $g(t)$ much faster than $f(t)$.

$g(t) \propto e^{-t/\tau}$ (compare with Figures 2.1 and 2.4). After this initial straight line, the vertical logarithmic scale also allows to appreciate the decaying tail, hidden in the linear plot, Figure 2.23. This tail is dominated by random noise, because we used a finite population P_0 and also the statistics corresponds to a finite number A of samples. The expected order of magnitude for this random noise is $1/\sqrt{AP_0} = 10^{-3}$, in agreement with the plot in Figure 2.24. Would we adopt a larger population and more samples, this noise would decrease as well. However, the important feature is the straight line behaviour observed before the system is dominated by statistical noise. In spite of the coalescence of the whole population into a common ancestor, the genetic features of this single individual are rapidly forgot, according to an exponential rate. No genetic trace of Eve can be found after a finite number of generations. Similarly, suppose some catastrophe kills almost all individuals of a real population, leaving only a few remaining founders for the next generations. After this bottleneck, the genetic diversity is restored according to a fast exponential rate. How fast? Depends on the slope of the quoted straight line, the inverse of which provides the characteristic decaying time τ. It is essentially determined by the mutation rate.

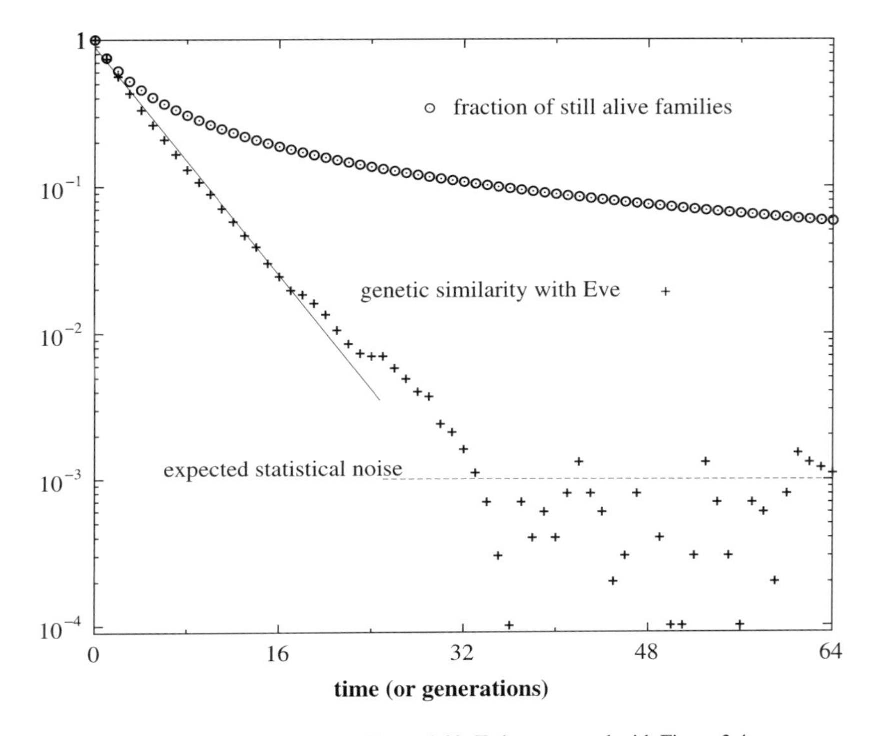

Figure 2.24. The same as Figure 2.23. To be compared with Figure 2.4.

Also, after a bottleneck, the genetic diversity is restored much faster for sexual reproducing species, compared to the asexual case. This behaviour is due to the crossing and recombination processes, which are much more effective than mutations alone to promote diversity, or to increase the entropy in physicists' jargon. Perhaps this is the explanation for the emergence of sexual reproducing species in a world where only asexual species existed at beginning (see, for instance, Section 3.2.2 or Martins (2000)).

Which is the mathematical functional form of $f(t)$? Both Figures 2.23 and 2.24 are aborted at generation $t = 64$, not enough to appreciate the decaying tail of $f(t)$. Figure 2.25 shows again the same data, now with logarithmic scales in both axes (compare with Figure 2.5). The horizontal scale is now much larger, and goes up to $t = 10^4$. The straight line observed for $f(t)$ denotes a power-law behaviour, asymptotically $f(t) \propto t^{-1}$, in complete agreement with the mathematical theorems of the coalescence theory. Indeed, the slope measured in Figure 2.25 gives a critical exponent of -0.98.

Finally, we have also measured the size of each family, i.e., the total number of individuals sharing the same family name during the whole history, since $t = 0$.

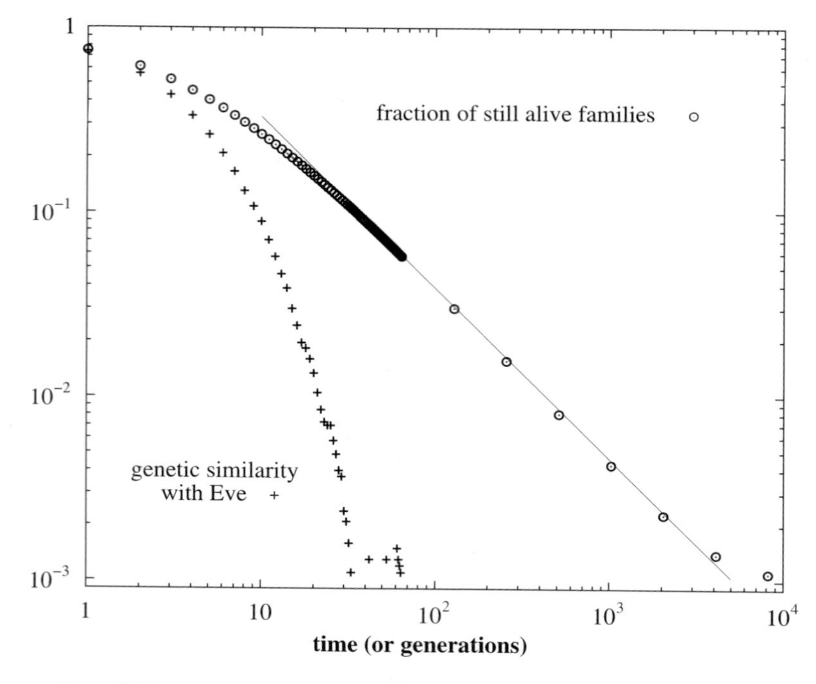

Figure 2.25. The same as Figures 2.23 and 2.24. To be compared with Figure 2.5.

Then, we have grouped these families according to their sizes and counted how many families belong to each group (size). The result is shown in Figure 2.26, another power-law with exponent $-1/2$.

The model treated in this section is an example of how Nature deals with slow power-laws versus fast exponential decays, in order to provide biological diversity. This is a crucial ingredient without which the evolution through natural selection cannot proceed. In this case, the genetic diversity (or entropy) is created according to an exponential fast rate. Its main source is the continuous appearance of mutations, but crossing and recombinations accelerate and enhance very much the process of visiting different parts of the genetic space.

On the other hand, it is equally important to preserve the diversity already obtained during the past history, avoiding the extinction of genetic characteristics which could be useful in the future. Let's show an example of this, concerning recessive diseases like phenulketonuria, falciform (sickle cell) anemia, etc. (see also Jacquard (1978), Cavalli-Sforza (1996)). These genetic diseases are caused by the inheritance of a defective gene p from both parents, instead of the wild, functional allele N. In early days, when no medical treatment was available, homozygous pp individuals died before they had the chance of breeding, therefore

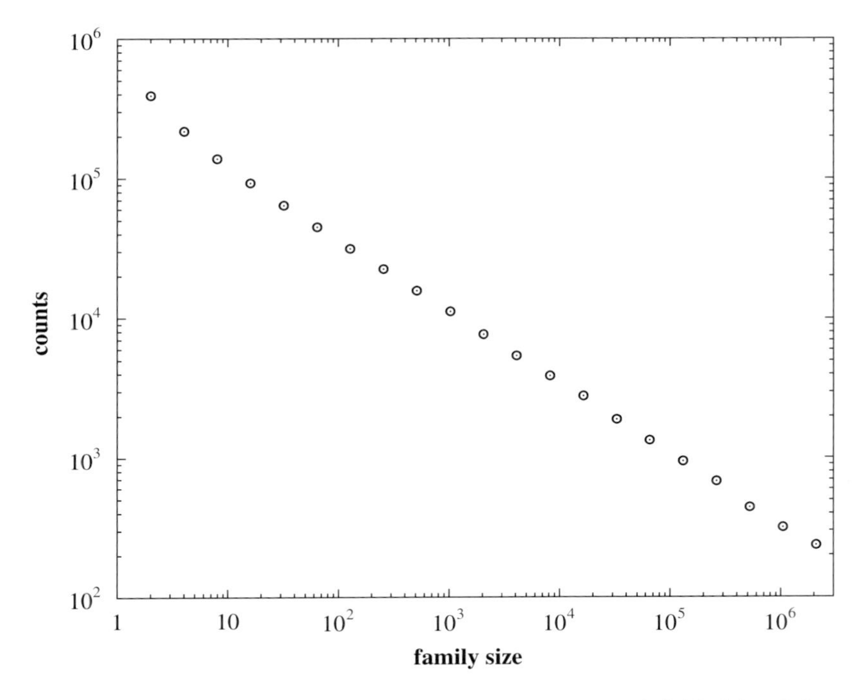

Figure 2.26. The whole set of families is classified according to how many individuals have belonged to each one, during the whole history since $t = 0$.

they were unable to pass the "bad" gene p to the next generations. Heterozygous Np individuals are the only source of this gene for the next generations. The obvious conclusion is the unavoidable extinction of the defective allele p, in the absence of any other evolutionary pressure, i.e., only homozygous NN individuals will be found in some future generation. How fast will be this extinction process? According to Albert Jacquard, the phenulketonuria defective gene occurs nowadays in France with frequency 0.95%, and will decrease to 0.90% within 6 generations, i.e., a century and a half from now, again supposing the absence of any other evolutionary pressure. It will further decrease to 0.50% only after 95 generations, i.e., 20 centuries! Indeed, a very slow process.

Let's return to our genetic space, Section 2.8, where each position along the bit-strings corresponds to a possible gene/allele. The dimension of this space is the number of bits along each individual bit-string, the number of different genes/alleles available among the whole population. The extinction of a gene corresponds to the definitive drop of a whole direction of this space, thus decreasing its dimension. It seems that Nature avoids this kind of gene extinction, postponing it forever, in the case of recessive diseases. The corresponding gene is kept alive among the population, although within a very low frequency.

Why Nature behaves like this? A possible answer is advanced in the quotes used above, when we refer to the "bad" gene. Is it really bad? Even in the possibility of being bad now, will it remain bad in the future, when another still unknown environment will be set? Falciform anemia is widespread in some African regions, the frequency of the corresponding p gene is much larger than in other regions of the world. Coincidentally, in these same regions malaria is endemic. Malaria is not a genetic disease, it is transmitted by the bite of a mosquito (a small flying insect). It was found that heterozygous Np individuals (for the falciform anemia) are more resistant against malaria than homozygous NN individuals. Thus, we cannot simply consider the p gene as bad. It is bad in what concerns falciform anemia, but it is not bad in what concerns malaria. Maybe another new disease, transmitted by the bite of a mosquito or by other means, will appear in France within the next 20 centuries. Maybe heterozygous Np individuals (for phenulketonuria) are more resistant to it than NN individuals. In this case, the French health authorities would profit from the still existing p gene at this time, within a very low (but non zero) frequency of 0.50%.

How Nature deals with the mathematics behind the recessive disease gene extinction? How is it so slow? Let's consider $x_t = x$ the frequency of the defective allele p among the current generation t. Therefore, $1 - x$ is the frequency of the normal allele N. The purpose is to calculate the corresponding frequency $x_{t+1} = \bar{x}$ for the next generation $t+1$. Table 2.1 shows random couplings between NN and Np individuals belonging to the current generation. No pp individuals are considered, because they do not survive enough in order to breed. Np individuals do not suffer any handicap, compared to NN individuals. The same table shows also all their possible offspring and the corresponding relative frequency.

The symbol † indicates individuals which will die soon, due to the disease. Because they are pp, the frequency according to which they appear in generation

Table 2.1

couples	offspring	frequency
$NN + NN$	NN	$(1-x)^4$
$NN + Np$	NN	$2x(1-x)^3$
	Np	$2x(1-x)^3$
$Np + Np$	NN	$x^2(1-x)^2$
	Np	$2x^2(1-x)^2$
	pp	$x^2(1-x)^2$ †

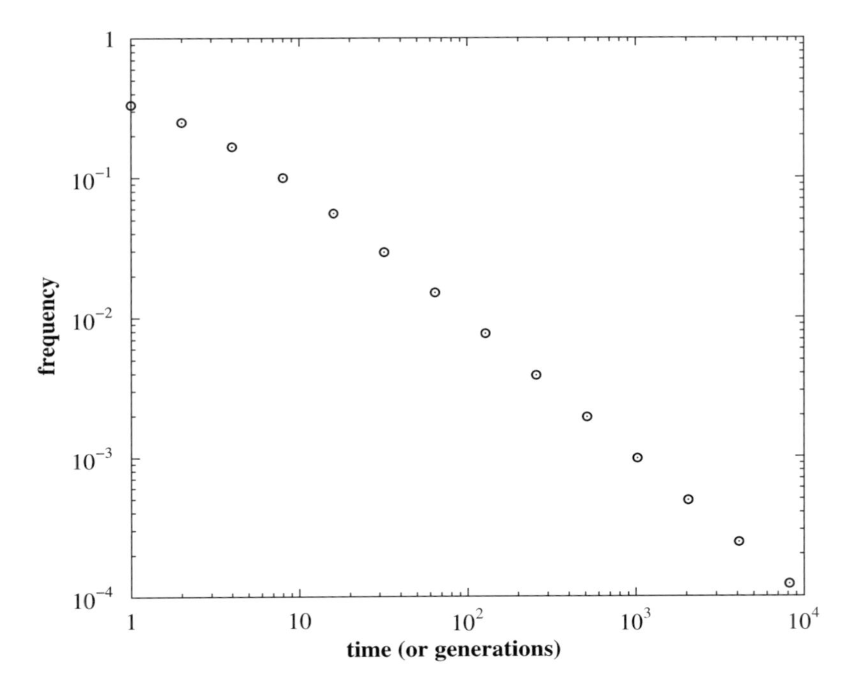

Figure 2.27. Recessive disease gene frequency decaying towards extinction. The data points follow a straight line signifying a slow power-law decay, with its characteristic infinite lifetime. This is the strategy adopted by Nature in order to avoid the unavoidable extinction: to postpone it forever.

$t + 1$ is $\bar{x}^2$, i.e.,

$$\bar{x}^2 = \frac{x^2(1-x)^2}{(1-x)^4 + 4x(1-x)^3 + 4x^2(1-x)^2}.$$

By simplifying the common factors to both the numerator and the denominator and taking the square root, we finally obtain

$$x_{t+1} = \frac{x_t}{1 + x_t}.$$

By successively iterating this last equation, starting for instance with $x_0 = 0.5$, one can determine the whole evolution of the frequency, x_1, x_2, x_3 etc., generation after generation. The result (plotted for clarity only at generations which are integer powers of 2) is shown in Figure 2.27. The straight line observed on this double logarithmic plot is the signature of a power-law.

Contrary to the fast creation of genetic diversity, exemplified with our simulational model, and described by the exponential function $g(t)$ plotted in Figures 2.23, 2.24 and 2.25, now Nature adopts the opposite strategy, i.e., a slow power-law decay in order to preserve the same diversity.

2.10. Another simple model

The title above is not completely fair. Better than this would be "The same model, with some minor modifications, but under completely different questions", a too long sentence. The minor modifications, compared with last section, are not important, and soon the reader will verify them. The new questions are related to the behaviour of the dynamic system for different genome lengths. Specifically, how should the mutation rate vary if one increases the genome length.

In real organisms, mutations are due to "errors" when the DNA is copied in order to produce the offspring. A point mutation, the simplest case, corresponds to one chemical base T, A, C or G being wrongly copied into another among the same set, for instance a T on the parent's genome transformed into a G on the offspring's copy. In our bit-string model, a point mutation is represented by a 1-bit transformed into a 0-bit, or vice versa, during reproduction.

The first, naïve idea about this process is to assume the number m of point mutations being proportional to the genome length L, or, in other words, the same mutation rate per bit, m/L, independent of L. Unfortunately, this simple reasoning does not work, as we shall see.

The first evolutionary bit-string model is that invented by Eigen (1971), after he got a Nobel prize, see also Eigen, McCaskill and Schuster (1989). He was interested in replication of molecules, the origin of primitive life on Earth. Thus, his bit-strings represent these molecules, not the individual genomes of an evolving population. However, the particular interpretation is again not important. Eigen himself invented the name "quasi-species" to denote the various possible bit-strings of his model, indicating that his own interpretation is much more general than the mere coexistence of self replicating molecules on some primordial chemical soup, supposed to have founded life on Earth 4 billion years ago. We are also interested in the replication mechanism of bit-strings and its consequences under Darwinian evolutionary rules, not in what precisely these bit-strings represent.

Within the Eigen quasi-species model, each individual is represented by a single bit-string, which corresponds to our haploid, asexual individuals, Figures 2.19 and 2.21, not our diploid, sexual population model treated in Section 2.9. The selection ingredient, however, is similar: the larger the number of 1-bits, the smaller the fitness of the corresponding individual face to the current environment. Thus, there is a master sequence, all bits set to 0, representing the best possible fitness. Even starting the dynamic evolution with a population where all individuals correspond to that master sequence, other sequences containing 1-bits will appear, due to mutations. Entropy increases, diversity appears. After many generations, some sort of dynamic steady state is reached, where different quasi-species coexist. One important feature we should require for this dynamic steady state is to keep alive the quoted master sequence. The selection mechanism should be strong enough to avoid what is called *error catastrophe*, where all individuals present a lot of 1-bits along the genome, in spite of the selection pressure. In other words, the genetic diversity of the steady population, after many generations, should be distributed *near* the master sequence, all individuals with relatively *few* 1-bits, and a finite fraction of them with only 0-bits.

The preservation of the master sequence is important because the model supposes a *fixed* environment. As discussed in Section 2.8, one cannot study the whole set of all real and potential forms of life at once, an artificial borderline should be adopted separating the population under study (a restricted set of individuals, a single species, a set of species, a whole genus, etc.) from the rest. This "rest" is the environment, which is not fixed at all, but is supposed to evolve within a much slower time scale than the population under study. An eventual change in this environment would change the Eigen master sequence to another neighbouring configuration, say with only one 1-bit, a single gene/allele which was well adapted to the former environment but which is now replaced by a new form better adapted to the slightly modified environment. If, in the steady population, all individuals are *near* the former master sequence, then they are as well *near* the slightly modified master sequence, and evolution can proceed with no further troubles. The dynamic evolution of the model itself does not need to be changed. Instead, one can re-define the genomes of all currently alive individuals by flipping the bit corresponding to the specific position where the environment modification acts. After that, the new master sequence has again all bits set to 0, and the evolution proceeds with the quoted modification unnoticed.

Some models, as the ones presented in this chapter, keep the population constant (the total number of individuals). It is a very useful artifact but rules out the possibility of extinction. Some other models allow the population to fluctuate, and can therefore exhibit the extinction phenomenon, normally called *mutational meltdown*: the genetic capacity of the individuals to face the current environment gradually deteriorates, and the whole population is eventually extinct. Instead, within the constant-population models, extinction is represented by the quoted error catastrophe, as follows. One can interpret the constant population as a representative sample of a much larger population, for which the number of individuals varies. When mutational meltdown occurs within this larger population, the smaller sample taken from it contains only genetically deteriorated individuals, far from the master sequence. Extinction could be the next step. See Lynch and Gabriel (1990), Bagnoli and Bezzi (2000).

Let's introduce now the asexual version of our model. We start with $P_0 = 100\,000$ individuals, the genome of each is a single bit-string with L 0-bits, the

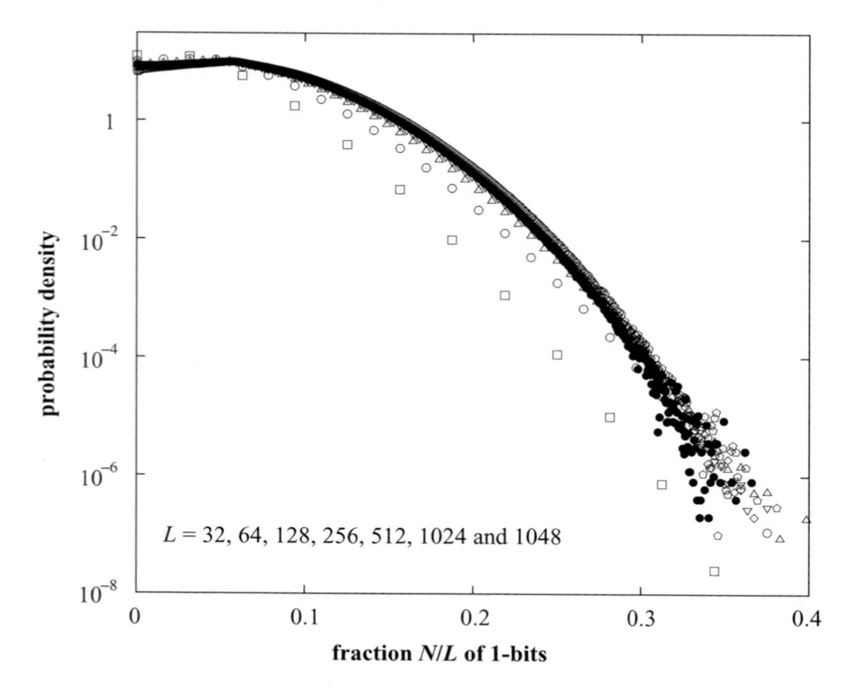

Figure 2.28. Frequency histogram $h(N)$, plotted against the fraction N/L of 1-bits, after many generations of the asexual model. Different symbols correspond to different genome lengths L. The curves collapse into a single one for large enough values of L (the largest length $L = 2048$ is represented by the full circles).

master sequence. We have tested $L = 32$, 64, 128, 256, 512, 1024 and 2048. The number of 1-bits along the genome of every individual i is $N_i = 0$, at beginning. Due to mutations, Figure 2.21, after many generations one will find also individuals with $N_i = 1, 2, 3$, etc., as the reader can see in Figure 2.28 commented later. The selection rule is similar to the one introduced before, Section 2.9: the survival probability for each individual i exponentially decreases with its number N_i of 1-bits. Let's call x the survival probability for individuals with $N_i = 0$. First, in order to obtain its value, we need to solve the polynomial equation

$$\sum_{i=1}^{P_0} x^{N_i+1} = P_0(1-b)$$

where $b = 0.02$ is the death rate, i.e., 2% of the current population die after each generation (and is restored with newborns, as explained soon). Equivalently, one can solve

$$\sum_{N=0}^{L} h(N)x^{N+1} = 1 - b \tag{2.18}$$

where $h(N)$ is the frequency of individuals with just N 1-bits along the genome.

At the beginning of each new time step (a new generation), after solving one of the above equations in order to find the precise value for x, the death roulette scans all the population. Individual i survives with probability x^{N_i+1}, and is kept in the population in this case. After the whole process, only 98% of the former population remains.

Let's open a parenthesis, returning to the equivalence between *error catastrophe*, the run-away from the master sequence, which may occur for a constant population, and *mutational meltdown*, where a fluctuating population becomes extinct due to genetic deterioration. Depending on the mutation rate, this run-away may occur during the evolution of a given population when the frequency $h(N)$, initially something like Figure 2.28, leaves the origin $N = 0$ and becomes sharply distributed inside a narrow range around a certain average mutation load $\overline{N}$. The larger the genome length, the larger this value $\overline{N}$. In this case, the solution x for equation (2.18) approaches the maximum possible value $x = 1$, meaning that all individuals survive with probability 1. Selection no longer holds, no deaths, no births, no evolution at all, hence the denomination *error catastrophe*. The degree of mutations was chosen too strong, and could not be controlled by the selection mechanism. The simplest interpretation is that this particular population would become extinct, the computer simulation does not need to proceed, and can be aborted.

However, a real extinction could be easily obtained by imposing an upper limit to the value of x, as follows. The limit $x_{\max}$ should be slightly smaller than 1 but larger than $1 - b$, for instance $x_{\max} = 0.999$, in our case. This procedure will not change anything while the master sequence $N = 0$ and its neighbouring forms $N = 1, 2, 3$, etc. are still present in the population, because in this case the root x obtained from equation (2.18) satisfies the condition $x < x_{\max}$. However, as soon as the run-away occurs, this condition fails and the killing process is performed according to probabilities $x_{\max}^{N+1}$ instead of x^{N+1}. Soon, all the population will be killed, showing the equivalence between the concepts of *error catastrophe* and *mutational meltdown*. This shows also the adequacy of keeping a constant number of individuals, in order to model a real population, provided the frequency $h(N)$ does not run-away from the master sequence. Closed parenthesis.

Let's return to our constant population model, after deaths were already implemented. Now, the missing 2% will be restored, i.e., just $(1 - b)P_0$ newborns will be included. Each one is the offspring of a randomly chosen individual, among the 98% which remain. The parent's genome is copied, and random mutations are set according to a rate of m/L, where the number m is fixed since the beginning, not necessarily an integer value. First, we toss a random number M of mutations to be performed in between 0 and $2m$. Then, we perform int(M) random mutations, i.e., we flip int(M) bits randomly chosen along the genome, where the symbol int(...) means the integer part of the argument. Then, with probability frac(M), where the symbol frac(...) means the fractional part of the argument, we perform a last mutation.

The steady state histogram $h(N)$, after $T = 1\,000\,000$ time steps is shown in Figure 2.28, averaged over the last 100 000 generations, for each genome length L. Here, we adopted $m = L/32$, i.e., a mutation rate of 1/32 per bit. The observed collapse of all curves into a single one, for large enough values of L, is the indication that both actors, mutation and selection, play their roles. Neither of them dominates the other, both together define the final destiny of the population, in an equilibrated dispute where mutations tend to increase the diversity while selection breaks the eventual explosive error catastrophe. Physicists would prefer to say that mutations create entropy, and selection minimises the free energy.

In order to better understand this competition between mutations and selection, let's resort to the extreme cases. First, imagine we have a too strong selection, overwhelming the role of mutations. In this case, the histogram would be a single point at $N = 0$, all individuals sharing the same master sequence. Of course this uniform scenario, without diversity, does not correspond to any evolutionary process. On the other hand, if mutations dominate the process, with no selection, the curve would be a narrow distribution around $N/L = 0.5$. In this case, the genomes would represent only a random noise, with 0 and 1-bits randomly distributed. This is just the maximum entropy situation characteristic of a chaotic system as described in Section 2.2. Again, this scenario would not correspond to any kind of evolution, it would correspond to the extreme case of the already quoted error catastrophe.

Furthermore, Figure 2.28 shows the class of $N = 0$ individuals (the master sequence) surviving, not extinct. Error catastrophe (or mutational meltdown) is avoided. This asexual version model is the basis for a more complete study on lineage branching, the speciation process for asexual reproduction, Section 4.3.2.

Here, we are interested in the sexual version, with a diploid population, as follows. For haploids, the diversity (or entropy) created by mutations alone was controlled by selection, by keeping the same mutation rate m/L independent of the genome length L. What will occur for diploid populations with sexual reproduction, for which crossing and recombination, Figure 2.22, are a further source of diversity besides mutations?

The sexual version of our model starts with $P_0 = 10\,000$ diploid individuals, each of them with two parallel bit-strings containing only 0-bits at beginning. After each new generation, they reproduce with crossing and recombination plus mutations, Figure 2.22, producing also individuals with 1-bits in various configurations along the parallel bit-strings. At each new time step, we start by the death roulette, where the genetic load N_i counts the total number of loci containing at least one 1-bit along the genome of individual i (both dominant and recessive genetic diseases are counted). The same polynomial equation, i.e., equation (2.18), is solved in order to obtain the survival probability x for $N = 0$ individuals, and then each individual i is killed with probability $1 - x^{N_i+1}$. The global death rate is $b = 0.02$. The initial population P_0 is then restored by including newborns. They are offspring of random couplings tossed among the 98% survivors.

We have adopted different values for the genome length, $L = 32, 64, 128 \ldots$ up to 16 384, and the evolution was followed up to $T = 10\,000\,000$ generations (for $L \leqslant 1024$) or $T = 1\,000\,000$ (for $L \geqslant 2048$). The number m of mutations is fixed for each run, and our purpose was to study the behaviour of the steady state population as a function of m. The first observed behaviour is the inexorable run-away from the master sequence, i.e., the error catastrophe for large enough genome lengths, when we kept the same mutation rate per bit, m/L, for different values of L. Contrary to the haploid, asexual case, now this quantity m/L cannot be considered intensive, the same constant for any genome length L. We have tested also other sub-intensive scalings, namely $m \propto L^e$, for diverse values of the exponent $e < 1$, but the run-away from the master sequence always appears. Only with $e = 0$, i.e., by keeping the same *absolute* number m of mutations independent of the genome length L, we succeed in avoiding the error catastrophe. In other words, sex with crossing and recombination turns the number m of mutations into an intensive quantity, when related to the genome length L, i.e., it should not vary when L increases.

This surprising behaviour, if extrapolated to real life, can be interpreted as follows. First, a larger genome length presents the advantage of allowing the storage of more genetic information. However, there is a price to pay: the chemical machinery responsible for the DNA copying process during reproduction should be improved in order to keep the same number of “errors”, independent of the genome length. In this case, at least part of the further information capacity should be used for this, to store the larger genetic information required by the more sophisticated chemical machinery. Why we, humans, have 23 pairs of chromosomes, not just a single, long one? Maybe the answer to this question keeps some relationship with the intensivity exhibited by our simple model.

Returning to it, for each L and m, we have measured the final distribution of N, among the whole population, averaged over the last 100 000 generation. Figure 2.29 shows the mean genetic load $\langle N \rangle$ (divided by L) obtained from these steady population distributions as a function of m, for $L \leqslant 1024$. It shows a clear phase transition, see Sections 2.5 and 2.7, in particular Figure 2.10. Due to our laziness, we will not present the plot equivalent to the asexual Figure 2.28, with the intensive variable N/L along the horizontal axis. Instead, we describe it in words.

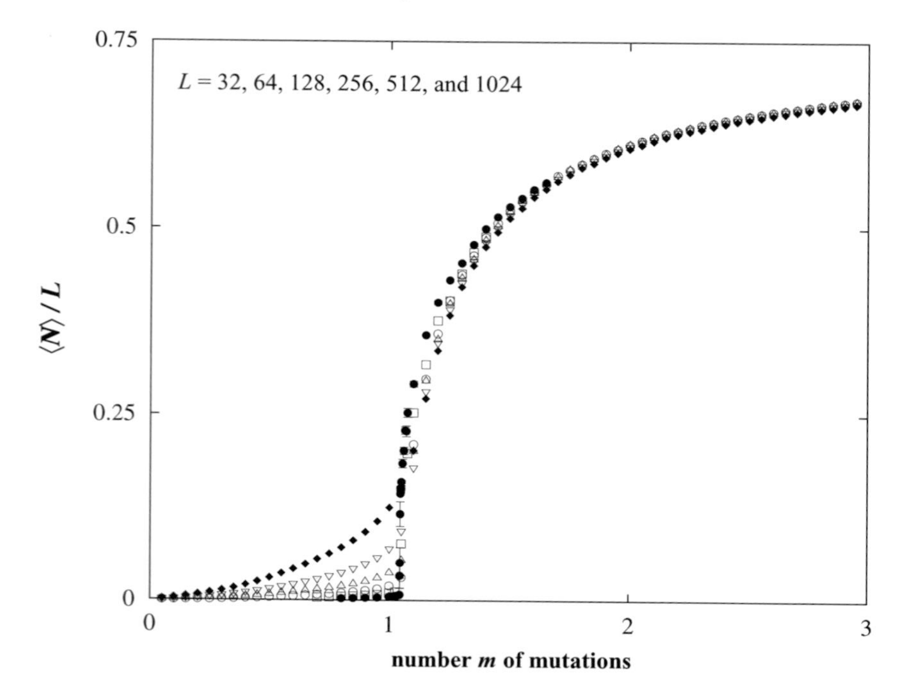

Figure 2.29. Phase transition observed within the sexual version model, measured after the steady population is reached. The *absolute* number m of mutations appears along the horizontal axis, not the mutation rate per bit. For m smaller than a certain critical value (here $m_c \approx 1.043$) all individuals keep their genomes near the master sequence with genetic load $N = 0$. On the right-hand side of m_c, however, one observes the error catastrophe. For each fixed m on that side, the population distribution runs away from the master sequence, all genetic loads $N \neq 0$ remain close to the average $\langle N \rangle$ which can be read on the vertical axis. The smallest and largest genome lengths $L = 32$ and $L = 1024$ are represented by full diamonds and circles respectively.

First of all, for the sexual case these distributions do not collapse any more into a single curve for different genome lengths, as they do for the asexual case, Figure 2.28. Let's take a fixed m below the critical point m_c, Figure 2.29. By increasing the genome length L, the steady population distribution (similar to Figure 2.28, but not collapsed) becomes more and more narrow, compressed near the leftmost point $N/L = 0$. Its fate is to converge into the uninteresting distribution where only the master sequence survives (within fluctuations proportional to $1/L$, as we shall see). On the other side, for a fixed m above m_c, the steady population distribution runs away from the master sequence, and becomes a bell shaped curve centred around the non-vanishing value of $\langle N \rangle / L$ displayed in Figure 2.29 for the corresponding value of m. The larger the genome length L, the narrower this bell shaped curve as a function of N/L.

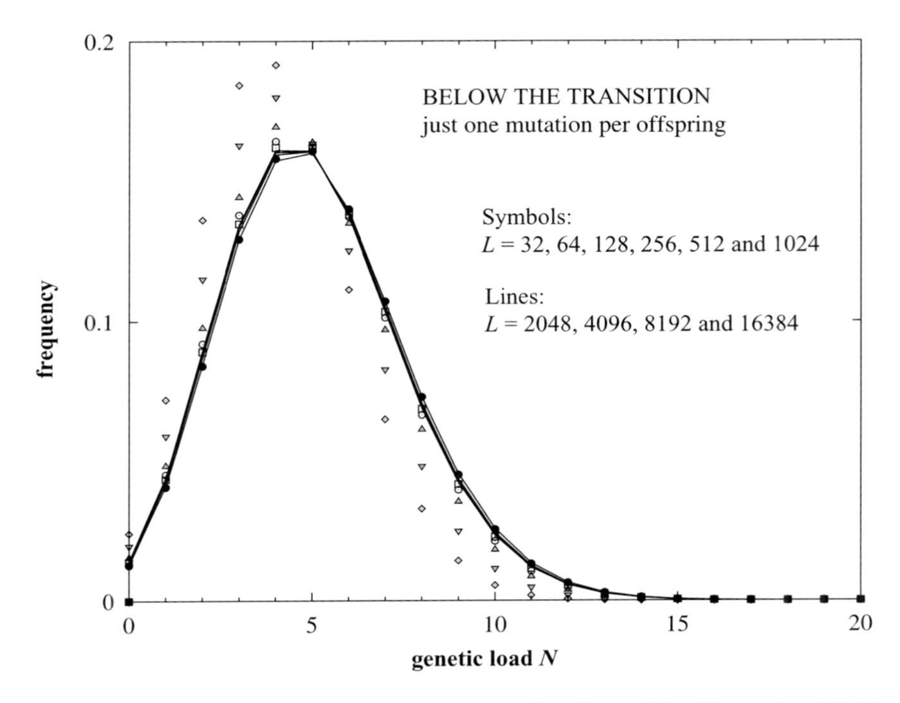

Figure 2.30. Steady population distributions for the sexual version model, with $m = 1$, a little bit below the phase transition, Figure 2.29.

The procedure of plotting the steady population distributions against the intensive variable N/L is no longer good for the present sexual case. Instead, it is better to plot the distributions against the extensive genetic load N itself. One of these plots is shown in Figure 2.30, corresponding to a value of m just below the critical point m_c. Now, the curves collapse again into a single one, for large enough genome lengths L, as expected. This behaviour shows once more the effect of the crossing and recombination sexual process into the dependence between genetic load and mutation rate: it was an extensive dependence within asexual reproduction, but becomes intensive with the introduction of sex. Because Figure 2.30 was taken very near the critical situation, the collapsed distribution is also very near the run-away process, as one can verify by looking to the small (but non-vanishing) presence of the master sequence.

On the other hand, Figure 2.31 shows the steady population distributions for m just above the phase transition. Now, one can clearly see the run-away from the master sequence occurring for the largest size $L = 1024$ (note the big interval on the horizontal axis).

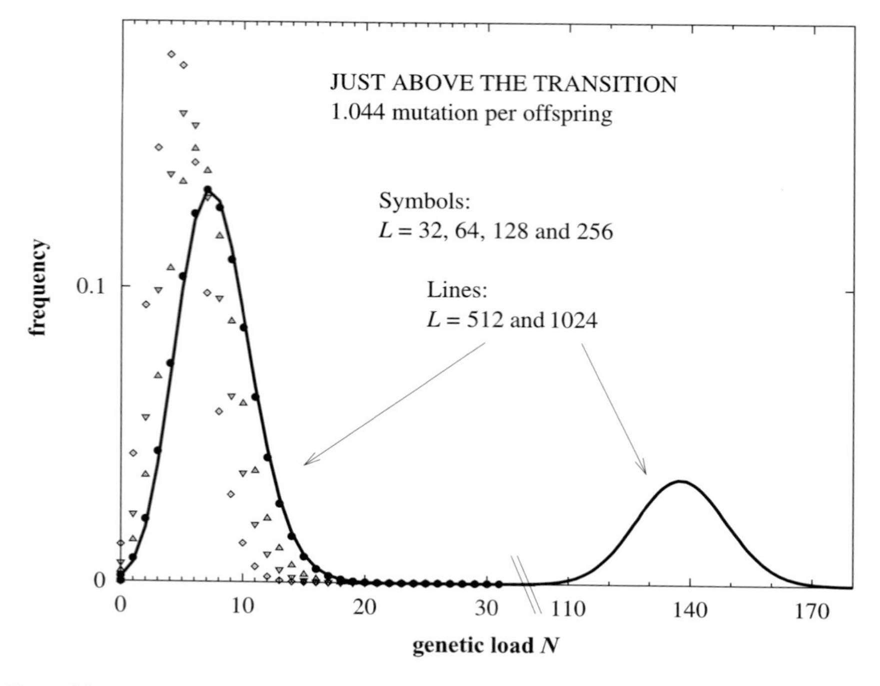

Figure 2.31. Steady population distributions for the sexual version model, with $m = 1.044$, just above the phase transition, Figure 2.29. The largest genome lengths $L = 512$ (left) and $L = 1024$ (right) are displayed by continuous lines.

For a low enough degree of mutations per offspring, $m < m_c$, all individuals become similar to the master sequence. In average, everybody differs from it by a limited genetic load $\langle N \rangle$ which *does not increase* if one increases the genome length L, Figure 2.30. Under an evolutionary point of view, this situation is not interesting, because it lacks the required diversity. Thus, Nature's task is to tune a larger value for m, in order to enhance diversity. However, it cannot be so large, because the error catastrophe waits on the right-hand side of m_c, Figure 2.29, as one can verify by the run-away from the master sequence observed in Figure 2.31. Nature should tune m near the critical point m_c. We have run other simulations with other values of the parameters, and observed that m_c indeed changes a little bit. This means that m should be adapted to the environment conditions, for instance by tuning the proper birth (or death) rate b in order to keep the population always near the critical situation.

We have also implemented a mean field version of this same model, see Section 2.6 and the last two paragraphs of Section 2.8. It is defined hereafter, according to the same general approach widespread among biologists (see, for instance,

Redfield (1994), Dieckman and Doebeli (1999), Kondrashov and Kondrashov (1999)). Instead of keeping in the computer memory the genomes of all alive individuals, we keep only a single L-entry array with the frequency distribution $h(N)$. It is the equivalent to Figures 2.30 or 2.31. Because this distribution will vary from generation to generation, we will denote it by the symbol $h_t(N)$, where t refers to the generation. Following what we have done with the simulations, we also start the dynamic evolution of this mean field approach by taking all genomes identical to the master sequence, i.e., $h_0(0) = 1$ and $h_0(N \neq 0) = 0$.

Each time step starts with deaths. They are implemented by first solving equation (2.18) for x, and then multiplying all entries of the array $h_t(N)$ by $x^{N+1}/(1-b)$, where the denominator $1-b$ is necessary in order to restore the normalisation condition $\sum_N h_t(N) = 1$.

Then, we construct an auxiliary array $h_0(N)$ for the offspring which later will be included into the population. Near the transition point, the genetic loads N are small if compared with L, supposed a large length. Thus, we can neglect the few instances of homozygous 11 loci. As a consequence, the genetic load N is just the double of the total number of 1-bits along both bit-strings of a diploid individual. As a further consequence, we can consider the genetic load N_0 of the offspring, after crossing and recombination performed on the parents' genomes, as the simple average $(N_m + N_f)/2$ of the parents' genetic loads N_m and N_f. Thus, apart from mutations, the offspring frequency distribution is obtained by the convolution

$$h_0(N_0) = \sum_{N=0}^{L} h_t(N) h_t(2N_0 - N)$$

As a technical point, N is always an integer number in the interval $[0, L]$. However, $N_0 = (N_m + N_f)/2$ can be a half-integer inside the same interval. In order to restore the restriction to only integer numbers, all $h_0(N_0)$ obtained by the above convolution for half-integer values of N_0 are re-distributed half-to-half between the two neighbouring integers $N_0 \pm 1/2$. Therefore, after this process, $h_0(N_0)$ is another frequency distribution defined at the *integer* genetic loads $N_0 = 0, 1, 2, \ldots, L$.

Now, we will introduce one mutation (a single one, for a while) on the offspring distribution, as follows. A fraction of $(1 - N_0/2L)^2$ is subtracted from $h_0(N_0)$, and then transferred to $h_0(N_0+1)$. Analogously, a fraction of $(1-N_0/2L)N_0/2L$ is subtracted from the same $h_0(N_0)$ and transferred to $h_0(N_0 - 1)$. We do that for all values of N_0 in parallel. If the number m of mutations is an integer, then we perform the whole parallel procedure just m times, sequentially. Otherwise, we repeat it int(m) times and then repeat it once more with the above factors multiplied by frac(m).

Finally, the frequency distribution $h_{t+1}(N)$ of the next generation is obtained by mixing the parents with their offspring, i.e.,

$$h_{t+1}(N) = (1 - b)h_t(N) + bh_0(N)$$

Note that no random numbers are needed during the whole process, which is then completely deterministic. The frequency distribution of each generation is completely determined by the previous one, no fluctuations, no contingencies.

The whole procedure (deaths, convolution, mutations and generation mixing) is iterated sequentially, up to equilibrium, i.e., until we obtain a no longer varying frequency distribution $h_t(N)$, after a sufficiently large time t. The result is twofold: (1) if m is less than a critical value m_c, the equilibrium distribution is concentrated near $N = 0$; or (2) if m is larger than m_c, it is equally concentrated near $N = L$. Figure 2.32 shows this behaviour for $L = 1024$. The symbols correspond to data really obtained from the computer, the step line is a guide to the eye.

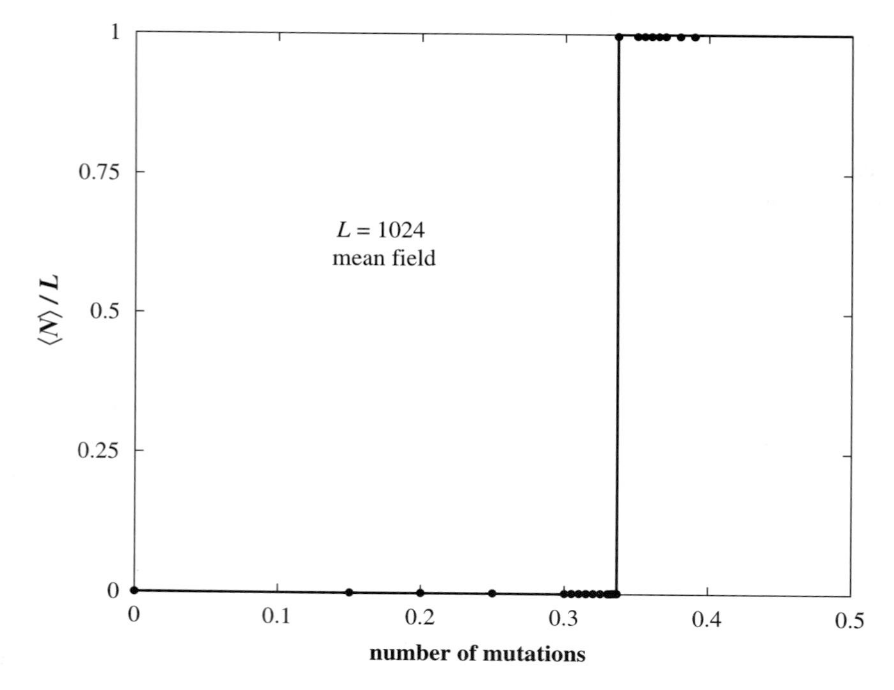

Figure 2.32. Mean field approximation for the sexual version model, genome length $L = 1024$, to be compared with the simulational result Figure 2.29. This is a good example of the danger of using mean field approaches near phase transitions.

We have already commented in Sections 2.6 and 2.8 the usual drawbacks of these mean field approximations. In particular, we have detected the problem's source: no fluctuations at all are taken into account. As we have also commented in Section 2.5, fluctuations are of fundamental importance in defining the critical behaviour near phase transitions like that exhibited in Figure 2.29. Here, the mean field approximation predicts a wrong value for the transition point, $m_c \approx 0.336$ instead of the correct value $m_c \approx 1.043$. Nevertheless, this drawback is not the most important. Worse yet in Figure 2.32 is the absence of the curvature observed in Figure 2.29. The run-away from the master sequence, if one just surpasses the critical number of mutations m_c, is not so drastic as the mean field approach indicates. The wrong jump from $\langle N \rangle = 0$ directly to $\langle N \rangle = L$ would forbid any kind of evolution, due to the complete lack of diversity in both sides of the transition, leaving no space for the action of selection.

2.11. Conclusions

In this chapter, we have introduced a lot of subtle concepts originally used in the study of critical phenomena and phase transitions, a branch of Statistical Physics. In particular, the ubiquitous appearance of power-laws, instead of exponential decays, describing the behaviour of various quantities as functions of both distance and time. We have also shown that this behaviour is not restricted to these physical studies, it appears also in many other systems outside Physics, in particular evolutionary dynamics through natural selection. The power-laws impose a strong difficulty, forbidding to separate a small piece of the system under study from the rest, or a small slice of time. All scales of both time and length should be considered, one can neglect neither the larger distances nor the remote past. The correct approach for these studies are the successive scaling transformations described in Section 2.7, which nevertheless are very difficult to be implemented by analytical mathematical treatment. The alternative is the computer simulation, the main object of this book. Fortunately, the length and time scale-free character of these systems gives to them a remarkable feature: the behaviour of distinct systems, which are very different from each other on the microscopic or short term scale, could be the same. Completely different systems belong to the same universality class, including some very simple computer toy models which nevertheless describe very well the critical behaviour shared with their partners. There are some few universality classes: inside one of them there is a real, very complicated system, but perhaps also a simple computer toy model which allows the researcher to study the critical behaviour of the whole class.

The last two sections were dedicated to two particular very simple evolutionary computer models, I used as introductory examples for the many others my three co-authors will describe in the following chapters.

Chapter 3

Biological Ageing

We are born, learn to walk, to speak, to read and write, and some of us even become professors. Then ageing sets in: we get more fat and wrinkles, become less original, retire and die. Why do we not stay youthful and healthy until we die for whatever reasons? Actually, Pacific Salmon has achieved this aim: It survives for years until it produces children once in life, and then rapidly deteriorates until it dies. Why do we not win football championships at an age of 100 years, and die of "old age" a few months later?

Beauty and scientific originality are difficult to measure, and thus we restrict the discussion of ageing to the most objective and reliable quantity, the mortality. If you become 30 years old in a peaceful rich country, then it is likely that you also reach 35; someone who celebrates the 100th birthday should not rely on reaching the 105th. Thus the probability $\mu(a)\,da$ to die within the next small age interval da, after having reached age a, increases with age for adults. This effect we call ageing and μ the mortality function. Before we go into computer simulations, we review some facts and theories.

3.1. Facts and theories

3.1.1. Facts

The medical progress of the last two centuries has not only increased human life expectancy, Figure 3.1, but also got humans closer to the above ideal of living healthily until a quite sudden death: Figures 3.2–3.4. There we see that child mortality was decreased by a large factor during the 20th century, while the mortality function near 90 years did not change much (in relative terms; note our logarithmic scale for μ). If this trend would continue over a long future (according to Yashin, Begun, Boiko, Ukraintseva and Oeppen (2001) it has already stopped) then the mortality over most of our life would fall to nearly zero, but stay constant near 100 years, as dreamed above. The probability of survival would be close to unity up to about 100 years and then drop towards zero rapidly; thus this trend is also called rectangularisation, or compression of mortality, Figure 3.4.

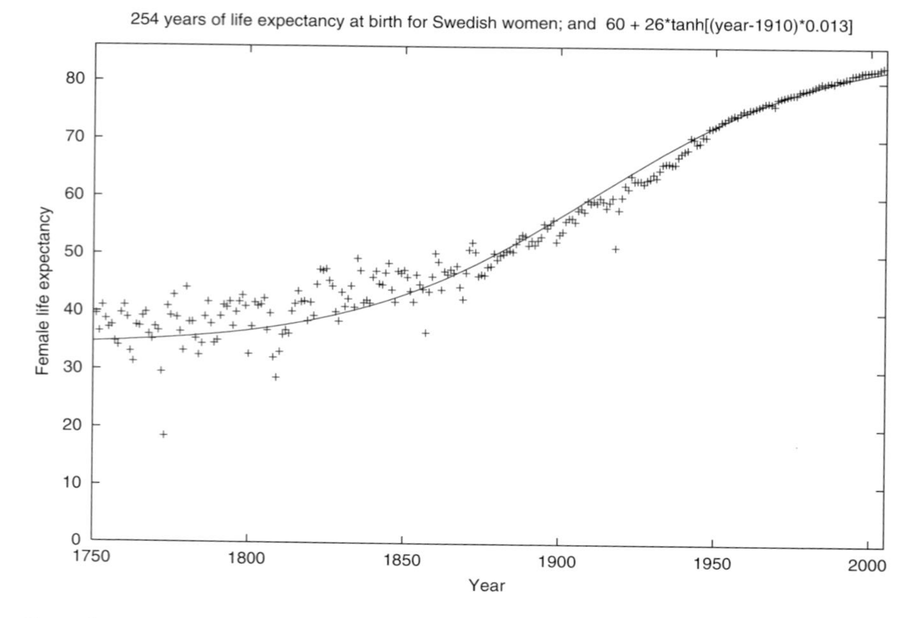

Figure 3.1. Life expectancy of Swedish women; from Wilmoth's mortality data base and the Swedish Statistical Central Bureau: www.scb.se.

Looking at present reality in Figure 3.2, we see a complicated behaviour in childhood, a shoulder near 20 years, and then a transition to a straight line in this semilogarithmic plot, i.e., an exponential increase of the mortality function $\mu(a)$ with age a

$$\mu(a) \propto e^{ba} \tag{3.1}$$

known as the Gompertz law of 1825. (The original paper is terrible to read and thus not cited here.) The straight line in Figure 3.2 indicates this exponential increase. The Gompertz slope b increased during the 20th century, Figure 3.3, and is nearly 0.1 per year, i.e., the probability to die increases each year by about 10 percent of its previous value.

The Gompertz law applies presently to many industrialised countries; Figure 3.5 shows the male mortality function of Sweden, quite similar to Figure 3.2 except that young Swedish men seem to live more reasonably than Germans. There besides the straight line for the Gompertz law, equation (3.1), we also show the Makeham modification, also going back to the 19th century, which adds a small constant to the Gompertz exponential increase. Now ages between 20 and 30 years fit better due to this extra parameter, and Gavrilov and Gavrilova (1991) have emphasised the Makeham correction. However, the mortalities be-

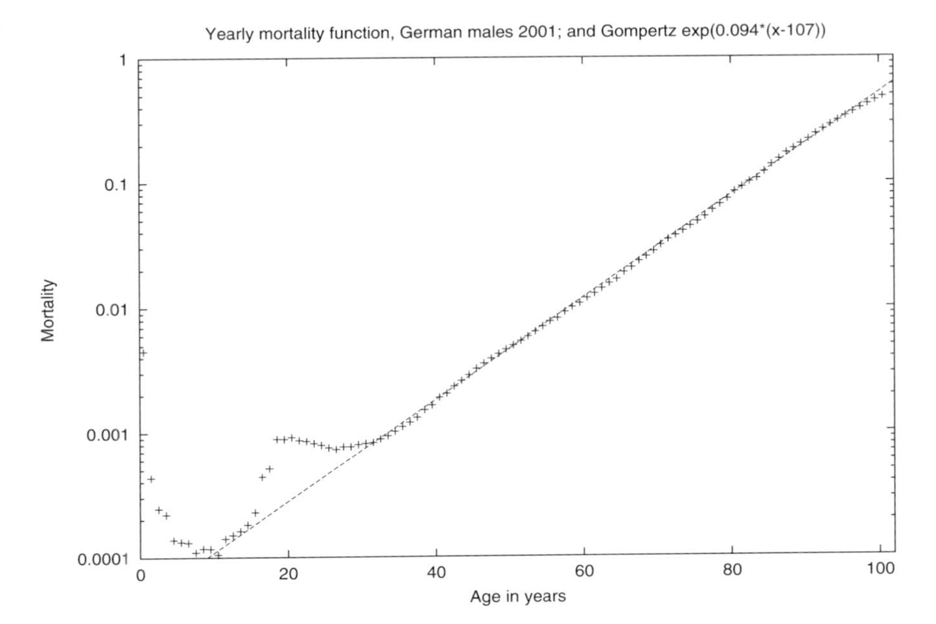

Figure 3.2. Recent mortality functions from www.destatis.de for German men.

low 20 years are far from being described by either the original Gompertz law or the Makeham corrections. (Figure 3.3 suggests that the Makeham correction was more important in the 19th century when it was suggested.) We mostly ignore here these childhood deaths and deal with adult mortalities. The possible downward deviations for the oldest old, about 100 years and above, are also discussed later for social simulations in connection with retirement demography, Section 6.1.3; they are visible in the upper right corner of Figures 3.2 and 3.5. (This later Section 6.1.3 will also complain about lawless women disobeying Gompertz.)

If $S(a)$ is the probability of surviving from birth up to age a, then the mortality function is $\mu = -d \ln S(a)/da$. Usual human life tables are given in yearly intervals, only for babies also weekly and monthly tables for their first year are widespread; with only yearly data we have to approximate $\mu = \ln[S(a)/S(a+1)]$. Except for the oldest old this mortality function (or "force of mortality", Thatcher, Kannisto and Vaupel (1998)) roughly agrees with the mortality $q = [S(a) - S(a+1)]/S(a)$ which is the fraction of people alive at age a and dying within the next year. Of course, $q \leqslant 1$ cannot obey the Gompertz law, since for large a the exponential function goes to infinity, and thus we will not use it here.

Thus far we only talked about humans, but also most (perhaps not all: Finch (1998)) other animals age. To find their intrinsic mortality we have to put them into protected environments like laboratories or zoos; in the wild, they are eaten

Figure 3.3. Older mortality functions for German men, for 1905, 1933 and 1950 (West Germany only for 1950). From the Statistical Yearbooks of Germany.

by predators or die from starvation, thirst, or bad weather. Then also the Gompertz law is fulfilled, except that some flies at old age have a mortality plateau (Carey, Liedo, Orozco and Vaupel, 1992; Curtsinger, Fukui, Townsend and Vaupel, 1992; Vaupel, Carey, Christensen, Johnson, Yashin, Holm, Iachine, Kannisto, Khazaeli, Liedo, Longo, Zeng, Manton and Curtsinger, 1998). Also single-cell yeast ages, if you don't kill it earlier by producing beer or bread. Even some bacteria age in the sense that they produce less offspring after many divisions (Ackermann, Stearns and Jenal, 2003). Mayflies (ephemerals, Carey (2002)) disobey Gompertz and have a mortality function increasing linearly with age; they are thus the counterpart of Pacific Salmon where the deviation from Gompertz is in the opposite direction of rectangularisation.

3.1.2. Theories

In a recent collection of reviews on ageing from the biological, not the computational point of view (Cell, 2005), Kirkwood on page 437 overviews many theories; we list here only some of them and refer to Kirkwood for the historical references, except for the last two theories not mentioned by him. He gives *Hydra* as an ani-

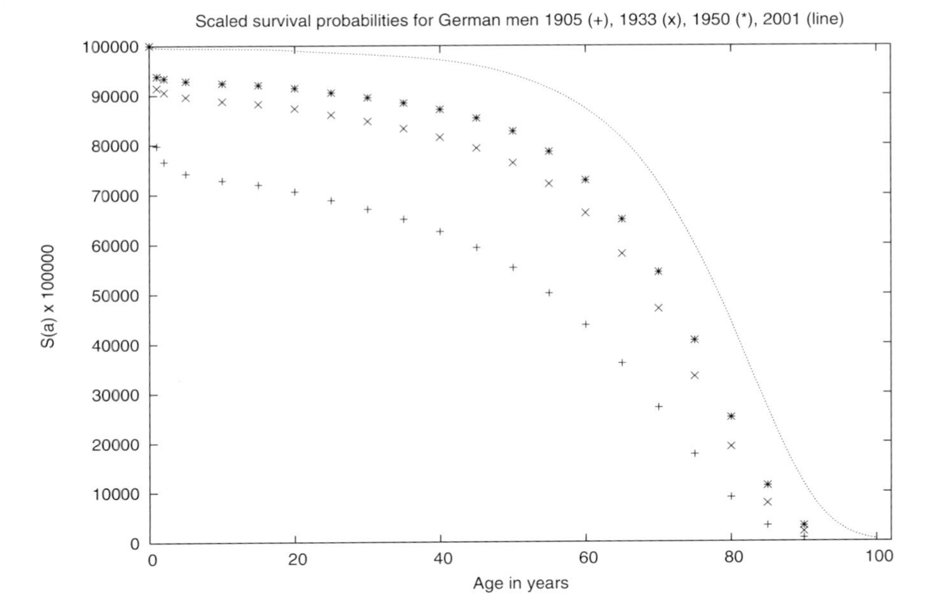

Figure 3.4. Survival probabilities $S(a)$ from birth to age a, scaled by 10^5, for German men in 1905, 1933, 1950 and 2001, as used for Figures 3.2 and 3.3. The curves show the possible approach towards a rectangle: Survival up to about 100 years followed by sudden death.

mal without ageing. A review much closer to what we will present here was given by Cebrat and Łaszkiewicz (2005) from a genetic institute.

At the end of the 19th century Weissmann suggested that we die to make place for our children; this death could be due to external reasons like for most animals in the wild, or due to ageing like for today's humans and for animals in a zoo. We will discuss at the end of Section 3.4 a simple simulation giving some (very late) justification to Weissmann.

More successful is the half-century old mutation accumulation theory of Medawar which is the foundation of most of the simulations. It assumes that ageing is due to inherited life-threatening diseases, each of which starts to act at a certain age. If such a disease kills an individual at young age before any offspring is produced, then this disease is not inherited by anyone and vanishes from the population. A hereditary disease killing us, after we got all the children of our life, was given on to them but is less detrimental for the species as a whole, since it does not prevent reproduction; if we die after finishing this book, the publisher does not have to pay us millions in royalties, our colleagues can use our offices, and the government saves on retirement funds. Williams's antagonistic pleiotropy happens if an inheritable mutation has good effects in the youth and bad effects

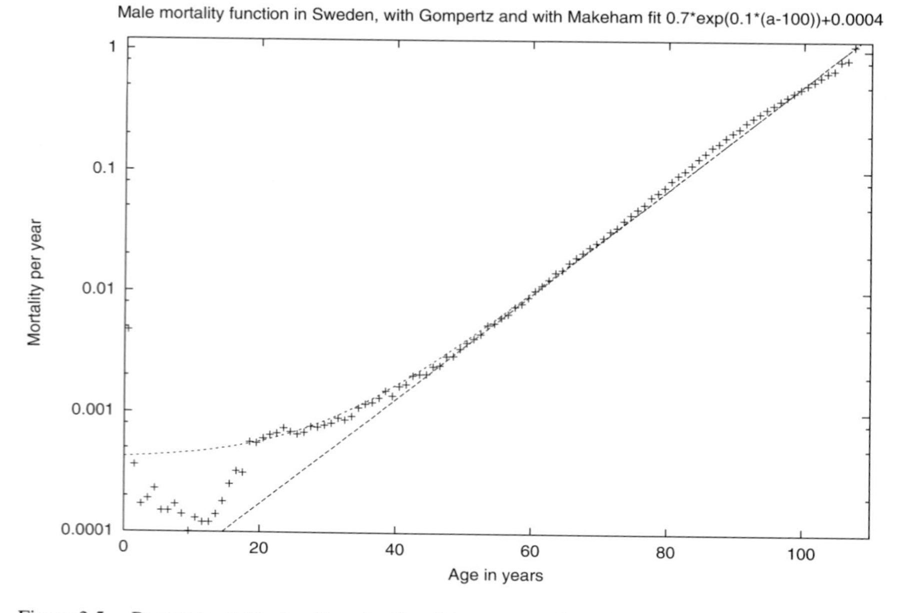

Figure 3.5. Recent mortality functions for Swedish men, together with a Makeham curve. The Gompertz law corresponds to a straight line in this semilogarithmic plot.

at old age (or the reverse); for example, a mutation increasing the amount of calcium in the body is good for children to build bones but bad at old age due to arteriosclerosis.

Kirkwood's disposable soma theory has similar effects: The body has a limited amount of resources to produce offspring and to ensure its own longevity. Some compromise is made by Darwinian evolution which optimises overall survival. Similarly, you may balance your money between spending it all in the youth, or saving it for retirement. One cannot have *everything* in life. Thus both Williams and Kirkwood lead to a trade-off between longevity and youthful strength.

Damage to bones, wings and other body parts may accumulate during life, like for professional athletes. We include in this effect also cancerogeneous substances produced by human pollution. One particular aspect of this damage are new bad mutations created during the life of one individual by oxygen radicals which damage the genome (DNA). Then the affected cells no longer work properly, endangering the life of the whole organism.

Programmed cell death happens during the development of each individual from the original single cell (zygote); otherwise we would be spherical with 2^t cells after t cell divisions. Perhaps there is a similar death gene hidden somewhere which kills us at old age. Recent experiments point in this direction: Genetic modification has increased enormously the life expectancy of nematode worms *Caenorhabditis elegans* (but see Davenport (2004)) and to a lesser extent for some other animals.

Telomeres are parts of the DNA strands at their end, and at each cell division some telomeres are lost if they are not restored later by the enzyme telomerase. For cells in vitro this leads to the Hayflick limit known since nearly half a century (Hayflick, 2003): After a few dozen cell divisions the cells no longer divide.

Most recently, Shklovskii explained ageing as a result of an exponentially rare escape of abnormal cells from immunological response; this lead to the Gompertz law (Shklovskii, 2005).

Finally, also cars and other non-living objects age, similarly to humans (p. 17 in Wachter and Finch, 1997; Vaupel group, 1998). This can be explained by reliability theory (Gavrilov and Gavrilova, 1991) where each important function may be fulfilled by many similar components of the system. First one function, and then the whole system, fails if all the components fulfilling this function have failed.

Not all these competing theories are mutually exclusive: The mutation accumulation theory obviously is based on mutations, which could be produced by oxygen radicals, against which the body has built defences which could be strengthened by suitable genetic modifications. The situation has been compared before by others to the difference between hardware and software in computing: Mutation accumulation theory is the software, and it is realised by mutations from oxygen radicals playing the role of the hardware. Genetic manipulation to increase the defence against oxygen radicals then corresponds to making the program more efficient.

It is not the aim of this book on simulations to determine which is the best theory; perhaps all of them are true and ageing is a multi-causal phenomenon. We discuss now those which have been used for simulations, and that is mainly the mutation accumulation theory (sometimes including antagonistic pleiotropy) and to a far lesser extent its combination with Weissmann's idea as well as some telomere simulations.

Many more older biological references are given in our previous book (de Oliveira, de Oliveira and Stauffer, 1999) and thus not listed here.

3.2. Penna model: asexual

3.2.1. Basic model

The Penna model (Penna, 1995), invented in September 1994 by a computational physicist at age 30, is by far the most widespread way to simulate ageing. It implements the mutation accumulation theory and was the first model to give the

Gompertz law. (With "model" we denote a description of individual elements and their interactions, as is customary in physics; other sciences may denote any mathematical formula, like equation (3.1), as a model.) It uses bit-strings, that means chains of zeroes and ones, as were introduced in biology long time ago (Eigen, 1971). Zero means health, one means a life-threatening inherited disease. This bit-string forms the genome, is inherited, but undergoes mutations at birth. In this Section 3.2 we review the asexual version, in the following Section 3.3 the sexual version.

More precisely, each position of the bit-string of typically 32 bits (also lengths from 8 to 4096 were simulated) represents one time interval in the life of the individual, like a day for flies or several years for humans. We will denote this time interval simply as "year", being accustomed to German Shepherds. A bit set to one means that from that age on until death this inherited disease reduces the health of the individual. If the number of active diseases reaches a threshold T, the individual dies. If it survives and is not younger than the minimum reproduction age R, it produces at every iteration B offspring, each of which inherits the genome apart from M bad mutations each of which randomly selects one bit and sets it to one independent of what its previous value was. A program is given in our appendix, Section 9.2.

If all diseases (bits set to one) are given on to the offspring, and only new diseases (bits switched from zero to one) may be created by mutations, without any back mutations (bits switched from one to zero), why does the population not die out? If the genome gets worse at each generation and never better, mutational meltdown (Lynch and Gabriel, 1990) is unavoidable: The mortalities rise while the birth rates stay constant; thus finally more individuals die than are born, and the species becomes extinct. Traditional theoretical biology assumes a constant population without checking first for mutational meltdown and thus is closer to biblical creationism than to Darwinian evolution. In the Penna model, the explanation for a stable population came from the analytical solution of the salmon model (Penna, de Oliveira and Stauffer, 1995). If a mutation hits a bit which is already mutated to one, then nothing happens. Thus with appreciable probability some offspring escape the mutational meltdown and have the same genome as the mother, not a worse genome. These children keep the species alive. If the mutation rule is modified such that it always searches for a zero bit to mutate it to one, then mutational meltdown happens and the population dies out quickly.

The parameters should be chosen such that the population neither dies out nor reaches an age of 32; typical values are $T = 3,\ R = 8,\ B = 3$. To prevent the population size $N(t)$ from growing to infinity, one applies an additional death probability $N(t)/K$, as in the logistic equation suggested by Verhulst in the 19th century. The denominator K is an input parameter limiting the population size. The equilibrium population then is determined through this K. These Verhulst deaths are due to the lack of food and space if the population size gets closer to

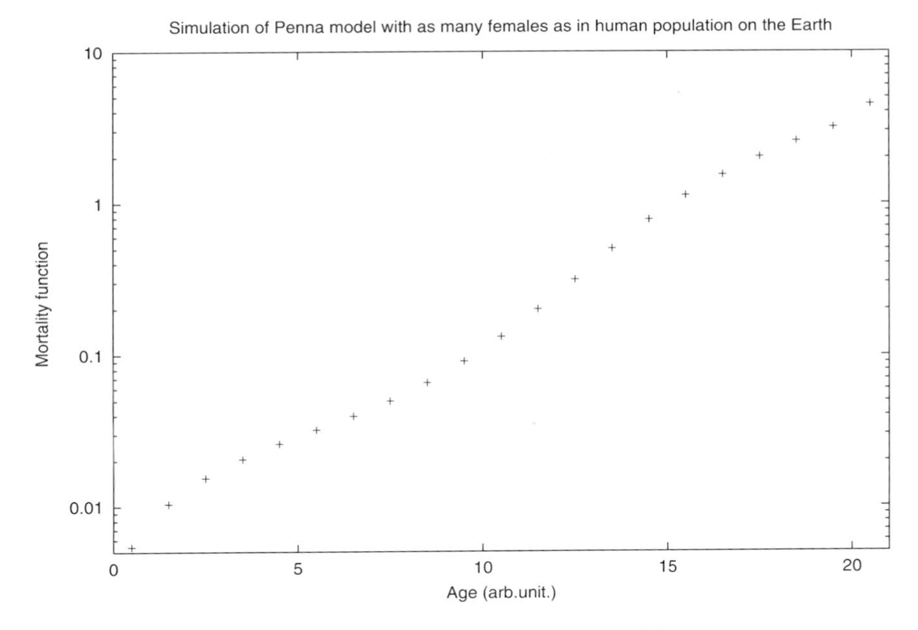

Figure 3.6. Mortality function μ for the standard asexual Penna model. The parameters do not matter much within reasonable limits. From Stauffer (2002d).

the carrying capacity of the environment. (They may be applied either to all ages, which is computationally better, or to the babies only as introduced by Martins and Cebrat (2000), which is biologically more realistic.) At each age, or at birth, the individual is killed if a random number is smaller than $N(t)/K$. About 10^4 iterations are usually needed to get a good equilibrium. The many young people reach their equilibrium population sooner than the few old people; thus watching the total $N(t)$ may give the wrong impression that already after 10^3 iterations an equilibrium is reached.

The semi-logarithmic plot of Figure 3.6 shows that the mortality function roughly obeys the Gompertz law, equation (3.1), which corresponds to a straight line. Deviations exist at young ages below $T = 3$ where the number of active bits cannot yet reach the dangerous threshold T, and at old age where μ diverges giving a maximum lifespan, according to an analytic calculation of de Almeida, de Oliveira and Penna (1998). (We hate to admit that this prediction of a maximum age occurs at ages barely reachable in simulations, a situation quite similar to reality.) The mortality μ_g in this figure describes only the genetic deaths and does not include the Verhulst deaths μ_V which are age-independent in these simulations; since for $T > 1$ they are equal to the total mortality μ_t in the first year, we can eas-

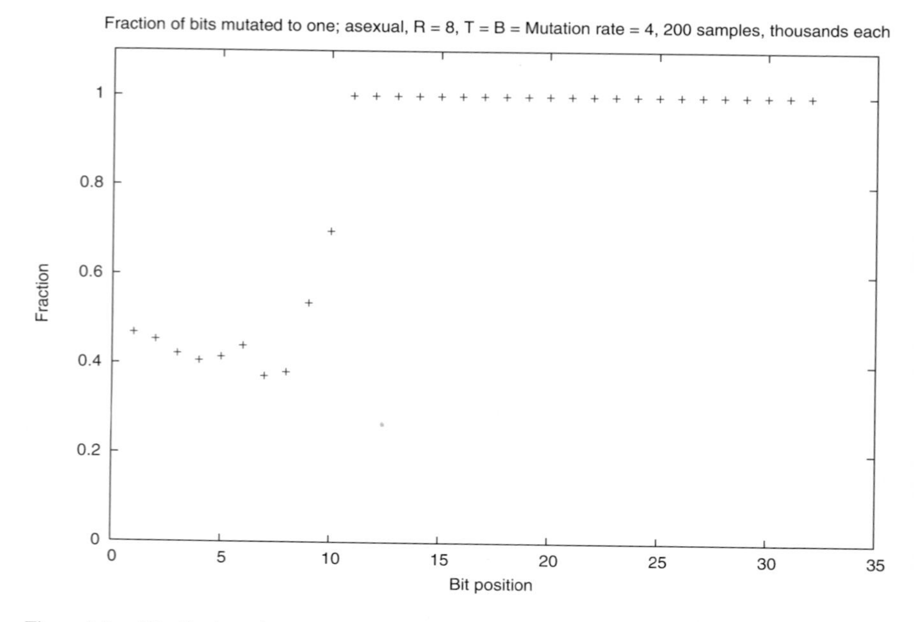

Figure 3.7. Distribution of mutated bits, averaged over 200 samples, each with thousands of individuals having four mutations at birth and $R = 8,\ T = B = 4$.

ily subtract them from the total mortality at older ages: $\mu_g(a) = \mu_t(a) - \mu_V(a)$ with $\mu_V(a) = \mu_t(a = 0)$.

If one population of size N is simulated over long times, with the number of iterations much larger than N, then all survivors have one common ancestor, which the Bible called Eve. The genome then has typically $T - 1$ bits set at ages younger than the minimum reproduction age R, and the positions of these bits are the same for nearly the whole population. (To avoid this effect one can average over many independent populations.) For ages after R the fraction of mutated bits rises rapidly and reaches unity at an age which then gives the maximum age of the population, Figure 3.7.

3.2.2. Applications and modifications

If one assumes that reproduction occurs only once in life, then the Penna model agrees with the reality for Pacific Salmon: everybody dies after reproduction (Penna, de Oliveira and Stauffer, 1995). In this case, the survival probability instead of decreasing smoothly with age as in Figure 3.4 jumps abruptly to zero at age $R + 1$. This is a clear success for the mutation accumulation theory and difficult to reproduce in other theories (e.g., Meyer-Ortmanns, 2001;

Jan, 1994). Kirkwood (p. 437 in Cell (2005)) feels that his disposable soma theory also explains the salmon effect but offers there no mathematical or computational evidence.

Brigitte Bardot has been blamed for the vanishing of Northern Cod off Newfoundland and the eastern coast of Canada in 1993. For in the 1980s she led a protest boycott against what some called the brutal slaughter and others the professional harvesting of baby seals on the ice floats there each spring, to produce white pelts commercially. She succeeded, pelts were no longer bought and thus baby seals were no longer killed. Therefore, the accusation goes, the baby seals survived and grew by eating cod fish, bringing it to extinction there. Men feel it entirely appropriate to blame women for their errors; Adam did so already at the beginning of the Bible. In this case, however, baby seals are now killed again as much as before, but the cod has not returned. And before her boycott, both seals and cod fish lived there together. Simulations of human over-fishing with the Penna model (de Oliveira, Penna and Stauffer, 1995) showed that it can lead to a rapid extinction, while sparing the young fish would help both the fishermen and the fish species. For lobsters, it is better to spare also the old ones (Penna, Racco and Sousa, 2001), since their fertility increases with age. It seems doubtful that European governments have learned from the Canadian experience; they still try to save short-term employment in fishing, endangering both the herring and long-term employment.

Both the number of births per iteration, and the number of offspring born in one birth, could be diminished by a large number of active mutations and thus decrease with age. This effect was studied thoroughly by Desai, James and Lui (1999).

Child mortality in reality is much higher than the extrapolation of the Gompertz law to the first year of human life would predict, Figure 3.2. The group of the geneticist Cebrat (Łaszkiewicz, Szymczak, Kurdziel and Cebrat, 2002; see also Magdon-Maksymowicz, Sitarz, Bubak, Maksymowicz and Szewczyk, 2002) explained enhanced child mortality by adding housekeeping genes to the Penna model. These are genes regulating the basic mechanism of life, not yet the type of complex organism which these mechanism are supposed to support. Failure of household genes causes early death during the development; in fact, most human zygotes never become a baby leaving the mother. In this sense, child mortality is simply the tail of the high death probability right after the formation of the first cell (zygote).

On the opposite end of the age spectrum we have the mortality plateau for flies (Carey, Liedo, Orozco and Vaupel, 1992; Curtsinger, Fukui, Townsend and Vaupel, 1992) for the oldest old. Several modifications of the Penna model were proposed (de Oliveira, de Oliveira and Stauffer, 1995; Sousa and de Oliveira, 2001) to reduce the old-age mortality. The latest success comes from Coe, Mao and Cates (2002) who applied an old idea (Thoms, Donahue and Jan, 1995) that the genes do not kill us deterministically when the number of active diseases reaches the threshold T. Instead they kill us with a smooth probability which increases from zero to unity when the number of active diseases increases from far below T to far above T. Then both analytically and in simulations they found a mortality plateau for the oldest old, using the Fermi function from statistical physics for this probability. Other reasonable choices gave similar results (Schwämmle and de Oliveira, 2005).

Downward deviations from Gompertz at old age are also obtained if one does not equilibrate long enough, both in the Penna model and elsewhere (Mueller and Rose, 1996). And the mortality maximum in de Oliveira, Stauffer, de Oliveira and Martins (2004) vanished when a programming error was corrected; obviously those authors should never be trusted. (See Sousa and de Oliveira (2001) below in Section 3.5.2 for more reliable simulations.)

Human genomes vary a lot, and human mortality is described by the Gompertz law which is found also in the Penna model. In contrast, inbred laboratory animals may all have approximately the same genome. If all individuals have exactly the same genome in the Penna model, their genetic death age is the same except for the probabilistic modification of Thoms, Donahue and Jan (1995) and Coe, Mao and Cates (2002). Geneticist Pletcher did not like this, and in a nice example of productive criticism of the Penna model, Pletcher and Neuhauser (2000) incorporated elements of reliability theory (Gavrilov and Gavrilova, 1991) into the Penna model to repair this defect.

Thus far no spatial structure was taken into account, as is appropriate for highly mobile individuals. For plants, civil servants, or other barely moving individuals it is more appropriate to put them onto a square lattice. Offspring can survive only if put on an empty lattice site besides the mother (Sousa and de Oliveira, 1999a). The random Verhulst deaths now can be avoided completely (Makowiec, 2001). The main results of the Penna model without lattice are confirmed by these lattice simulations. He, Pan and Wang (2005) put wolves, sheep and grass onto a square lattice, with the animals ageing according to the Penna model. The wolves eat sheep, and the sheep eat grass. Three possible equilibrium states were found: coexistence of wolves, sheep and grass, extinction of the wolves and survival of sheep and grass, or extinction of both animals and survival of grass only. Strong oscillations were found similar to the Lotka–Volterra differential equations for prey and predator.

3.2.3. *Plasticity*

At the Atlantic coast of the USA, the Virginia Opossums live partly on the continent when their life is endangered by predators, and partly on islands where their predators are missing. The island mammals not only live longer than those on the continent, which is plausible, but also mature later (Rose, 1991; Austad, 1993).

Thus the environment influenced over thousands of years through selection and mutations the ageing process (the minimum reproduction age). And this happened not in an artificial laboratory experiment but in reality. Non-evolutionary theories of ageing, like those built on oxygen radicals damaging the DNA, have difficulties to explain such a change of genetic properties, which is called "plasticity" by experts. Recent experiments on other animals, including the cod fish simulated earlier in Section 3.2.2, confirmed this reduction of reproductive age through higher predation (Olsen, Heino, Lilly, Morgan, Brather, Ernando and Dieckmann, 2004; Branco and Sherman, 2005). This plasticity requires many generations and is therefore not the simple genetic change in children through traits acquired by their parents, on which Lamarck and others built their theories two centuries ago, Section 2.8.

For guppies, however, Reznick, Bryant, Roff, Ghalambor and Ghalambor (2004) found the opposite effect: The minimum reproduction age went up for higher predation risks. This was thought to contradict evolutionary theories. Computer simulations of Altevolmer (1999), published long before the surprising new results on guppies, show that these new results are fully compatible with the Penna model. He assumed that the risk of predation is not constant over the whole age, but is particularly high for the weak, i.e., for the young and for the old. At middle age, animals can defend themselves better against predators, for example by running away fast. He also assumed self-organisation for the minimum reproduction age, which can mutate by ± 1 from parent to offspring. If in the computer simulations, only the old animals were killed by predators, then the minimum reproduction age went down, as for the Virginia Opossums. If, on the other hand, the predators kill only the young, then the minimum reproduction age went up, as for guppies. Thus the mutation accumulation theory is compatible with both effects: The sign of the change depends on the ages at which animals are eaten by predators.

How is this Penna-type simulation made? If R, the minimum age of reproduction, is kept as a constant input parameter, then by definition one cannot see a change in R. Thus (Ito, 1996) one needs mutations in R such that the fittest R emerges by self-organisation. Therefore at the time of birth the R of the mother is given on to the child, apart from a random mutation by one unit, up or down. If this is the only effect, then R would diminish to zero or one, such that even at the lowest age offspring can be produced. This limit is biologically nonsense for all species which require growth of the offspring before it reaches maturity; we are not dealing now with bacteria. Thus a counter effect has to be introduced, like a risk $(1 - x)^2$ for mother and child to die at birth, where x is the age divided by the maximum age (= length of bit-string.) Thus attempting to give birth at age one would lead almost certainly to death and thus would hardly represent a high fitness leading to lots of offspring. These methods (Ito, 1996; Altevolmer, 1999) are standard to achieve self-organisation of biological parameters and correspond closely to half a century of Monte Carlo simulations for Boltzmann distributions in statistical physics.

3.3. Penna model: sexual

3.3.1. Basic model

Hundreds of million years ago Mother Nature invented sex by copying Holland's genetic algorithm: Children get half of their genome from the father and the other half from the mother. Genetic algorithms assume that mixing the lines of two different versions of a program could make it more efficient. Similarly, one could program sexual reproduction by still giving everyone one string of 32 bits, and the child gets on half of the positions the paternal bits and on the other half of the positions the maternal bits. This, however, is not how nature works and how the program (appendix in de Oliveira, de Oliveira and Stauffer (1999)) was written. To formulate it simply, Nature distinguishes between asexual haploid (e.g., bacteria) and sexual diploid (e.g., humans) living beings. Haploid means that the genetic information is stored only once, as in the single string of 32 bits in the asexual Penna model. Diploid means that the genetic information is stored twice, one set coming from the father and the other set from the mother. Thus in the sexual version of the Penna model each individual has two bit-strings of the same length, typically 32. Now we call these two bit-strings the upper and the lower one and imagine them arranged horizontally:

0 0 0 1 0 0 0 1 0 1 0 1 1 0 1 1 1 1 1 1 1 1 1 1 1 1 1 1 1 1 1 1
0 0 1 0 0 1 0 0 1 0 1 1 0 1 1 1 1 1 1 1 1 1 1 1 1 1 1 1 1 1 1 1

To get a child with the same number of bit-strings as each of the parents, and not twice as many, Nature as well as the computer program produces haploid gametes: sperm and egg cells. Two gametes, one from each type, are then combined in ways not described in this book, to form a combination: the diploid zygote. This zygote then starts dividing and becomes a new individual. The production of these gametes is programmed as follows: One position along the bit-string is determined randomly as crossover point, and one gamete consists of the bits to the left of the crossover point x from the upper bit-string plus the bits to the right of the crossover point from the lower bit-string. The other gamete combines the remaining bits: lower left plus upper right. The two gamete genomes in the above example, if the crossover point x is between bits 4 and 5, look like:

0 0 0 1x0 1 0 0 1 0 1 1 0 1 1 1 1 1 1 1 1 1 1 1 1 1 1 1 1 1 1 1
0 0 1 0x0 0 0 1 0 1 0 1 1 0 1 1 1 1 1 1 1 1 1 1 1 1 1 1 1 1 1 1

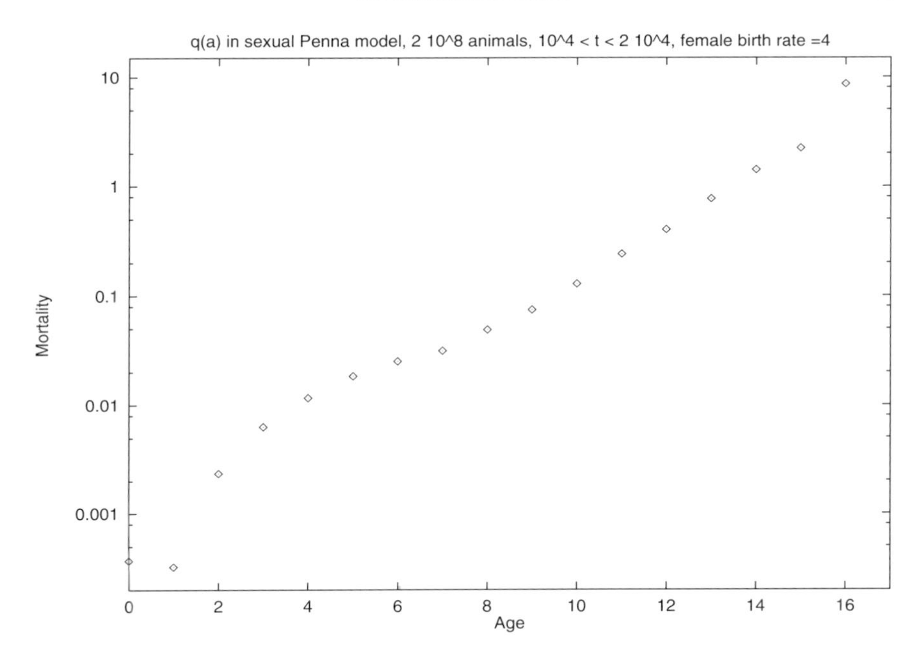

Figure 3.8. Semilogarithmic plot of mortality function versus age for the sexual Penna model. From de Oliveira, de Oliveira and Stauffer (1999).

One of these two gametes from the father is then selected randomly for the zygote, and the same happens in the mother. In this way the sexually produced offspring differs genetically from both mother and father, but combines features of both, just as is the end result in the simpler genetic algorithm. (Results barely change if instead of recombining left and right parts one recombines the bits randomly.) Then, random mutations are introduced in both gametes. The sex of the child is determined randomly.

How do we now count the life-threatening diseases in these bit-strings? Typically six of the 32 bit positions are selected randomly as dominant and all others as recessive. On dominant positions one bit set to one suffices to produce a disease, on recessive positions both the maternal and the paternal bit have to be one to produce a disease.

How do the individuals find partners? We know that women think that all men are alike. Thus they select a partner randomly, check if he has reached the minimum reproduction age, and if not they look for another randomly selected man, up to 20 times. Do you select the oxygen atoms which you breathe?

The simulation in Figure 3.8 shows somewhat more curvature than in asexual reproduction but roughly the mortality function there still obeys the Gompertz law which corresponds to a straight line in this semilogarithmic plot.

3.3.2. Applications and modifications

Women have menopause, and the same cessation of reproduction at middle age applies to many animals, even a fly species (Austad in Wachter and Finch (1997), Figure 5 in Novoseltsev, Novoseltseva and Yashin (2003)). (The old claim, that only humans and pilot whales show it, was based on observation in the wild when animals die long before their genetic death age; we humans are not so special.) According to the above mentioned simulations of Pacific Salmon, mutation accumulation theory kills individuals after their last act of reproduction. Also in the sexual Penna model one may introduce a maximum reproduction age for the females, but then both males and females can survive beyond that age. Why this difference to asexual reproduction?

It is the presence of the males which helps females to survive menopause or its analogs. Men can reproduce until old age; on the other hand the sex of the child is determined randomly. If all the bit positions above the menopause age would be equal to one, they would not only kill all the post-menopausal females but also the males above that age, and thus would reduce the number of births.

This argument can be generalised. For a given individual, to have a 0-bit at a given age a is a selective advantage, compared to other individuals, only if this age falls into the reproductive period of its descendents. This advantage is the more pronounced the larger is the number of descendents which will be still reproductive at that same age a.

In a population where all individuals stop reproduction at a given maximum reproductive age M, individuals with a 0-bit at age a have selective advantage over others only if $a \leqslant M$. There is no selective advantage at all for $a > M$. Thus, genetic drift will sooner or later populate all individual genomes with 1-bits at ages beyond M. The consequence is the catastrophic senescence, like Pacific Salmon.

In populations where the maximum reproduction age is not the same for all individuals and is not given on unchanged to all offspring, the above assertion continues to be valid. Namely, to have a 0-bit in some age a is a selective advantage for the individual. Even if the maximum reproduction age of this particular individual falls below a, the 0-bit still configures an advantage, provided some future descendents can reproduce beyond a. If, for instance half of its descendents are supposed to do so, the 0-bit is a "half advantage", compared with another individual for which all descendents will be able to reproduce at age a, a "full-advantage". The important point, here, is to realise that the 0-bit at age a is a selective advantage for *any* individual, independent of its own maximum reproduction age being smaller or larger than a.

Survival after menopause is just a particular realisation of the above mentioned case. Both males and females have the quoted "half-advantage", because in both cases the descendents are half-to-half divided into male and female offspring. There is absolutely no advantage at all of males over females. Taking into account the slight male-female difference due to X-Y chromosomes for male and X-X chromosome for female mammals, evolution could kill the females after menopause only by putting in a deadly gene into the X chromosome at $a = M$ and at the same time putting into the (small) Y chromosome a gene counteracting this deadly gene. That seems more improbable than a victory by the first author at the Olympic Games in London.

S. Cebrat pointed out that for men staying faithful to their wives the argument does not apply; the men then would also stop reproducing after their wives reach menopause. Humans are known to live beyond 50 years; we do not have here the space to discuss what this tells the wives about their husbands.

But why did menopause arise? Again some people like to give reasons which make us humans special, like grandmothers helping to teach the children, Section 3.5.3 (Voland, Chasiotis and Schiefenhövel, 2005). Indeed, some correlations between the survival of grandmothers and grandchildren were found for humans (Lahdenperä, Lummaa, Helle, Tremblay and Russell (2004), and the accompanying News and Views comments of Hawkes), but not for lions. However, without any such cultural traits the sexual Penna model gave a self-organisation of menopause if the risk of giving birth increases with the mother's age and the offspring needs the mother in order to stay alive during the first few years (de Oliveira, Bernardes and Martins, 1999). Thus starting without menopause, the survival of the fittest simulated by the program with these two modifications creates menopause, Figure 3.9, the other big success of this model apart from reproducing the Gompertz law.

Also the correlations between children and their grandmothers, observed empirically by Lahdenperä, Lummaa, Helle, Tremblay and Russell (2004), are reproduced by this menopause Penna model without any human traits, Figures 3.10 and 3.11, with and without considering maternal care and the risk of giving birth. (In Figure 3.10 the upper data correspond to a higher birth rate than the lower ones.) Good genomes are inherited and are therefore partly in common for each individual and its grandparents (de Oliveira, de Oliveira, Bernardes and Stauffer, 1998). Of course, we do not deny that *in addition* particularly for humans the grandmothers help, following the principle: "Marry early so that your parents take care of your children while you go to the movies."

Returning to the question whether husbands are faithful to their wives, Sousa and de Oliveira (1999b) were afraid of their spouses and claimed that, under certain circumstances, marital fidelity is better genetically, even though a Brazilian

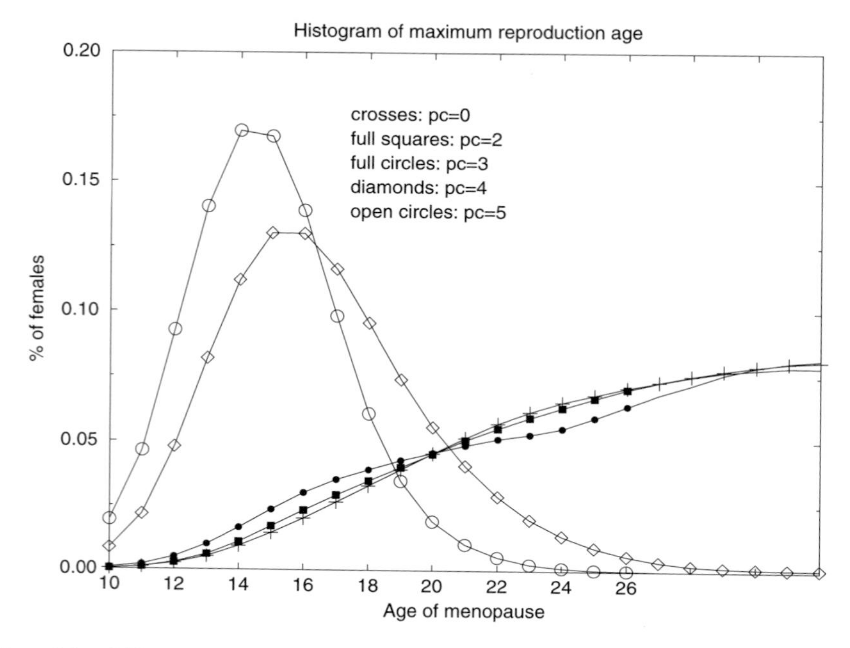

Figure 3.9. Self-organisation of menopause or its analog (de Oliveira, Bernardes and Martins, 1999). The variable pc is the number of iterations over which parental care is required for the children. With no or little parental care, the preferred menopause age equals the maximum age of 32, i.e., there is no menopause. If longer parental care is needed, the menopause age self-organises far below this maximum lifespan.

movie shows how one woman was married successfully with four men (simultaneous updating, not sequentially like Liz Taylor).

Why do men die sooner than women? The forces of evil (wives, medical doctors, ...) claim that men eat too many steaks and drink too much alcohol. The Penna model blames Mother Nature (Schneider, Cebrat and Stauffer, 1998) and with a larger diploid genome distinguishing between X and Y chromosome could reproduce typical human reality: Mortalities for men are twice as high as for women up to very old age where the ratio of the two gets close to unity. The reason is that an "error" in one X chromosome can be counteracted by the other healthy X chromosome for women, but not for men since they have only one X chromosome. In this way, men have something else to blame for their weakness. They may not be entirely correct, however. Figure 3.12 shows the difference between female and male life expectancies from Sweden over two centuries: Women always lived longer but their advantage changes more rapidly than the species changes. Thus, both male behaviour and their discrimination by nature seem to

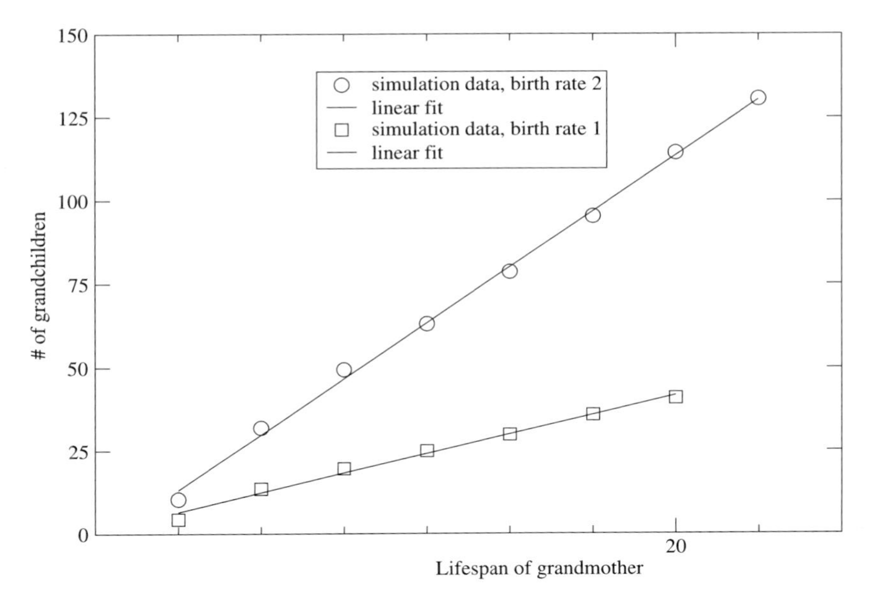

Figure 3.10. Correlations between the age at death of grandmothers and the number of her grandchildren, in the menopause Penna model without human traits, but with child care and risk of giving birth. These simulations follow the empirical observation of Lahdenperä, Lummaa, Helle, Tremblay and Russell (2004).

play a role. (Mortality functions for Swedish men and women will be presented later in Figure 6.1.)

A crucial test for this XX-XY chromosome explanation are the life expectancies of birds, since there the females have two different and the males the same chromosome, opposite to the X and Y chromosomes for mammals. Indeed Paevskii (1985) found in general male birds to live longer than female birds but Austad (2001) in later independent work found no reliable difference between mammals and birds on male-female life expectancies; see also Carey and Judge (2000). Understanding this difference could led to medical treatment prolonging at least men's lives by years, Figure 3.12. But while lots of money is spent on cancer research, which if successful may lead to a similar prolongation of human life expectancy, letting numerous birds live and die in a zoo or laboratory is apparently not "sexy". Outside of Russia, the book of Paevskii was cited in journals of the Science Citation Index only by us physicists who got it from geneticist Cebrat. And for Austad's review of 2001 we could not find any journal citation.

If men are merely faulty versions of women, why did nature invent them? Why did it not continue with asexual cloning of haploid individuals, like the bacteria? Thus we do not discuss here why today sex is useful; the asexual individuals

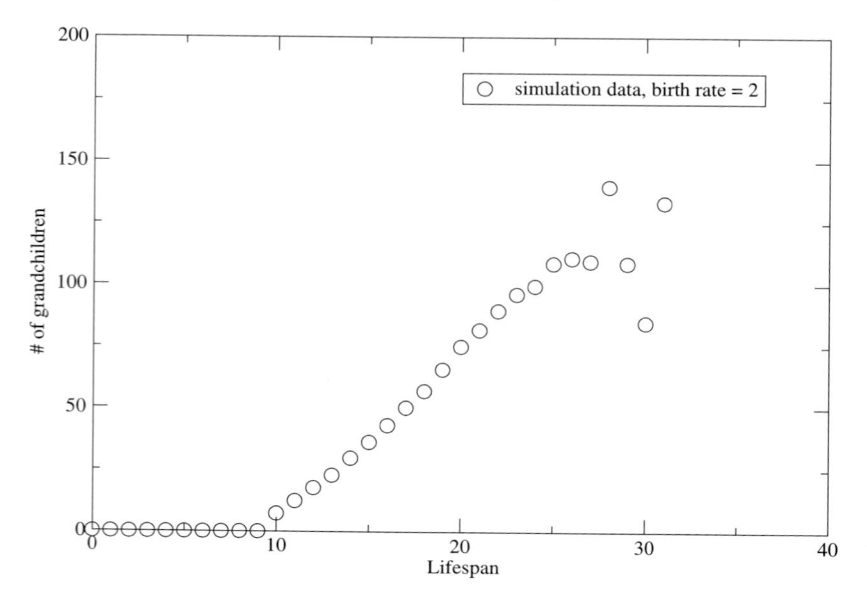

Figure 3.11. As in Figure 3.10 but for the standard sexual Penna model, without child care and without maternal risk at birth.

of old times might not have been convinced that they should switch to sexual reproduction in order that the city of one of the present authors can establish a sex tax hundreds of million years later. Sex must have been useful already after a few generations since otherwise sex would have died out again.

Sexual reproduction gives offspring different from parents, and so sex hinders parasites to adjust well to a host (Howard and Lively, 1994; Martins, 2000). But were there already parasites important when sex was just invented? Sex also helps survival after a catastrophic change of the environment (Martins and de Oliveira, 1998; see also He, Ruan, Yu and Yao, 2004) and recent experiments on yeast by Goddard, Godfray and Burt (2005) confirmed this; but the catastrophe may come too late to allow the first generations with sexual reproduction to compete. Are there intrinsic advantages for sex which counteract the loss of births by a factor two? Males (mostly about half the population) do not get pregnant (with the exception of His Honor the present governor of California) and compete with females for food and space. Much of the sex simulations of physicists were triggered by zoologist Redfield (1994) who found the cost of sex for females too high. It lead to later propaganda with article titles "Why sex—Are men useful for anything" and even "On the uselessness of men—Comparison of sexual and asexual reproduction". Men were saved from this feminist danger by Martins and Stauffer (2001) who modified the Penna model such that sexual reproduction overcame

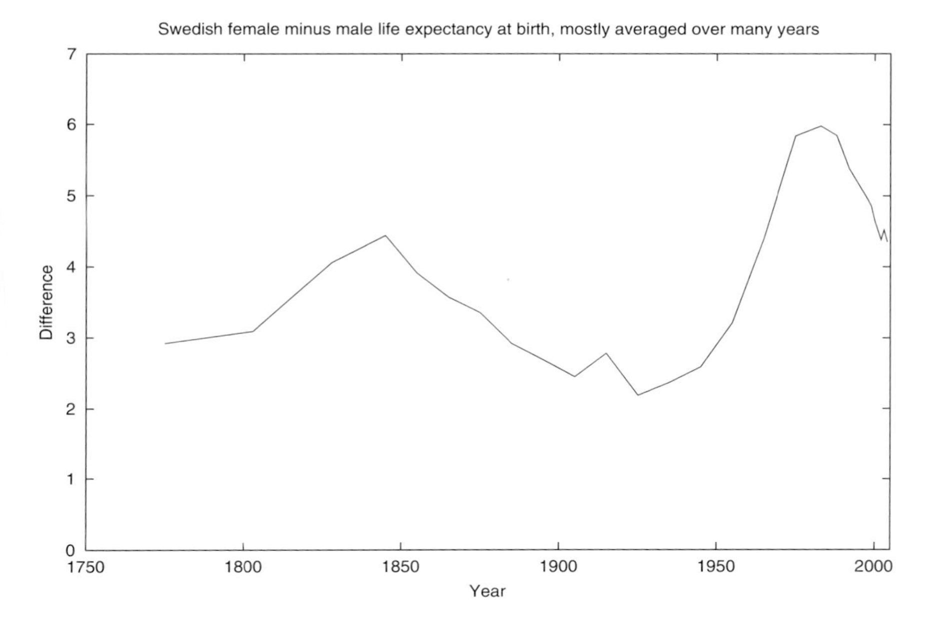

Figure 3.12. Difference in life expectancy between women and men in Sweden, versus calendar year. From www.scb.se.

the loss of a factor two: There, each deleterious mutation diminishes the survival probability in every time interval by a small percentage, in addition to the usual lethal effect if the threshold T is reached, Figure 3.13.

They were supported by Scharf (2004) who simulated pre-selection: many bad mutations may already reduce the ability of, e.g., the sperm cells to swim towards the egg cell. Thus only the fitter males produce offspring in his model and help the species to survive. For asexual cloning, neither selection of a partner nor this pre-selection exists to improve the offspring fitness. Depending on parameters, sex was preferred or rejected, Figure 3.14. Maybe men are useful, or we follow Hitchcock and claim: The trouble with Eve is over.

A technical remark: Evolutionary selection of the fittest acts on individuals and their genes, not on populations as a whole. A proper computer simulation comparing two populations (in the present case one sexual and one asexual one) would have to put both into one common simulation in one common environment. This requires many changes in the computer program and thus has a high probability for programming errors. Instead, one may approximate the truth by simulating both populations separately, but with the same carrying capacity K for both. Then the population with the larger stationary population in the comparison of the two separate simulations uses the resources better and is fitter. This trick was tested

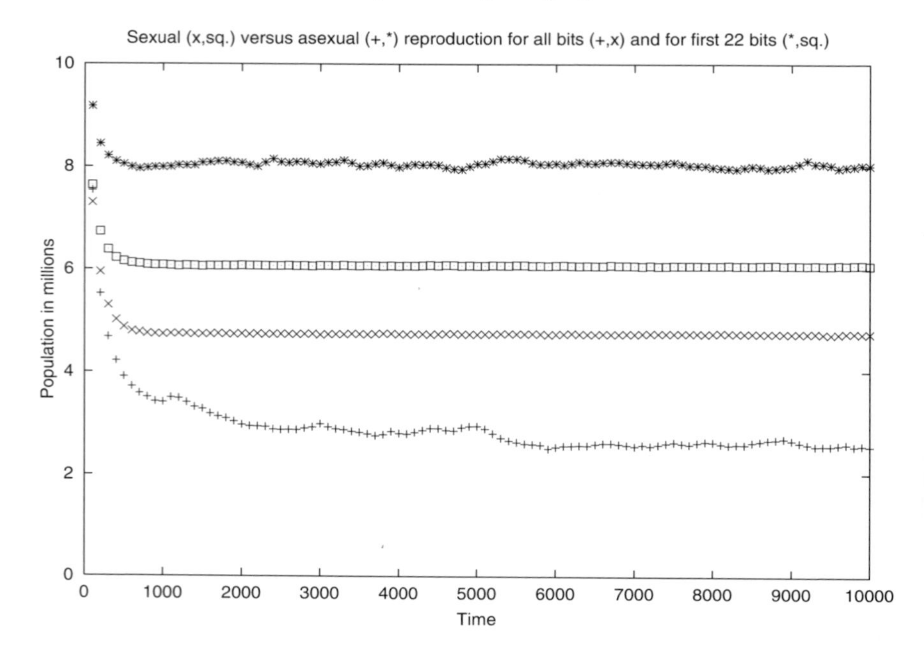

Figure 3.13. Fitness difference, measured through the population size in the same environment, between asexual and sexual Penna model, with the modifications of Martins and Stauffer (2001). If all bit positions are relevant, sex wins; if only the first 22 of 32 bits may reduce survival probabilities, sex loses.

successfully in some simple cases; it may be incorrect if one population size fluctuates much stronger than the other.

The same conclusion that sexual reproduction is better than asexual one for some parameter values and not always, was also found by Bagnoli and Guardini (2005) in a different model and agrees well with reality: Bacteria are not divided into males and females (though they practice parasex: de Oliveira, de Oliveira and Stauffer (2003)), and live on Earth since a much longer time than sexual species.

The crucial Redfield (1994) paper was more or less repeated by Siller (2001) while Otto and Nuismer (2004) from the same department as Redfield propose a more complicated model; both papers do not cite the Redfield paper. And all three papers were ignored in the review of sexual reproduction by Partridge, Gems and Withers on p. 461 in Cell (2005). High impact factors of journals do not necessarily ensure quality for every article in them.

Finally, why are there only two types, male and female, with diploid genomes? As can be seen from Danny de Vito in the film "Twins", having more than one father may cause problems, as was later simulated by Sousa, de Oliveira and Martins (2003).

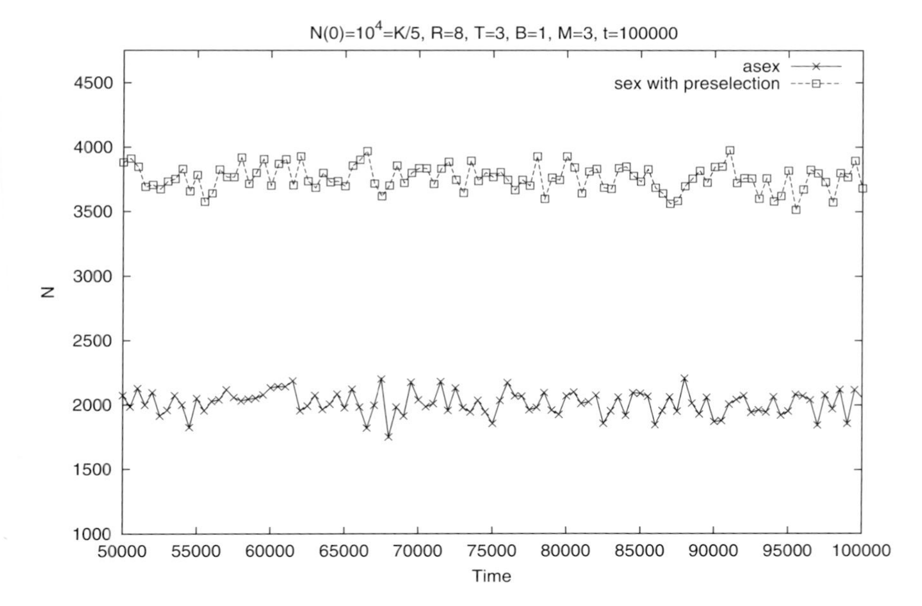

Figure 3.14. Comparison of sexual (top) versus asexual (bottom) populations under otherwise corresponding conditions. The mode of reproduction which gives the higher population in a fixed carrying capacity of the environment is the better one. From Scharf (2004).

3.3.3. Scaling

In all these simulations, time is discrete and increases in unit steps. This makes sense for animals and plants with lives strongly depending on seasons. But even then, the time unit is not one year except if the typical life expectancy is about a dozen years, since the Penna model gives a typical life expectancy of about a dozen iterations. In general, we need a continuous time, just as for the mortality function μ one should let the time interval go to zero to get a derivative. Therefore a proper ageing model would have the number of bits going to infinity, and the age interval associated with each bit would go to zero, such that the total life expectancy (product of number of bits and size of age interval) remains the same if measured in years.

This is easily said but not so easily done (Malarz, 2000; Brigatti, Martins and Roditi, 2004; Łaszkiewicz, Cebrat and Stauffer, 2005) but the last of these three papers, for the sexual Penna model, gave good agreement, Figure 3.15, for bit-string lengths between 32 and 512 if the age associated with the bit position is suitably normalised depending on the length of the bit-string. There was one mutation per genome (as in Section 2.10, i.e., not one mutation per bit), and one birth

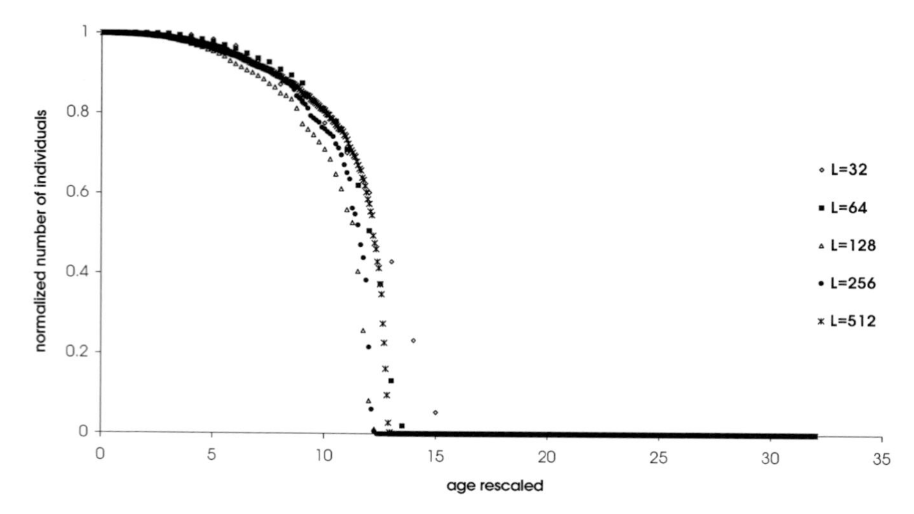

Figure 3.15. Probability to survive birth to age a, plotted versus scaled age. The scaled age is the age, multiplied by 32, and divided by the length of the genome which varied here between 32 and 512 bits. From Łaszkiewicz, Cebrat and Stauffer (2005).

per iteration. For the asexual case no such scaling was found yet; thus sex helps to get proper scaling.

3.4. Other models

Shklovskii (2005) argues that dangerous cells (cancer, ...) are removed by our immune system with a reliability decreasing linearly with age. The immune system has n random opportunities to remove these cells, and this number n follows a Poisson probability distribution $n_0^n \exp(-n_0)/n!$ about its average $n_0 = \langle n \rangle$. Thus the immune system fails, i.e., $n = 0$, with probability $\exp(-n_0)$ and leads to a mortality function $\mu \propto \exp(-n_0)$. If n_0 decreases linearly with age a by an amount ba, then the mortality function $\mu(a) \propto \exp(ba)$ obeys the Gompertz law, as desired. Since n_0 cannot become negative, this linear decrease cannot continue forever and presumably is replaced by $n_0(a \to \infty) = \text{const}$, leading to a mortality plateau as in Gavrilov and Gavrilova (1991). This Shklovskii theory requires no simulations but is included here because of its simplicity. It can be tested by quality experiments on species without immune system, like the unicellular yeast; the yeast mortalities given by the Vaupel group (1998) seem too unclear for a test.

Telomeres are sections at the end of each DNA strand in the genome, and at each cell division, which requires duplication of the DNA, telomeres are partly

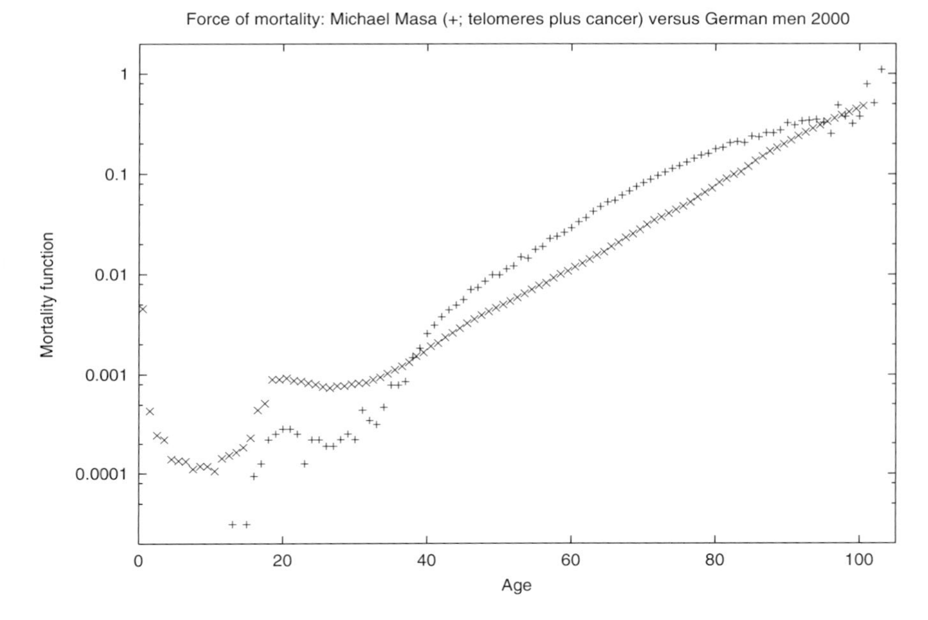

Figure 3.16. Comparison of telomere model including cancer and telomerase (+), with the same German data (x) as in Figure 3.2. Simplified from Masa, Cebrat and Stauffer (2005).

lost. Thus *in vitro*, i.e., in a laboratory vessel outside the living body, cells can only divide a few dozen times; then their telomeres are exhausted: Hayflick limit. This simple counting does not require any computer simulations; Tan (2005), Proctor and Kirkwood (2002) and Aviv, Levy and Mangel (2003) published more sophisticated simulations.

In a living body, the enzyme telomerase restores the telomeres, and thus cells can divide longer. Can we live forever by just getting more telomerase and thus more telomere restoration? Actually, some cells are indeed immortal, and these are the cancer cells. But instead of making us living forever, cancer kills us. Thus real life is a delicate balance between cell death (which leads to our death if the cell is not replaced) and cell survival (which leads to cancer if the cells proliferate too much). If for little worms with no cell division in adult life one can prolong life enormously by genetic manipulation, it does not necessarily mean that reliable anti-ageing medication for humans will appear in the new future.

Masa, Cebrat and Stauffer (2005) took into account telomere attrition and assumed that at every iteration 10 percent of the adult body cells die, and the body tries to replace them by "asking" a randomly selected cell of the same type to divide. If that cell has reached the Hayflick limit it cannot divide, and one cell is lost for the body. If the body has lost too many cells, it dies. This simple model gives a

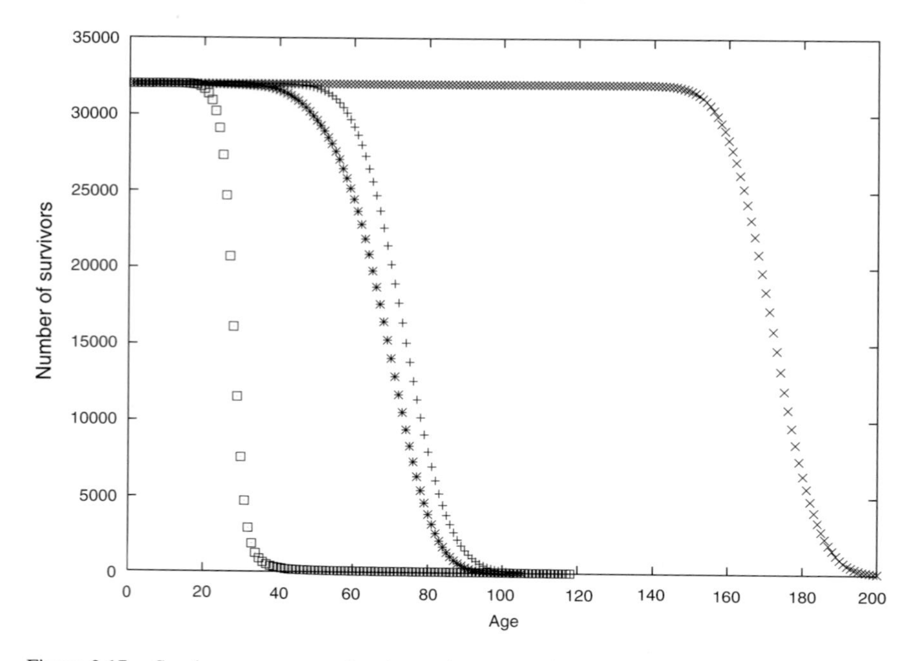

Figure 3.17. Survivors versus age for shorter (+,stars) and longer (×,squares) initial numbers of telomeres. Without the probability to develop cancer (+,×), the larger number is better, with the possibility of cancer (stars, squares) the smaller number is better. Artificial telomere enhancement may be bad.

mortality increasing with age, but not strong enough to follow the Gompertz law, equation (3.1). Then these authors included telomerase and cancer, resulting in a much more realistic mortality function, Figure 3.16. Increasing the amount of telomeres increases the Hayflick limit but also the dangers to cancer. Perhaps real life has already obtained the optimal number of telomeres such that any artificial increase of their number reduces via cancer the life expectancy, as seen in Figure 3.17. Recent experimental facts how *cellular* senescence fights cancer were summarised by Sharpless and de Pinho (2005).

The Dasgupta (1994) model precedes the Penna model and was based on earlier work of Partridge and Barton, of Ray, and of Jan. It does not have an explicit bit-string as a genome but rather survival probabilities (see also Charlesworth, 2001). Originally it had only two age intervals, and the generalisation to many ages (Heumann and Hötzel, 1995) was made efficient only with modifications by Medeiros and Onody (2001). It was generalised to sexual reproduction by Sousa (2003b).

Similarly, also a Weissmann-type mutation accumulation model (Stauffer, 2002d; Stauffer and Radomski, 2001) does not have an explicit genome. Instead the properties inherited and mutated are the minimum reproduction age R and the genetic death age D. Both are given on to the offspring except that they both can be mutated by ± 1. If Weissmann's old idea, that we die to make place for our children, would be generally valid, then the death age D should self-organise to a stationary finite value. However, for a constant birth rate, the death age instead increases fluctuatingly towards infinity, which is unrealistic. Only if we take into account a trade-off between birth rate and longevity by assuming a birth rate $\propto 1/(D - R)$, i.e., by assuming that each individual has the same number of children during its genetic life span, do we get a reasonable stable death age of order 10^2. It was modified for lattices and for sexual reproduction by Sousa, de Oliveira and Stauffer (2001) and could explain menopause (Sousa, 2003a). However, except for additional assumptions (Makowiec, Stauffer and Zieliński, 2001), the mortality function increases more linearly than exponentially with age, and also the Pacific Salmon is not as well described as with the Penna model (Meyer-Ortmanns, 2001).

3.5. * Additional remarks

3.5.1. *Eve effect*

According to some religions and Section 2.9, all humans are offspring of Adam and Eve. Simulations of the Penna model, as well as older theories, confirm this assertion, as reviewed by de Oliveira, de Oliveira and Stauffer (1999). (However, the female and the male ancestor of all living humans according to the simulations in general never met each other, if one does not start the simulation with only one pair.) Initially all N individuals are equal, without any mutations. Then accidentally some become better than others genetically. They and their offspring will slowly overwhelm the others since the Verhulst factor keeps the total population roughly constant. If for asexual reproduction we call the whole offspring of one individual a family, then the number f of families decreases from its initial value N until finally there is only one family left, which remains "forever". Since men are so often oppressed by women, this reduction of ancestry was called the Eve effect, not the Adam effect, in the literature. It also exists for sexual reproduction and for real humans (Cann, Stoneking and Wilson, 1987). In the asexual case, recent simulations (Sitarz and Maksymowicz, 2005) showed that the decay of f with time t is not a simple power law (in the sexual case, f decays as $1/t$). At the beginning one sees a plateau, which extends over longer times if the Verhulst deaths apply only to the babies (Sá Martins and Cebrat, 2000). Then the decay sets in smoothly, reaching $1/t^2$. Afterwards the decay becomes less rapid, roughly as $1/t$, and finally approaches zero if $f(t) = 1$ is constant. Figure 3.18 shows the whole story for an equilibrium population of 22 million; if $f - 1$ instead of f is

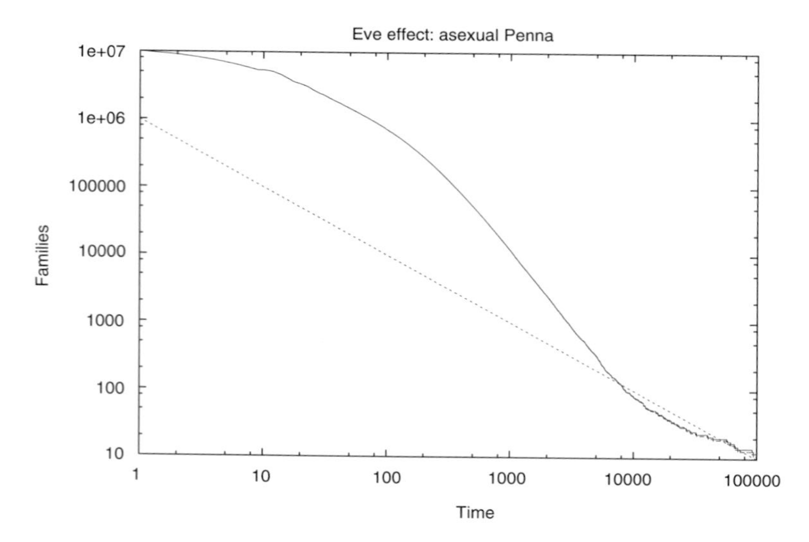

Figure 3.18. Decay with time t of the number f of families in the asexual Penna model. For sufficiently long times, $f = 1$ (Sitarz and Maksymowicz). The line has slope -1 in this log-log plot and fits badly the S-shaped curve.

plotted the figure barely changes, even in its right tail. It looks different from the simple sexual model shown earlier in Figure 2.25.

3.5.2. *Antagonistic pleiotropy*

Pleiotropy means that one allele (one version of a gene, one bit value in the Penna model) has different effects, which may become active at different ages. It is antagonistic when one effect is good and the other is bad for the body. As we mentioned at the beginning of this chapter in Section 3.1.2, calcium is helpful in the youth and dangerous at old age. And this is typical for the examples in the literature: first good, later bad.

Sousa and de Oliveira (2001), in contrast, simulated an antagonistic pleiotropy which is first bad and then good, and which is connected with bit positions 9 (maturity) and 16 (old age) in the 32-bit sexual Penna model. If bit 9 is mutated to 1, then it is detrimental starting from that age on. However, if the individual nevertheless survives up to age 16, then the positive effects of the mutation at age 9 become active. From age 16 on at each "year", i.e., at each iteration which makes one more bit visible, the number of active mutations is decreased by one, with cleaning probability p. Since we deal with the sexual case and two bit-strings, the mutation at bit 9 acts only if either both bit-strings have a 1 there, or the position is one of the dominant ones and at least one of the bit-strings has a 1 there.

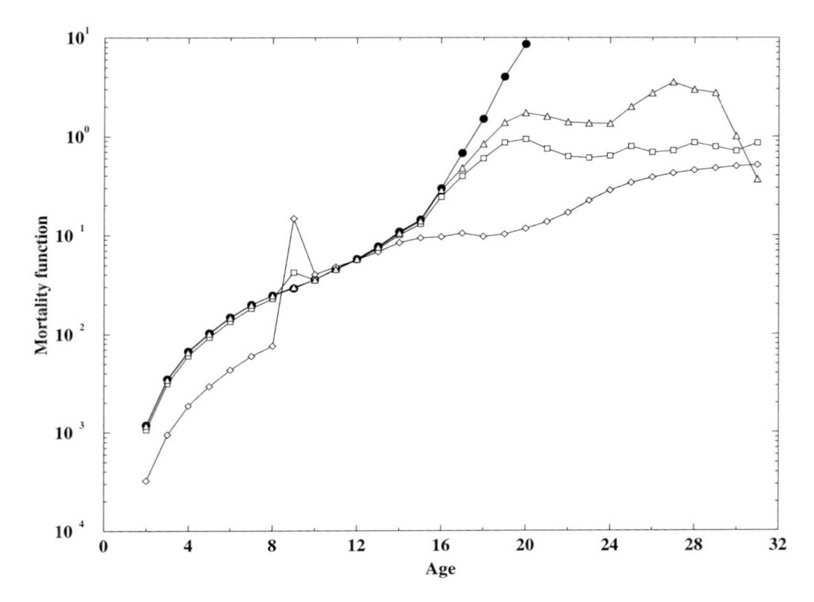

Figure 3.19. Improved survival chances for old age, at the expense of health at age 9, in a sexual Penna model with antagonistic pleiotropy. The cleaning probability p is zero (standard model), 0.2, 0.4, and 0.5, from top to bottom. From Sousa and de Oliveira (2001).

As a result, the mortality self-organises to have a small peak at age 9, but is reduced and even may have a maximum at old age, as Figure 3.19 shows, in agreement with fly experiments at old age. Such a mortality maximum is therefore compatible with mutation-accumulation ant antagonistic-pleiotropy theory. This result is analogous to people who save lots of money at young age in the hope to use it after their retirement (see Section 6.1).

3.5.3. *Grandmother effect*

We discussed near Figure 3.9 already the origin of menopause and mentioned the grandmother hypothesis of anthropologists, according to which a grandmother who no longer can give birth to children helps her daughters to raise the grandchildren (Voland, Chasiotis and Schiefenhövel, 2005). Figure 3.9 showed, however, that without any specifically human traits, menopause can self-organise if we assume a need for childcare, and a risk of giving birth increasing with the age of the mother. The results of Figure 3.10, corresponding to anthropological observations, can even be explained rather trivially: A long-living grandmother can produce more children and thus also have more grandchildren. Therefore the anthropological figure wisely starts only at age 45.

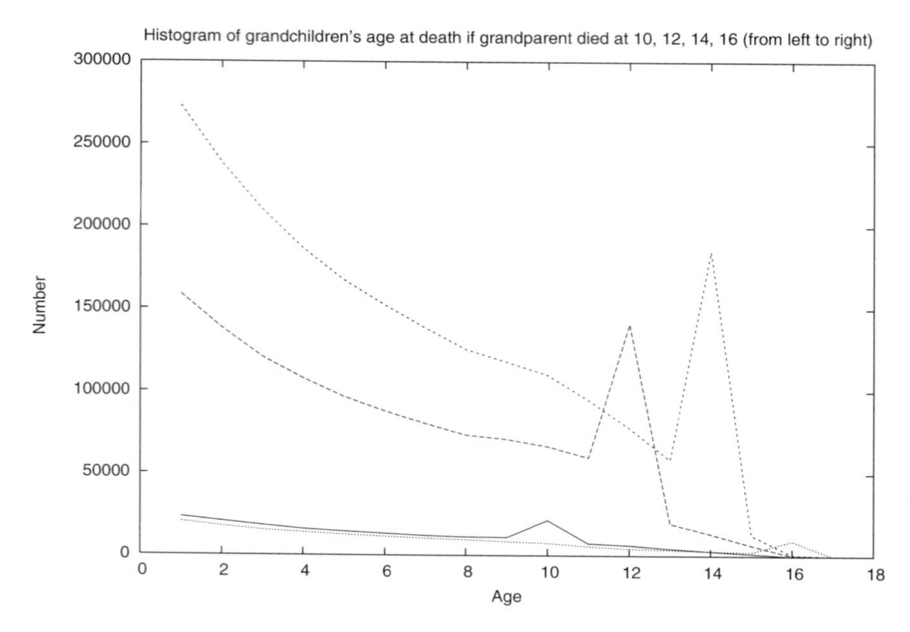

Figure 3.20. Correlation between death ages of grandparent and grandchild in standard asexual Penna model.

Less trivial is the genetic effect that long-lived grandmothers may have healthier genes than short-lived grandmothers, and give them partially on to their children and grandchildren. This effect was simulated by de Oliveira, de Oliveira, Bernardes and Stauffer (1998) for the relation between the ages of death of parents and their children, and is presented in Figure 3.20 for the relation between grandparent and grandchild, using the standard asexual Penna model. We separate all deaths of grandchildren into different statistics (shown as different curves in this figure) depending on the death age of the grandparent. We see little correlation for the normal death ages, but for particularly long lives we see a peak in the grandchild's distribution of genetic death ages. The position of this peak moves to longer life if the death age of the grandparent increases. These histogram show clearly that the Penna ageing model agrees with the general wisdom: A good method to live long is to select the proper parents. Longevity is partially hereditary and can explain correlations for real humans between grandmothers and grandchildren, without assuming any specifically human traits like transmission of culture by grandmothers, or their help in rearing their grandchildren. Menopause can be explained otherwise and its analogs may, and actually do, exist for other sexually reproducing species.

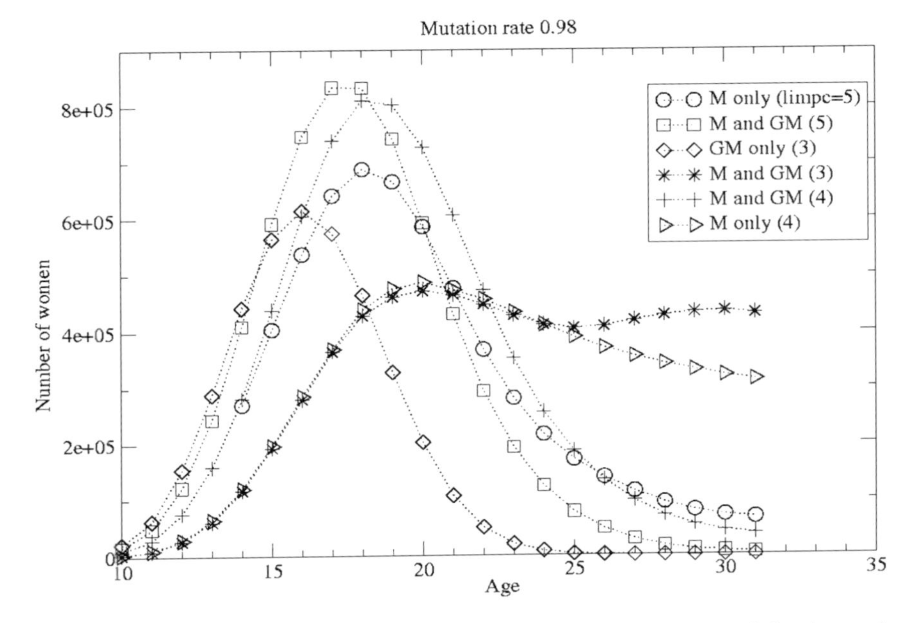

Figure 3.21. Self-organization of menopause age in case only the mother (M) and also the grandmother (GM) offers childcare up to the child age given in parentheses. This histogram complements the histogram of Figure 3.9 where only the mother helps the child.

Anthropologists could clarify the role of the grandmother effects by comparing it with the grandfather effect. Usually fathers contribute half of the genome but less than half of the household work. What correlations exist between the survival of grandfathers and grandchildren? If the grandfather effect is about as large as the grandmother effect, then genetics should be the main cause; if the grandfather effect is much smaller than what Lahdenperä, Lummaa, Helle, Tremblay and Russell (2004) found in their anthropological studies for the grandmother effect, then the help of grandmothers in rearing their grandchildren is really important and could enhance the self-organisation of menopause specifically for humans or other species with helping grandmothers and lazy grandfathers. Indeed, simulations similar to Figure 3.9 but including help from the grandmother show that the population is higher (in a fixed environment K), if also the grandmothers and not only the mothers help the children to survive, Figure 3.21.

3.6. Conclusions

A dozen years of ageing simulations by physicists, in cooperation with some biologist like Cebrat, gave good agreement with the Gompertz law of exponentially increasing mortalities, simulated Pacific Salmon, predicted that more predation can also decrease instead of increase senescence, explained menopause without needing any special human properties, and could explain the emergence of sexual reproduction. The future will show which other models have combined at least the same advantages.

Chapter 4

Biological Speciation

The common ancestors of today's humans and today's chimpanzees presumably lived several million years ago. Then, due to genetic mutations and/or changes in the environment, the population split into the ancestors of humans and the ancestors of chimpanzees. Such a separation of one species into two is called speciation. It involves the division of a species on an adaptive peak, so that each part moves onto a new adaptive peak without either one going against the upward force of natural selection. This process is readily envisioned if a species becomes subdivided, for example, by a river, whereby each part experiences different mutations, population fluctuations and selective forces. If they sufficiently diverge, then even if the river dries and the two parts can make contact again, inter-breeding between them will not occur, two new species have been formed. This kind of speciation is called allopatric and is currently accepted by the majority of the biologists. Besides a geographical barrier to prevent gene flow, it requires a long time to be completed. In contrast, conceiving the division of a single population and radiation onto separate adaptive peaks without geographical isolation, in what is called sympatric speciation, is intuitively more difficult (Tauber and Tauber, 1989). Through which mechanism can a single population of interbreeding organisms be converted into two reproductively isolated segments in the absence of spatial barriers or hindrances to gene exchange? Many evidences and experimental data have appeared in recent years giving support to the existence of a sympatric mechanism of speciation. The cichlid species living in volcanic lakes of western Africa (as well as in some lakes of Nicaragua) are probably the most studied examples. The main features of these lakes are the environmental homogeneity and the absence of micro-geographical barriers (Bagnoli and Guardini, 2005). It is interesting to mention that while this book was being concluded (under the pressure of the exhausted senile author), a paper appeared (McBrearty and Jablonski, 2005) showing that fossil chimpanzees were found in the African rift valley, the same region where fossil homo-species were found. So, perhaps we separated from them by *sympatric* speciation and *not* because the rift valley separated us.

The largest part of this chapter is devoted to computational models which correspond to an extension of the Penna model, just reviewed in the previous chapter, and that have been developed in order to study the origins and dynamics of this speciation process.

4.1. Sympatric speciation

According to Sara Via (2001) the idea that natural selection can lead to divergence and speciation of sympatric populations dates back to Darwin (1859). However, in the mid-1900s the pioneer work of Mayr (1963) on the allopatric mode of speciation shifted the focus of speciation research away from natural selection as the driving force and towards the role of geography in limiting gene flow and promoting genetic drift. The main point of Mayr's theory was that geographically isolated populations can diverge freely, while those found in sympatry (sharing the same habitat) can only escape from the homogenising effects of gene flow under very special circumstances. Such a point of view has been the dominant one for many decades, turning sympatric speciation an extremely controversial process. Recently, however, a variety of approaches and laboratory studies have provided increasing evidence that reproductive isolation may set in due to multiple selective forces, suggesting that sympatric speciation can be in fact a rather common phenomenon (Rice and Hoster, 1993; Tregenza and Butlin, 1999; Odeen and Florin, 2000).

One of the most important ingredients to obtain sympatric speciation is sexual selection, that is, mating partners are chosen assortatively instead of randomly. We may say that sexual selection plays the role of a geographical barrier preventing mating between individuals of too different phenotypes. But how does assortative mating develop? Consider for instance an ecology which presents a broad distribution of seed sizes such that birds that feed on these seeds can equally compete for them, independently of their beak sizes. Suppose now that due to oscillations in the rainfall regime, this distribution of seed sizes becomes bimodal, peaked at very large and very small seeds. In this case the number of birds with intermediate beak sizes rapidly decreases, since they lose out in competition for either resource. So the establishment of a bimodal distribution of resources provokes the so called *disruptive selection* which splits the population into two distinct ecological characters determining adaptation to the environment. Such a splitting of the ecological character is called polymorphism and may be considered as a first step towards sympatric speciation. However, with random mating, individuals with intermediate beak sizes would continue to be produced and the great majority would die of starvation before the minimum reproduction age. The development of assortative mating is then a kind of self-organisation of the population to guarantee the perpetuation of the species and to stabilise the two distinct ecological characters, concluding the speciation process.

This is the mechanism proposed by Kondrashov and Kondrashov (1999) to explain speciation in sympatry. At the same time, and in fact in the same issue of Nature, Dieckman and Doebeli (1999) presented a model of *evolutionary branching* which does not require a bimodal distribution of resources to obtain speciation. Starting from a Gaussian resource distribution, the population first concentrates around the phenotype with the highest fitness, that is, the phenotype for which the available resource distribution is maximum. Then, due to the strong competition, the population splits into two different groups that later become reproductively isolated due to selection through assortative mating. Anyway, assortative mating again evolves as a consequence of competition for resources, since less fit phenotypes suffer less competition than the most fit ones.

Many models have been proposed during the last ten years to explain sympatric speciation (for a review see Turelli, Barton and Coyne (2001) as well as Via (2001)). Some of these models (Panhuis, Butlin, Zuk and Tregenza, 2001) focus on the process leading to reproductive isolation, usually neglecting ecological divergence. Other models focus on ecological differentiation (Schluter, 2001) without giving much attention to the mechanisms underlying the evolution of mating structure. We prefer to follow van Doorn and Weissing (2001), who argue that both approaches present mutually dependent rather than conflicting explanations of sympatric speciation. Thus the results we are going to present were mostly obtained through computational models that consider both, competition and assortative mating, as the fundamental ingredients to obtain sympatric speciation.

4.1.1. *Minimal model: Speciation defined by a single bit*

The most simple strategy to obtain sympatric speciation using the sexual version of the Penna model (Chapter 3) was adopted by Luz-Burgoa, de Oliveira, Martins, Stauffer and Sousa (2003). In this case, one bit position of the bit-string pair that represents the individuals genomes was defined, which was taken as position 11, as an identifier of the species influencing mating. Each diploid individual has $n = 0, 1$, or 2 bits set at this position. A female with n such bits at position 11 selects only partners with the same number n of such "speciation bits" (assortative mating). Due to the randomness of mutations and crossover, its children do not necessarily have n speciation bits set to one, and this randomness allows the emergence of a new species out of the initial one where all n were zero. At every time step t three populations N_n, depending on the number $n = 0, 1, 2$ of speciation bits set to one, may coevolve, and each of these three sub-populations is half male and half female.

Coexistence is achieved by replacing the completely random Verhulst factor into three separate Verhulst factors for the separate populations $n = 0, 1, 2$ (intra-specific competition). We may imagine, for example, that the original population $n = 0$ is vegetarian, and that the second population $n = 2$ emerging out of it consists of carnivores. Both populations are limited by the amount of food, but their food sources are completely different; thus, there is no competition between the two different populations, but the meat-eating females will not select any herbivore males for mating, and vice versa. The population with $n = 1$ can be regarded as one that feeds in both niches. It is added half to $n = 0$ and half to $n = 2$ for the evaluation of the two intra-specific Verhulst factors $V_0 = (N_0 + N_1/2)/N_{max}$ and $V_2 = (N_2 + N_1/2)/N_{max}$ and has the arithmetic average of these two Verhulst factors as its own food-limiting Verhulst factor.

The simulations start with a single population of $n = 0$ individuals. Figure 4.1 shows how the new species N_2 emerges, within about a hundred iterations, from the old species N_0. The intermediate population N_1 is only about one percent of the total and is not shown.

Shifting the speciation bit position from 11 to 21 or to 1 does not change much the results. If the birth rate is changed from 1 to $1 + n$, where n is the number of bits set in the female's genome, then the new species $n = 2$ ends up with a larger population than the original one but still may both coexist (not shown).

Sousa (2004) obtained similar results using this same simple model but distributing the individuals on a square lattice. The simulation starts randomly distribut-

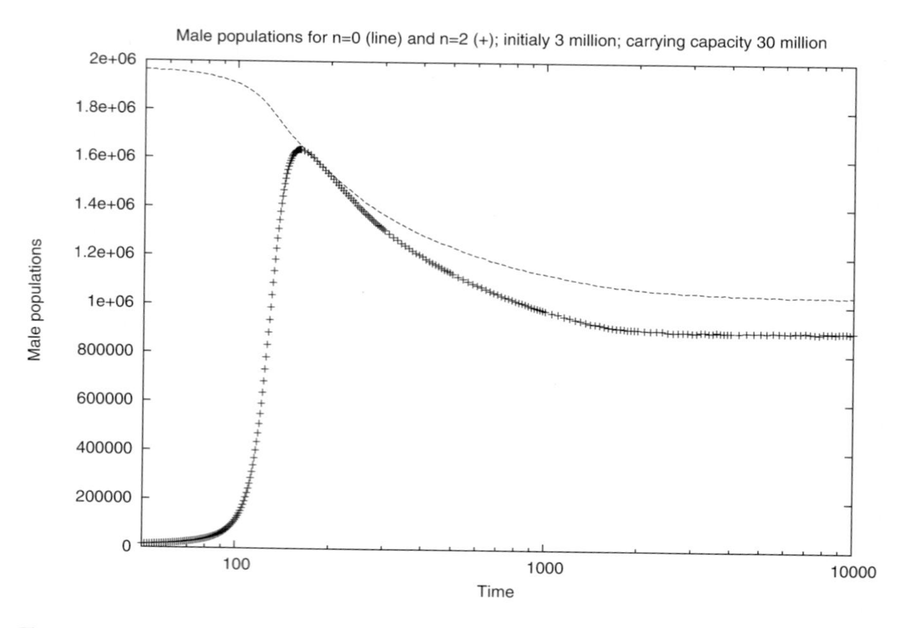

Figure 4.1. Variation in time of N_0 (line, original population) and N_2 (+, new species), with $N_{max} = 30$ million. The simulation started with 3 million males and as many females of the original species. The plots show only the males but the results for females are the same.

ing a single individual per site and at every time step each one has a probability to move to the neighbouring site that presents the smallest occupation, if this occupation is smaller or equal to the current individual's site. Again mating occurs only between individuals of the same species: A female already able to reproduce (age $\geqslant R$) chooses a neighbouring able male with its same value of n to breed. Each of the B offspring is placed randomly at one of the neighbouring sites according to the following rules:

(1) The selected site occupation must be $\leqslant 1$;
(2) A newborn with $n = 0$ ($n = 2$) can occupy an empty site or one already occupied by an individual with $n = 2$ ($n = 0$). This rule means that individuals with $n = 0$ and $n = 2$ do not dispute for the same food resources and so may share the same habitat.
(3) A newborn with $n = 1$ can occupy only an empty site, which means that the $n = 1$ population feeds in both niches.
(4) If it is not possible to find a place respecting the constraints above, the newborn dies.

Again a new $n = 2$ population emerges from the original $n = 0$ one, and the intermediate $n = 1$ population corresponds to 0.5% of the total population. The new feature of these simulations is the complete absence of random deaths, that is, the intra-specific Verhulst factors are replaced by the above occupation rules for the newborns.

As pointed by Sousa (2004), in spite of the simplicity of the model its results fit very nicely to the real situation of three different snake species inhabiting the Australian Fogg Dam Natural Reserve. They differ considerably in body sizes and dietary habitats. The water python species feeds almost exclusively on a single type of native rodents; the keelback species feeds primarily on frogs and the third species, the slatey-grey snake, has a broad diet consisting of reptile eggs, frogs, small mammals and lizards. According to Brown, Shine and Madsen (2002), the population of slatey-grey snakes is smaller than the other two during the whole year. Particularly from April to May, when neither the frogs nor the rats are very abundant, the water-python and the keelback populations are almost of the same size, while the population size of the slatey-grey snakes is around $1/7$ of this value.

4.1.2. *Speciation defined by a single phenotypic trait*

The more realistic computational model introduced by Martins, de Oliveira and de Medeiros (2001) was the first one to add a non-age-structured part to the original genome of the Penna model, to represent a given individual phenotypic characteristic. It can be interpreted as a blow-up of a certain region of the genome, where the genes of a particularly important characteristic are found. We will call this extra, non age-structured bit-string the "phenotype", for simplicity. The purpose of the authors was to study the genetic patterns generated by the order-disorder conflict between selective pressure and mutation accumulation in the presence of an environment that favours particular phenotype configurations. Their biological motivation was the evidence of a stable polymorphism observed in the population of ground finches inhabiting the Galapagos Islands, also known as Darwin finches. In fact, the assumptions that selection, mediated through rainfall and its effects on the availability of different sized seeds, can have a dramatic impact on the beak morphology of the finches that feed on these seeds (and that much of this morphological variation is genetically inherited), are reasonably well established since the field work of Grant and his collaborators (Boag and Grant, 1978; Boag and Grant, 1981; Grant, 1986; Grant and Grant, 1989).

In the computational model, the beak morphology is represented by a single pair of bit-strings (of 32 bits each) added to the genome of each individual. The dynamics of reproduction and mutations are the same for both the age-structured and the new pair of bit-strings – for the latter, a mutation that changes a bit from 1 to 0 is also allowed (see Figure 4.2). The beak size is computed by counting in this new pair the number of recessive bit positions (chosen as 16) where both bits are set to 1, plus the number of dominant positions with at least one of the two bits set. It will therefore be a number k between 0, meaning a very small beak, and 32, for a very large one. Its selective value is given by a fitness function $F(k)$. For a given value of the beak size k, $F(k)$ quantifies the availability of resources for individuals with that particular morphology. The probability of death

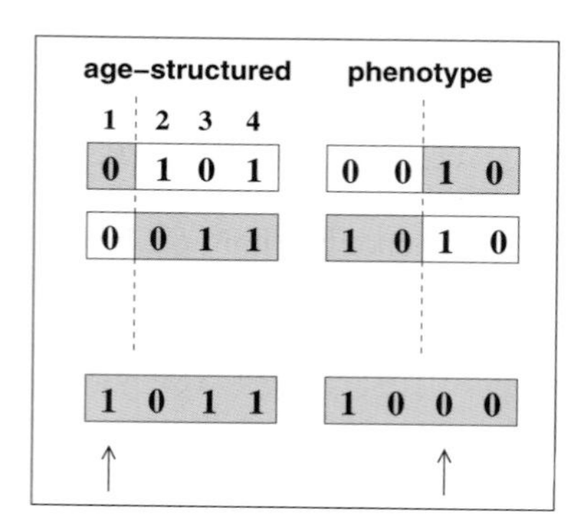

Figure 4.2. Schematic representation of gamete formation, using bit-strings of only four bits. The left side corresponds to the Penna age-structured part where only bad mutations can occur. The right side corresponds to the portion of the genome that encodes the beak morphology and which can suffer mutations in both directions. The arrows indicate the positions where random mutations were introduced. This process of crossing, recombination and mutation occurs with both parents genomes.

by competition at each time step is given by:

$$V(t) = \frac{N(t)}{N_{\text{max}} * F(k)} \tag{4.1}$$

where $N(t)$ accounts for the population that competes for resources available to individuals of beak size k, and N_{max} is a constant proportional to the carrying capacity, as already mentioned in Chapter 3, and related to the maximum number of individuals that the environment can support. Observe that if an individual's phenotype is perfectly adapted to the consume of the available resources, its $F(k) = 1$ and for this individual the probability to die due to competition for food is the same given by the standard Penna model. Less fit individuals have $F(k) < 1$, which increases the Verhulst factor given by equation (4.1) and consequently enhances their chances to die at every iteration.

A final addition refers to mating selectiveness. Martins, de Oliveira and de Medeiros (2001) introduced into the genome a single locus that codes for this selectiveness, also obeying the general rules of the Penna model for genetic heritage and mutation. If it is set to 0, the individual will not be selective in mating (panmictic mating), and it will be selective (assortative mating) if this locus is set to 1. When a female is ready to mate, she chooses a partner according to the expression of this gene. A randomly selected male in the population, to be accepted as a partner, has to either feed on the same niche, in which case the mating selection gene becomes irrelevant, or, if he feeds on a different niche, both parents have to be non-selective in their mating preferences for reproduction to occur. The offspring inherit the mating preferences of either the mother or the father, randomly selected at birth, and this gene can also suffer a mutation in either direction with probability 0.001.

At the beginning of the simulations all individuals are non-selective. Assortative mating following the establishment of a stable polymorphism is essentially equivalent to speciation in this context, and one of the purposes of the simulations was to follow the rising of the fraction of the population that becomes sexually selective as a result of the evolutionary conflict between selection and mutation.

In this model there is a single phenotypic trait, the beak size, and this trait acts both on the individual's fitness and on its sexual selectiveness. In fact, the sexual imprinting-like mechanism is apparently ubiquitous in Darwin's finches and is present in some species of all orders of birds examined so far (Grant and Grant, 1996). It has been shown that as a consequence of beak evolution there have been changes in the structure of finch vocal signals (Podos, 2001). Patterns of correlated evolution among morphology and song are consistent with the hypothesis that beak morphology constrains vocal evolution. Different beak morphologies differentially limit a bird's ability to modulate vocal tract configurations during song production. Data (Podos and Nowicki, 2004) illustrate how morphological adaptation may drive signal evolution and reproductive isolation, and furthermore identify a possible cause for rapid speciation in Darwin's finches.

Results without sexual selection

In order to study the effect of an abrupt change in the ecology alone, the program was first run considering only random mating. The simulations start with a small-sized beak population ($k = 0$ for all individuals) immersed in an ecology with a broad distribution of edible seed sizes available, peaking at middle-sized seeds. During this period the whole population competes for the same general food resource and $N(t)$, in equation (4.1), is equal to the total population. This ecology is represented by the fitness function:

$$F(k) = 1 - \frac{|16 - k|}{128} \tag{4.2}$$

where the denominator 128 ensures a mild selective pressure for middle-sized beaks ($k = 16$). The population evolves for 200 generations, when a snapshot of the phenotype distribution is taken. This distribution is bell shaped, with its peak located at $k = 16$, corresponding to middle-sized beaks. Because mutations can both increase or decrease the beak size, and because the number of positions where each allele is dominant is the same (dominance = 16), there is no inherent bias to the equilibrium distribution: Its position can sit at the same position as the one for the fitness function, as shown by the circles in Figure 4.3.

After 200 generations, for instance due to a variation in the rainfall regime whose effect is to decrease the availability of seeds, the function $F(k)$ changes to a two-peaked shape, with maxima at $k = 0$ and $k = 32$; the food resources concentrate on either small or large seeds, with a vanishing number of medium-sized ones. The fitness function that expresses this new ecology is:

$$F'(k) = \frac{|16 - k|}{16}. \tag{4.3}$$

Now only small(large)-beaked individuals – those with $k < (>)16$ – can compete for the small(large) seeds, that is, competition becomes intra-specific. For that reason, the death probability $V(t)$ of an individual with $k < (>)16$ is computed by assigning to $N(t)$ the number of individuals with $k < (>)16$ plus half of the population that has $k = 16$. An individual with $k = 16$ competes either for small or large seeds, and this choice is random. As a consequence, the beak-size distribution splits into a double-peaked one, centred on large and small beaks, as shown by the squares in Figure 4.3. However, since mating selectiveness has not been introduced this polymorphism is reversible: If, in a subsequent time step, the pattern of availability of edible seeds reverts to its original configuration, so does also the distribution of beak sizes, that becomes again unimodal. This reversibility is indicated by the small bump at $k = 16$: Since there is no reproductive isolation, mating between birds feeding on different niches generate offspring with medium-sized beaks. This is in complete agreement with the field observations in the Galapagos ground finches (Lack, 1983;

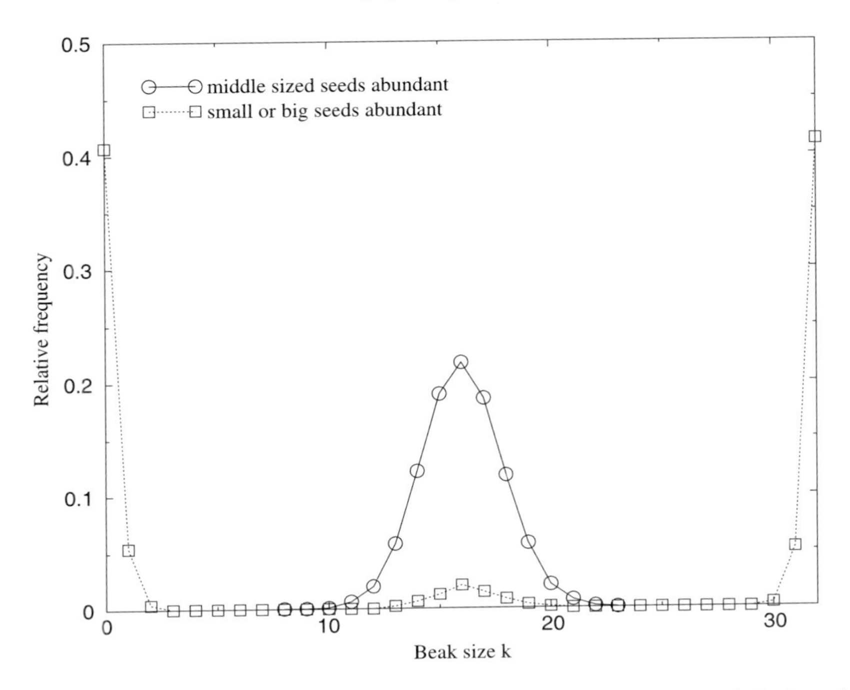

Figure 4.3. Distribution of the beak-sizes before and after a climate induced-change in food supply. The circles correspond to the equilibrium population when there is a broad distribution of seed-sizes available peaking at middle-sized seeds. The squares correspond to the equilibrium population after the change, when only small or large seeds can be found. In this simulation there is no assortative mating and finches with medium-sized beaks continue to be produced after the change, as shown by the small bump at $k = 16$ (squares).

Grant, 1986), whose beak sizes vary according to the seasonal amount of rain in a continuous and very fast process of adaptation.

Results with assortative mating

The simulation above was repeated now considering the gene for selectiveness, but starting with a completely non-selective population. In this case, before the splitting of the distribution of seed-sizes only 0.3% of the population becomes selective. However, after the splitting, this fraction rises to 94%: Now, even if the climate changes again, there will be no cross-mating between the two extreme beak-sized populations. Figure 4.4 shows the distribution of beak-sizes when sexual selection is considered. It can be seen that the medium-sized beaks completely disappear after the ecological change.

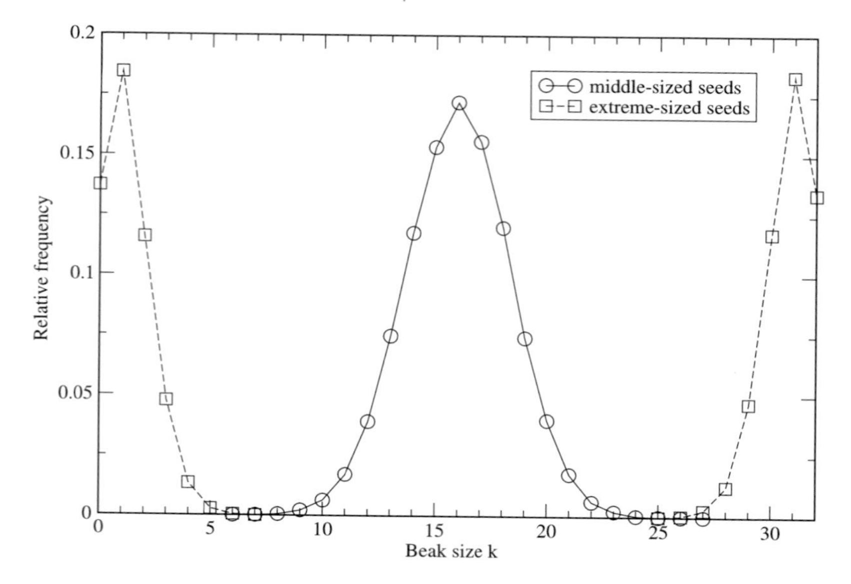

Figure 4.4. The same as in the previous figure, but now considering sexual selection; the small bump at $k = 16$, after the ecological change, disappears.

4.1.3. Speciation in a food chain

A model rather similar to the one just described was developed by Luz-Burgoa, Dell and de Oliveira (2005) in order to study sympatric speciation in a simple food web. Initially, the web consists of a primary food source and a single herbivore species that feeds on this resource. In this case the herbivore is the top species of a two-species food chain. Subsequently they introduce a predator that feeds on the herbivore, simulating a three-species food chain. As will be seen, sympatric speciation is obtained for the top species in both cases, and the speciation velocity depends on how far up, in the food chain, the focus population is feeding. The main difference of their model to the previous one is that competition is maintained constant during the whole simulation, instead of changing according to the ecology. That is, in the previous model, before the splitting of the seed-size distribution, competition was not intra-specific: All individuals disputed equally for the available resources. In the present model competition is intra-specific since the beginning of the simulation.

Two-species food chain

Their two-species food web consists of a basal resource (plants) and a consumer that feeds on this resource (herbivore). The herbivores have the same genetic properties as in the previous model, represented by the two pairs of bit-strings. The phenotypes are again characterised by the integer k, computed from the non-structured pair, and the death probability by intra-specific competition for extremal phenotypes is now given by:

$$V_{1(2)}(k,t) = \frac{P_{1(2)}(k,t) + P_m(k,t)}{N_{\max} * F(k,t)} \tag{4.4}$$

where $P_{1(2)}(t)$ accounts for the population with phenotype $k < 16$ ($k > 16$), respectively, and P_m accounts for the population with phenotype $k = 16$. The Verhulst factor for intermediate (m) phenotypes is:

$$V_m(k,t) = \frac{P_m(k,t) + \left[P_1(k,t) + P_2(k,t)\right] * 0.5}{N_{\max} * F(k,t)}. \tag{4.5}$$

Now individuals with extremal phenotypes (P_1, P_2) compete for small/large plants among the individuals with its same extremal phenotype, and also with the whole intermediate population (4.4). Individuals with intermediate phenotypes (P_m) compete among themselves and also with half of each population presenting an extremal phenotype (4.5). Again at every time step, and for each individual, a random number is generated; if this number is smaller than V, the individual dies.

Mating selectiveness is also encoded by a single locus introduced into each genome, but females that are selective choose mating partners according to one of the following mating strategies:

(1) If a female has phenotype $k < 16$ ($\geqslant 16$) it mates with the first randomly chosen male that presents the same phenotype $k < 16$ ($\geqslant 16$).
(2) The female chooses, among six males, the one with the smallest difference between its own phenotype k_F and the male's phenotype k_M.
(3) The mating of a pair occurs with probability $= |(k_F - k_M)|/32$, where k_F is the female phenotype and k_M is the male phenotype. The female tries to mate for at most six times; if it doesn't find a proper male, than it mates randomly.

The fitness function that expresses the individuals ability in using the available resources is now given by:

$$F(k,t) = \begin{cases} 1.0 - \dfrac{|16-k|}{20}, & t \leqslant 250 \text{ generations}, \\ 0.1 + \dfrac{|16-k|}{20}, & t > 250 \text{ generations}, \end{cases} \tag{4.6}$$

which is essentially the same as the one of the previous model, since it first favours herbivores of medium sizes and suddenly changes, favouring extreme-sized individuals.

The resulting distribution of phenotypes for the mating strategies (1) and (2) are the same, and shown in Figure 4.5.

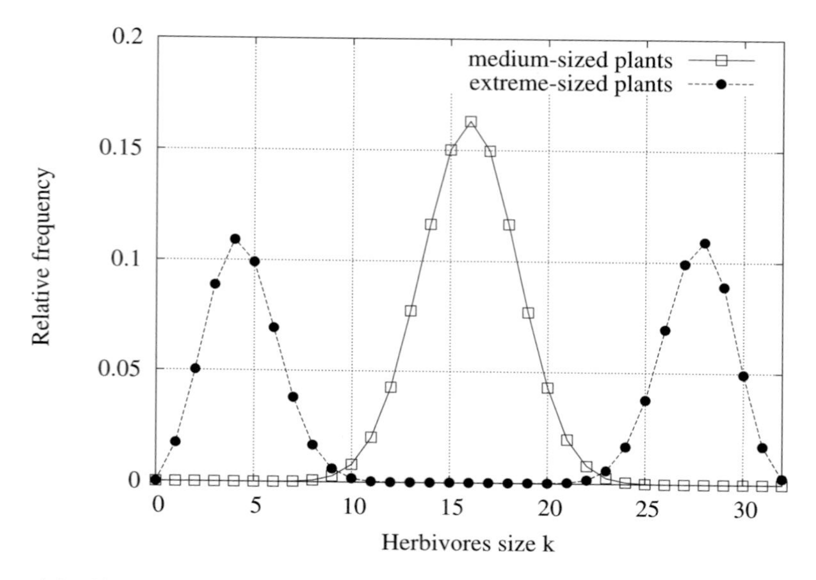

Figure 4.5. Phenotype distributions of the herbivores before (squares) and after (filled circles) the abrupt change of the basal resource species distribution, for mating strategies (1) and (2).

If we compare Figure 4.4 with Figure 4.5 we notice that in the latter the extreme-sized populations, after the ecological change, are smaller. The reason is that in the present model these populations compete with the whole intermediate population (4.4), while in the previous model each one competes only with half of the intermediate population. The crucial effect of the competition level between extreme and intermediate phenotypes will appear in a much clearer way in Section 4.1.4, where we will show that a phase transition may occur, depending on the value of this competition level between different populations.

Still concerning Figure 4.5, observe that in case of mating strategy (1) the female follows the drift direction of the ecological change and it is easy to understand why the final distribution of phenotypes is bimodal and how reproductive isolation has driven the elimination of all intermediate phenotypes. However, with mating strategy (2) females do not know this direction and even so, the ecological change drives their preferences in the same way as with strategy (1). With mating strategy (3) the intermediate phenotypes are not totally eliminated, as Figure 4.6 shows, since in this case random matings may occur.

Three-species food chain

Now there are predators that compete among themselves also according to equations (4.4) and (4.5), and that feed solely on the herbivores. All animals have

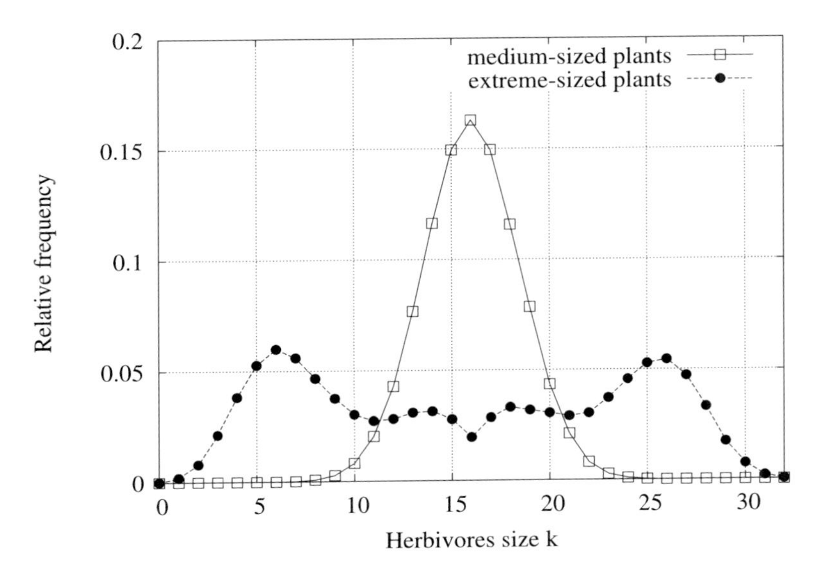

Figure 4.6. Phenotype distributions of the herbivores before (squares) and after (filled circles) the abrupt change of the basal resource species distribution, for mating strategy (3).

the genetic properties already presented, but only the predators have mating preferences: The herbivores mate randomly. The basal resource species distribution (plants) on which the herbivores feed is still given by equation (4.6). Before the change of the plant distribution, the unimodal phenotype distribution of the herbivores is stationary, represented by the open squares in Figure 4.7. However, after the ecological change, this distribution becomes bimodal and presents an oscillatory behaviour: Sometimes there are more individuals with one of the extremal phenotypes than with the other. The period of these oscillation was found to be equal to the minimum reproduction age R. This oscillatory polymorphism is represented by the circles and triangles in Figure 4.7.

In this three-species food web, speciation is not always obtained for the predators (top species), that is, for them to speciate it is necessary to have the herbivores polymorphism, but not sufficient. Observe that now the change in the plants distribution is directly felt by the randomly mating herbivores which feed on these plants, but acts on the predators only in an indirect way, turning this population very sensible to the fluctuations that may occur in the herbivores population. When speciation occurs, the predators populations with $k < 16$ and $k > 16$ also oscillates with the same frequency as the herbivores ones, as shown in Figure 4.8. However, the amplitude of these oscillations is small if compared to the herbivores one which makes the phenotype distribution of predators to remain stationary and

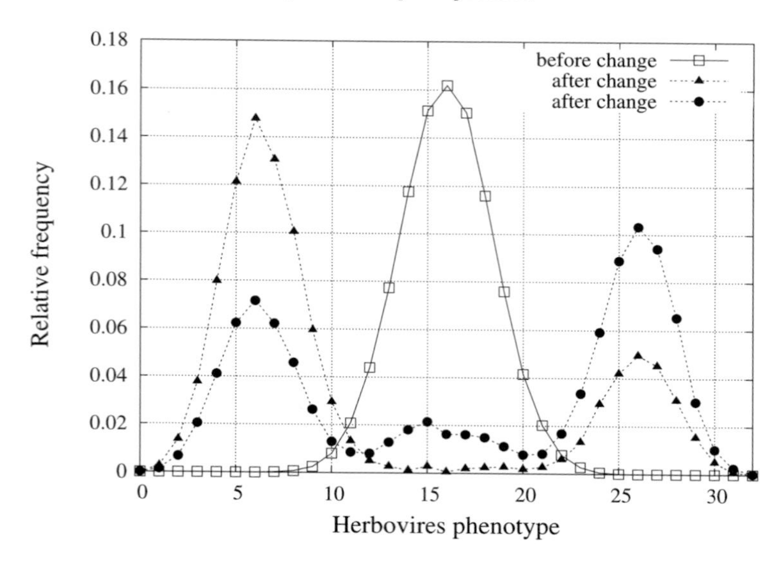

Figure 4.7. Phenotype distributions of the herbivores, which now are not submitted to sexual selection. Open squares: stationary distribution before the abrupt change of the basal resource species distribution. Circles and triangles: two different instantaneous snapshots of the oscillatory polymorphism observed after the basal resource change.

again almost equivalent to the one shown in Figure 4.5, for mating strategies (1) and (2). For mating strategy (3) predators with intermediate phenotypes do not disappear, as obtained for the herbivores in the two-species food web where they were the top species with mating preferences (Figure 4.6).

The most important difference between the two food chains is the speciation velocity, measured through the time evolution of the fraction of selective individuals in the populations. Figure 4.9 shows that intermediate phenotypes disappear faster in the two-species food chain than in the three-species one, which leads to the conclusion that higher level consumers take longer to speciate when the distribution of the basal resource is altered. A process of speciation that possibly fits into this model and respective results is the one that has occurred with one of the three lineages of the Darwin's finches, named the tree finches. There are six species in this group; all of them, except the vegetarian finch, *P. crassirostris*, are insect eaters. Inside this lineage, according to analysis made in mitochondrial DNA sequences (Sato, O'h Uigin, Figueroa, Grant, Grant, Tichy and Klein, 1999), the vegetarian tree-finch may have diverged from the ancestral stock before the divergence of the rest of the tree-finch group (probably due to some irreversible climate change that modified the existing distribution of plants).

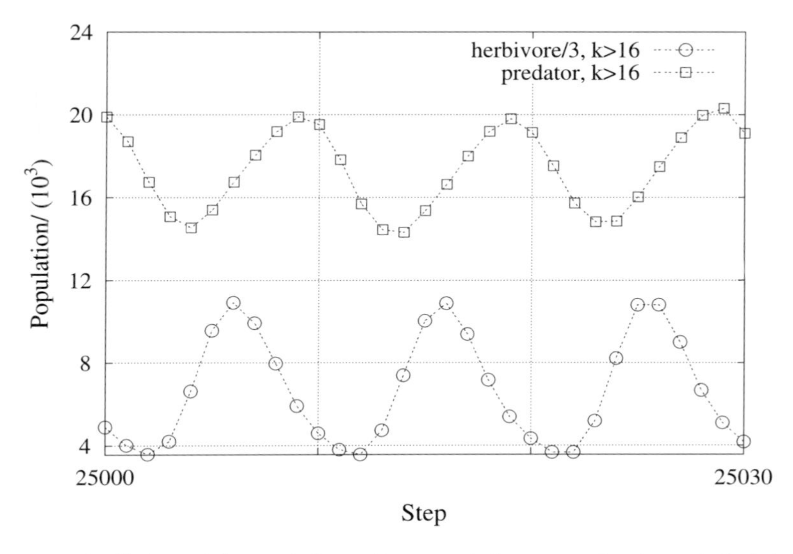

Figure 4.8. Oscillations of the populations with extremal phenotypes after the ecological change. Open squares correspond to predators and open circles to herbivores. Notice that the amplitude of the herbivores population oscillations is divided by three in the figure. The phase difference of three time-steps between them remains the same for different values of the model's parameters.

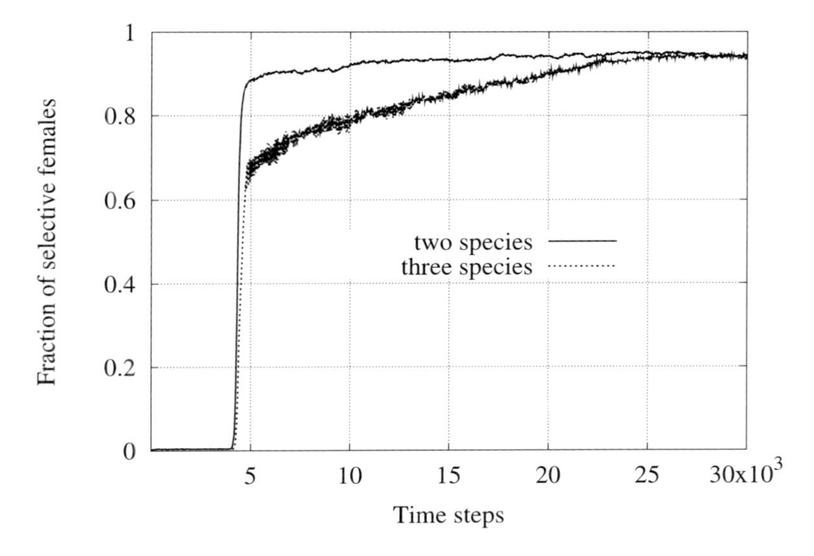

Figure 4.9. Comparison between the speciation velocities for the top species of the two food webs, using the mating strategy (1). The fraction of selective females increases faster for the two-species food web than for the three-species one.

4.1.4. Phase transition in the sympatric speciation process

As explained at the beginning of Section 4.1, in the Kondrashov and Kondrashov (1999) model an abrupt change in the ecology was introduced in order to provoke a disruptive selection which may lead to a stable polymorphism that, if followed by assortative mating, gives rise to speciation. The model we are going to present now has been recently developed (Luz-Burgoa, Schwämmle, Martins and de Oliveira, 2005), and does not require any splitting of the resources distribution to obtain speciation; in this way, it is closer to the model from Dieckman and Doebeli (1999) also mentioned at the beginning of Section 4.1. Anyway, the present model also has competition between common (intermediate) and extreme phenotypes and sexual selection as its main ingredients. As we will see, the degree of competition between these two classes of phenotypes plays the role of a control parameter in a phase-transition-like behaviour found in the speciation process.

The model considers a single initial species, let us say of birds, living in an environment where there is a constant supply of seeds of all sizes. All individuals have the same ability (fitness) $F = 1$ to use the resources, independently of their beak sizes. Competition is intra-specific since the beginning of the simulations and given by:

$$V(k,t) = \begin{cases} V_1(k,t), & 0 \leqslant k < 13; \text{ extreme phenotypes} \\ V_m(k,t), & 13 \leqslant k \leqslant 19; \text{ intermediate phenotypes} \\ V_2(k,t), & 19 < k \leqslant 32; \text{ extreme phenotypes} \end{cases} \tag{4.7}$$

where

$$V_{1(2)}(k,t) = \frac{P_{1(2)}(k,t) + P_m(k,t)}{N_{\max}} \tag{4.8}$$

and

$$V_m(k,t) = \frac{P_m(k,t) + \left[P_1(k,t) + P_2(k,t)\right] * X}{N_{\max}}. \tag{4.9}$$

Now intermediate phenotypes are not only those with $k = 16$, equation (4.7), and the competition degree X between the intermediate population and the extreme ones can vary between zero and one, equation (4.9), instead of assuming a constant value equal to 0.5, equation (4.5).

Another difference between this and the previous food web model is the mating rule, which is now stronger: A selective females chooses, among N_m males, the one that lies deepest into its phenotype range. That is:

- If $k_{\text{female}} < 16$ then it selects the male with the smallest k_{male};
- If $k_{\text{female}} > 16$ then it selects the male with the largest k_{male};
- If $k_{\text{female}} = 16$ then the female chooses randomly to act as one of the above.

The simulations start with non selective populations of random phenotypes.

The phenotypes distributions rapidly converge to a Gaussian centred at some intermediate phenotype k, independently of the value of X. However, for $X = 0$ the distribution remains unimodal for the whole simulation, while for $X = 1$ in some moment it splits into a bimodal one, centred at some opposing extreme phenotypes. Observe that now there is no ecological change: The splitting (speciation) occurs only due to the high degree of competition. The shape of these distributions are equivalent to those shown in Figure 4.5, where the squares would represent the situation for the non-speciation case $X = 0$, and the filled circles would represent the case $X = 1$ where competition drives the population to reproductive isolation.

For $X = 0.5$ the scenario is quite different. Figure 4.10, upper part, shows the phenotype distributions at three different moments. At the very beginning the distribution is also a Gaussian centred at some intermediate value of k, but the final distribution is not stationary: It remains oscillating between the distributions represented by filled circles and triangles, respectively. The lower part of Figure 4.10 shows the time evolution of the density of selective females, ρ_s, for the three values of X. It is nearly zero for the non-speciation case $X = 0$, goes very fast to one for $X = 1$ and presents strong fluctuations for $X = 0.5$.

Observe that $\rho_s = 0$ and $\rho_s = 1$ characterises two very different states of the population organisation: In one of them mating is completely random and in the other totally assortative. In order to analyse the behaviour between these two states the authors performed 10 long simulations (10 different initial random seeds) for each value of X; in each simulation they calculated the final density of selective females (averaged during the last 10^4 time steps). The ten final densities obtained were then averaged producing, for each value of X, the mean density of selective females $\langle \rho_s \rangle$. The upper part of Figure 4.11 shows the behaviour of $\langle \rho_s \rangle$ as a function of X; the lower part shows the log-log plot of the standard deviation of $\langle \rho_s \rangle$, $\sigma(\langle \rho_s \rangle)$, as a function of X.

The behaviours of the curves presented in Figure 4.11 are very typical of phase transitions in physical systems, where the average density $\langle \rho_s \rangle$ plays the role of the order parameter, as the magnetisation in a magnetic system, and X plays the role of the control parameter, like temperature. Within this analogy, $\sigma(\langle \rho_s \rangle)$ is equivalent to the magnetic susceptibility. In one "phase" ($X < 0.5$) there is a single species of intermediate phenotypes and in the other ($X > 0.5$) there are two species with extreme opposing phenotypes. However, for this "biological system" such a phase transition is, by definition, a non-equilibrium one.

In Figure 4.12 we show the behaviours of $\langle \rho_s \rangle$ and its fluctuations, $\sigma(\langle \rho_s \rangle)$, as a function of X for different values of the bit-strings lengths. In this case the bit-string length seems to be equivalent to the size in a magnetic system. To decide the order of such transition it would be necessary to increase the bit-strings up to

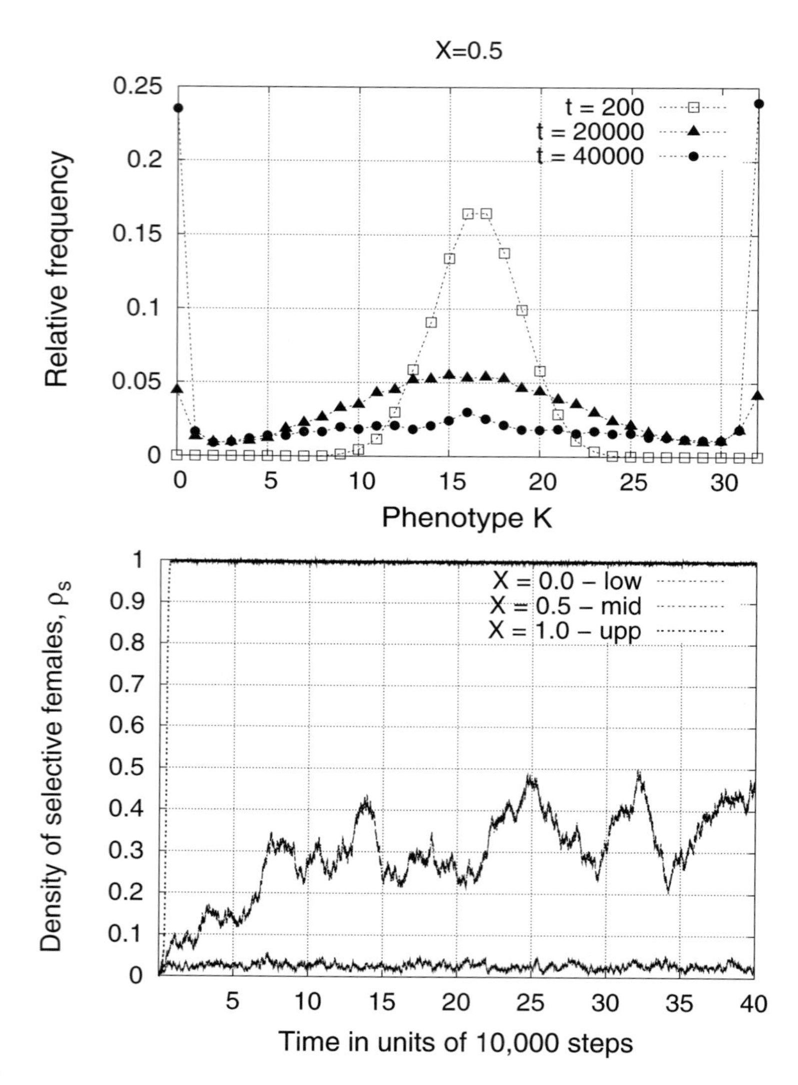

Figure 4.10. Upper part: Phenotype distribution of the whole population for $X = 0.5$, which after equilibration remains oscillating between the curve represented by filled circles and the one represented by filled triangles. Lower part: Temporal behaviour of the selective females density for three different values of the competition level, $X = 1$ (upper curve), $X = 0.5$ (middle curve) and $X = 0$ (lower curve).

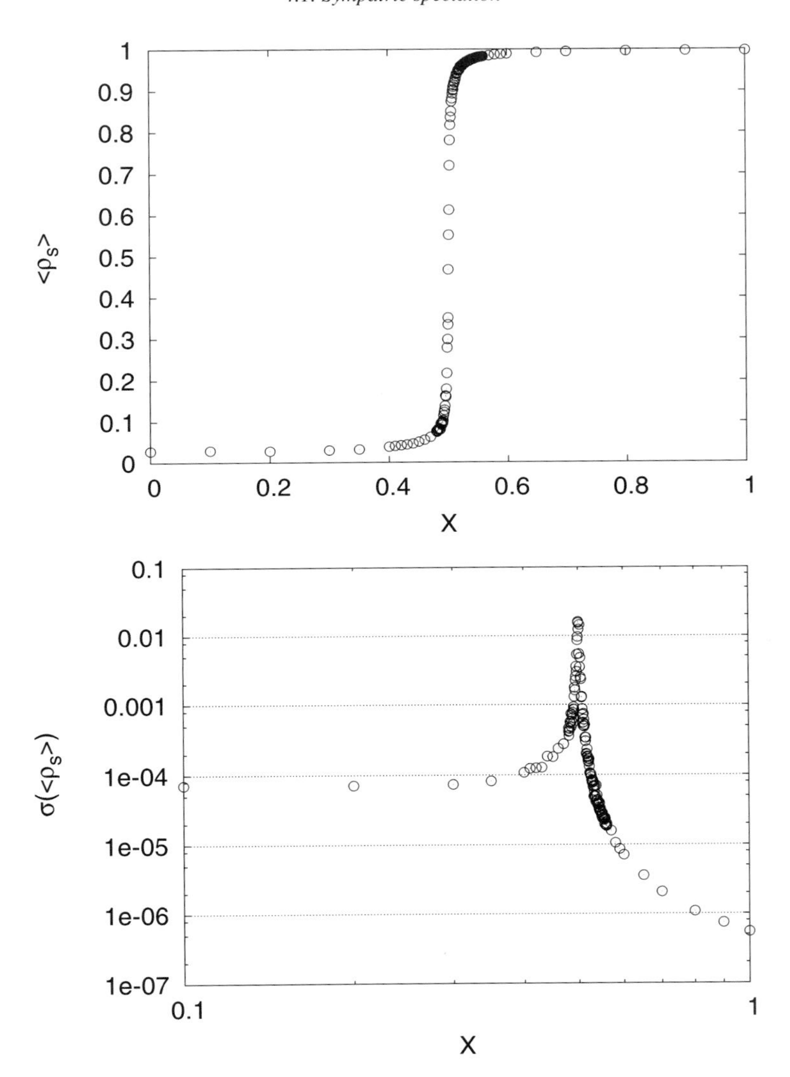

Figure 4.11. Mean value of the selective females density as a function of the competition level (upper part) and standard deviation of $\langle \rho_s \rangle$, $\sigma(\langle \rho_s \rangle)$, versus the competition level in a logarithm scale (lower part).

64 or more bits, and check whether the value of the susceptibility peak increases or not (in a second order phase transition of an infinite system this value goes to infinity, that is, the susceptibility diverges at the critical point). More simulations are being developed to clarify this point.

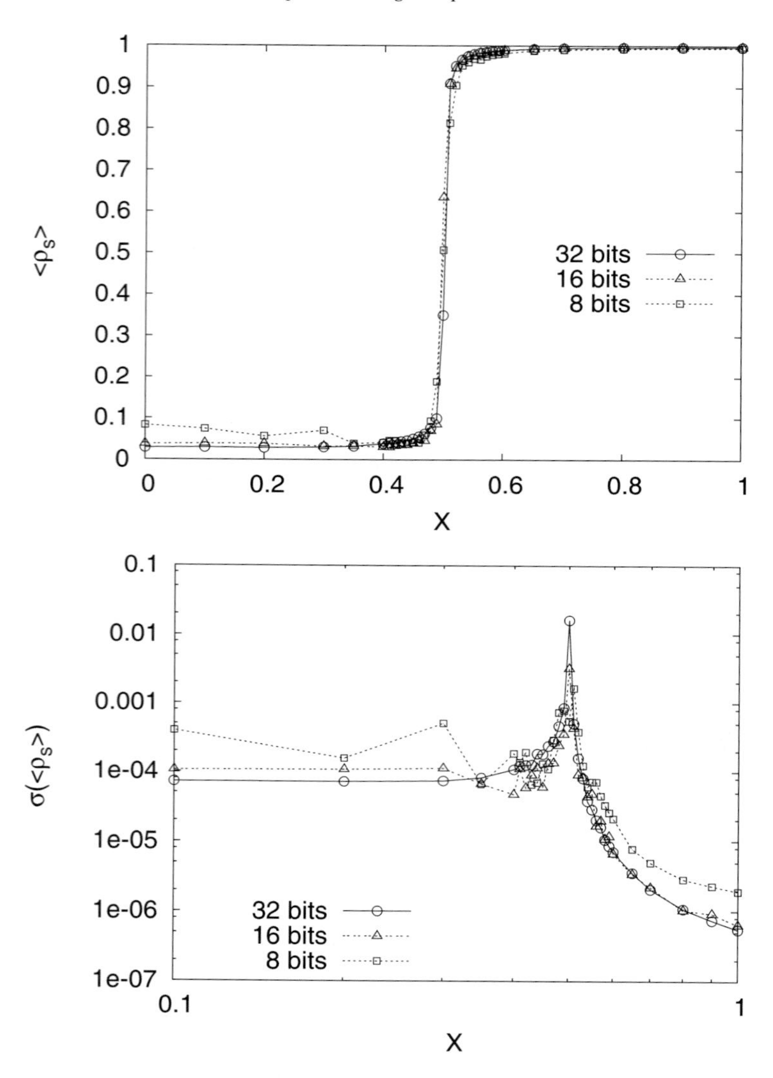

Figure 4.12. Mean value (upper part) and standard deviation (lower part) of the selective females density as a function of X, for different values of the bit-strings lengths.

All simulations presented above were done assuming that each female could choose to mate, among $N_m = 50$ males, the one whose phenotype lies deepest in the female's phenotypic group. When this number of choices is decreased to $N_m = 3$, the transition is destroyed, as can be seen from Figure 4.13. Continuing

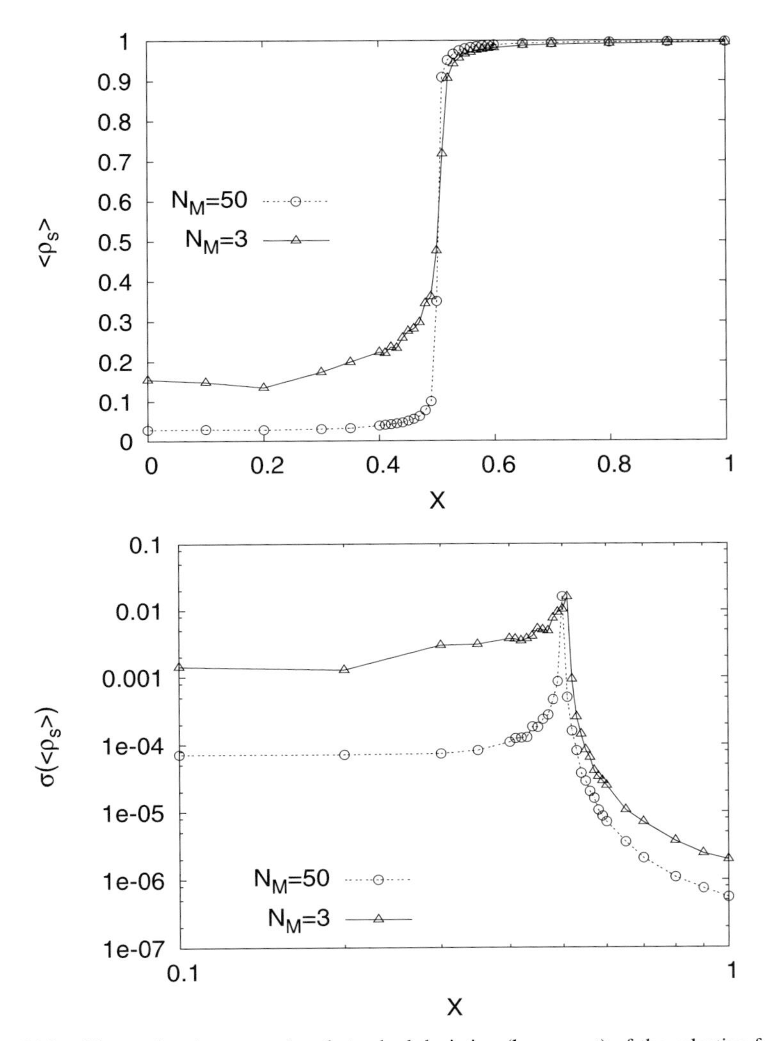

Figure 4.13. Mean value (upper part) and standard deviation (lower part) of the selective females density as a function of X, for two different numbers of mate choices. The simulation time needed for $N_M = 3$ was 8×10^5 with averages performed during the last 10^5 steps; for $N_m = 50$ the time requested to reach equilibrium is shorter, 4×10^4, with averages performed during the last 10^4 time steps.

with the same analogy, the number of choices N_M seems to play the role of the inverse of the magnetic field in a ferromagnetic/paramagnetic transition. Apparently, life is interesting if and only if women have a large number of men to choose from (something that only one of this book's authors finds perfectly reasonable).

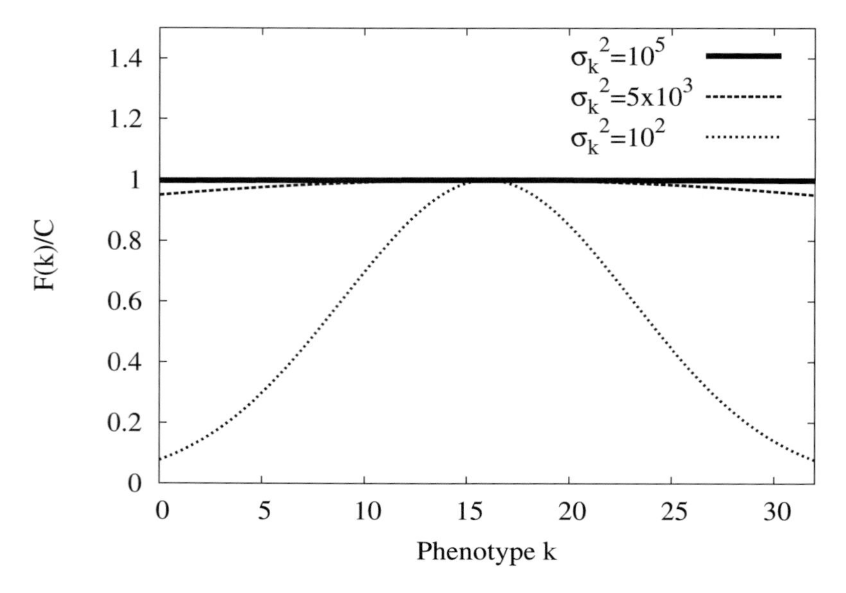

Figure 4.14. Fitness function according to equation (4.10), for different values of σ_k. For large values of σ_k all individuals have almost the same ability in using the available resources. The smaller the value of σ_k is, the smaller is the extreme sized individuals' fitness.

Luz-Burgoa, Schwämmle, Martins and de Oliveira (2005) also performed simulations where the individuals ability in using the available resources, F, is not constant but depends on the phenotype number k in the following way:

$$F_{\sigma_k}(k) = C * \exp\bigl(-(k-16)^2/\sigma_k^2\bigr). \tag{4.10}$$

Observe that for large values of the parameter σ_k, the value of $F_{\sigma_k} \sim C$, that is, all individuals present the same fitness, as has been considered until now; the smaller the value of σ_k is, the smaller is the fitness of the extreme-sized phenotypes, as shown in Figure 4.14.

Figure 4.15 shows the behaviour of $\langle \rho_s \rangle$ as a function of X, for different values of σ_k. From this figure we see that even for $X = 1$, when medium-sized phenotypes have to compete with the whole extreme-sized populations, equation (4.9), there is no speciation for $\sigma_k = 10$ (triangles) since the amount of food available for the extreme-sized populations is not large enough to compensate the high degree of competition to which the intermediate phenotypes are submitted to.

Figure 4.16 shows the final distributions of the phenotype numbers for different values of σ_k, when $X = 0.6$. From this figure we see that when σ_k is too small (triangles), the intermediate phenotypes completely dominate in spite of competition; on the other side, for large σ_k values (squares) speciation occurs and

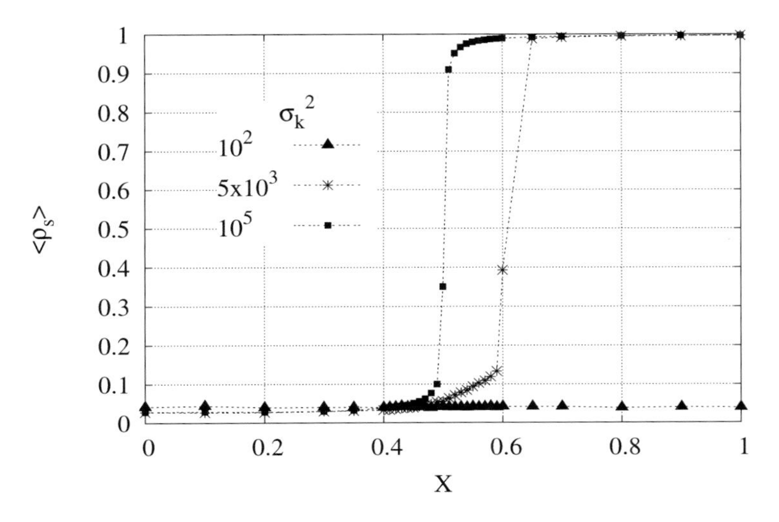

Figure 4.15. Behaviour of the mean selective females density as a function of the competition level X, for different values of σ_k.

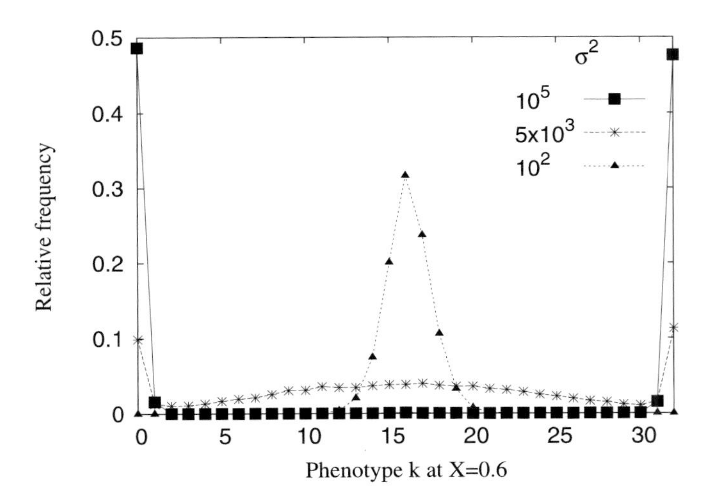

Figure 4.16. Final distribution of the phenotype numbers for different values of σ_k, when $X = 0.6$.

only the extreme phenotypes remain in the population. For intermediate values of σ_k (stars) we have the most interesting situation where almost all phenotypes coevolve, which signals that the phase transition may be a first order one.

A much simpler mean-field version of this computational model also predicts a phase transition, depending on the strength of competition between different phenotypes (Schwämmle, Luz-Burgoa, Martins and de Oliveira, 2005).

4.1.5. Models with two phenotypic traits

In the computational models presented until now the ecological character related to the ability in using the available resources was the same as used for sexual selection. However, it is also possible to obtain speciation considering these two characters as independent ones; in fact this is the case of the mean-field like models of Kondrashov and Kondrashov (1999) and Dieckman and Doebeli (1999). The first simulations based on the Penna model considering two phenotypic traits were performed by Luz-Burgoa, de Oliveira, Martins, Stauffer and Sousa (2003). They added to each individual's genome a third pair of non age-structured bit-strings, that suffers crossing, recombination and mutations (in both directions) in the same way as the other two pairs. The number of bits 1 in this new pair is also computed considering dominance (equal to 16), and the result is again an integer k' between zero and 32. In this way each individual is characterised by three pairs of bit-strings, the first one age-structured and related to genetic diseases, the second one related to some ecological trait (e.g., size) and the third one related to mating preference (e.g., colour). As before, sexual selectiveness is assigned by a single independent locus inherited with some mutation probability (in both directions). The mating rule is the same strong one used in the previous Section 4.1.4.

The authors followed Kondrashov and Kondrashov (1999) strategy and also considered an abrupt change in the ecology. The behaviours of the fitness function (ecology) and intra-specific competition are the same as those presented in Section 4.1.2, that is, $F(k)$ is single-peaked up to step 12 000 when it becomes bimodal. The distribution of the fitness trait (size) is single-peaked at $k = 16$ at step 12 000, as a consequence of the number of loci (16) where the 1 allele is dominant, and moves into a polymorphism after the ecology becomes bimodal. The sexual selection trait (colour) also shows an unimodal distribution at step 12 000 centred at $k' = 16$, and splits the population in two groups afterwards. But now a strong correlation develops between these traits, that is, a female chooses a mating partner because of his colour, and its correlation with size allow them to generate viable offspring.

The correlation between traits k (size) and k' (colour) is given by:

$$R(t) = \frac{\langle kk' \rangle - \langle k \rangle \langle k' \rangle}{\sigma_k \sigma_{k'}}$$

where

$$\langle k' \rangle = \sum_{i=1}^{N(t)} \frac{k'_i}{N(t)}$$

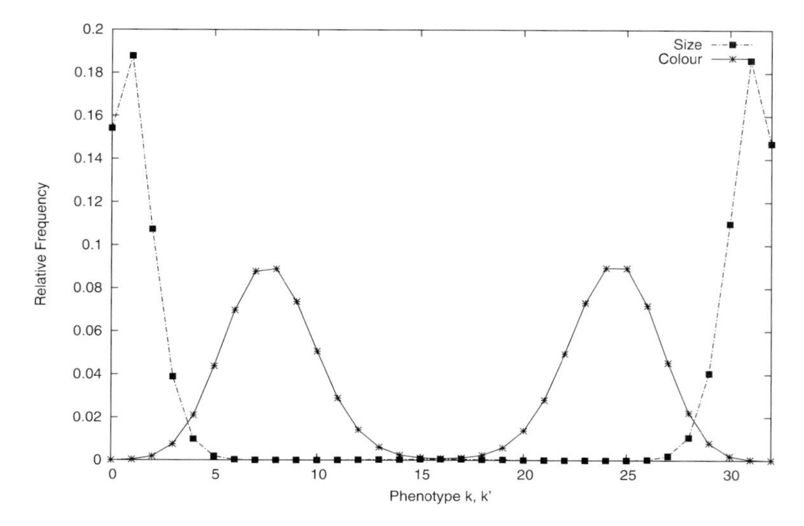

Figure 4.17. Distribution of sizes (squares) and colours (stars) at the end of the simulation. In this case the traits are positively correlated, and the population with fitness trait (size) to the left has its sexual trait (colour) also to the left of the plot.

is the mean value of the colours distribution of the whole population at time-step t, $N(t)$ is the population size and $\sigma_{k'}$ is the width of the distribution. The same calculation is done for trait k and the product kk'. Observe that this correlation can be positive or negative. When it is positive most individuals with $k > 16$ (< 16) have also $k' > 16$ (< 16) while when it is negative individuals with $k > 16$ (< 16) prefer $k' < 16$ (> 16). If for instance we consider that $k < 16$ (> 16) means small (large) fish and $k' < 16$ (> 16) means blue (red) fish, the simulations start with fish of all sizes and colours and finish, for instance, with all small fish blue coloured (positive R) or all small fish red coloured (negative R).

Figure 4.17 shows the distribution of both traits at the end of the simulation and Figure 4.18 presents the correlation between them. In the latter figure we also present a case where speciation failed. Sexual selectiveness also develops as a result of the evolutionary dynamics. At the end of the simulation all females are selective, and again assortative mating and reproductive isolation are the proxies in this model to the development of two separate species out of the single one that existed at the beginning.

4.1.6. Conclusions

Individual-based computational models have been successful in simulating the sympatric speciation process, using competition between different phenotypes

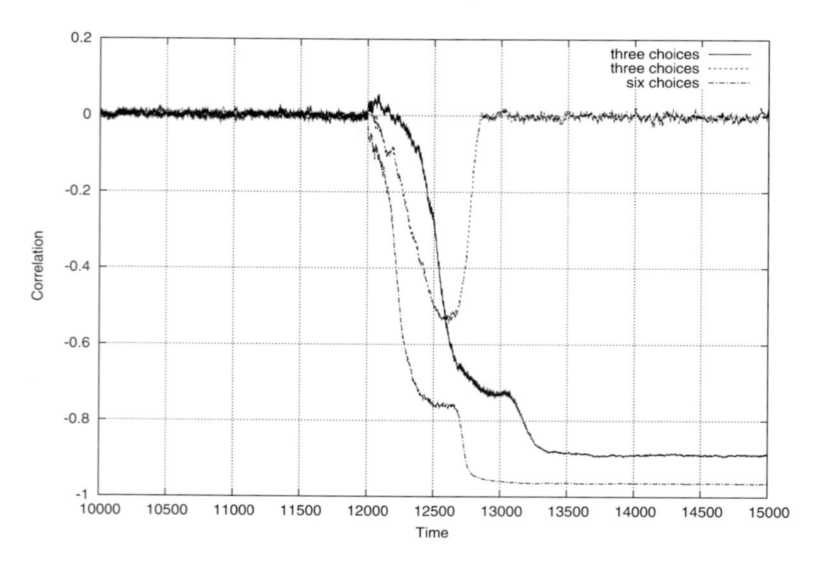

Figure 4.18. Correlation between the fitness trait and the sexual trait as a function of time, for two different numbers of available males a females has to choose in the moment of reproduction. A higher number of choices increases the speciation probability and the correlation between the traits.

and assortative mating as its main ingredients. When disruptive selection is induced by an abrupt change in the available resources, speciation is obtained even for moderate levels of competition (Section 4.1.3, equation (4.5)). When the environment is kept constant, it is necessary to have a strong competition between intermediate and extreme phenotypes in order to provoke disruptive selection, which, followed by assortative mating, leads to reproductive isolation (Section 4.1.4, equation (4.9)). In this case a phase-transition has been observed between a non-speciation state, where the females mate randomly, for low values of the competition level, and a state where all females mate assortatively, occurring for larger values of the competition level. If sexual selection is not included, competition will induce only a polymorphism (Sections 4.1.2 and 4.1.3) and the speciation process remains incomplete. The most important aspects of these computational models are that they offer the possibility of verifying whether speciation is likely to occur depending on different contingencies (which means that the outcome of two different simulations with identical parameters may depend strongly on the random number seed used in each one), and to measure the speciation velocity (Figure 4.9) which, as already emphasised in Section 2.6, is related to some critical exponent. The importance of fluctuations, absent in mean-field approaches, will appear again in the next section.

4.2. Parapatric speciation

As mentioned in the introduction of this chapter, allopatric speciation occurs when a physical barrier divides an original population into two geographically separated ones; in this case genetic drift and adaptation to the environment are the main ingredients for speciation. In case of sympatric speciation the two populations inhabit the same region, and its most important ingredients are competition for resources and assortative mating. Parapatric speciation is an intermediate case where there is no physical barrier but there is a gradient of temperatures or altitudes, for example, across the same region (for a review see Gavrilets (2004) as well as Coyne and Orr (2004)). The idea of traits being differentially adapted in different spatial locations is not new (Endler, 1973; Lande, 1982; Sanderson, 1989; Kirkpatrick and Barton, 1997), although still under investigation. The interesting question is how much does gene flow actually retard the development of geographic differentiation within a species. Here we concentrate on a numerical model which is, as far as we know, the first one to adapt the Penna ageing model to simulate parapatric speciation (Schwämmle, Sousa and de Oliveira, 2005). In this case the age structure is very convenient, since it allows a measurement of the life span of the individuals according to age and so to determine whether the hybrids are viable (survive until the minimum reproduction age) or not.

The model considers a single phenotypic trait, that is, a single pair of non age-structured bit-strings, and individuals survival probabilities are connected to this trait and to their geographic positions. Initially individuals are randomly distributed on a two dimensional square lattice of linear size L. They move at every iteration, with a rate m_m, to a randomly chosen less or equally populated nearest neighbouring site. If all nearest neighbours are more populated than the current individual's site, the movement is not carried out. This strategy guarantees a fast and balanced distribution of individuals over the whole landscape. The reproductive females select their mating partners randomly from the reproductive males localised at the same or at a nearest neighbouring site. Reproduction between different phenotypes, k, is allowed. Offspring are distributed into empty nearest neighbouring sites. If there is no empty site, the offspring is not produced, which means that the population size is controlled by the size of the lattice (Makowiec, 2001; Sousa, 2004).

The probability of an individual to die, at every iteration, depending on its x-position and phenotype number k is given by:

$$P_{\text{death}}(x, k) = S \times \left(1 - \left|g(x) - \frac{k}{32}\right|\right). \tag{4.11}$$

In this equation S is a parameter between zero and one representing the environmental selection pressure and $g(x) = x/(L - 1)$, where the coordinate x is an

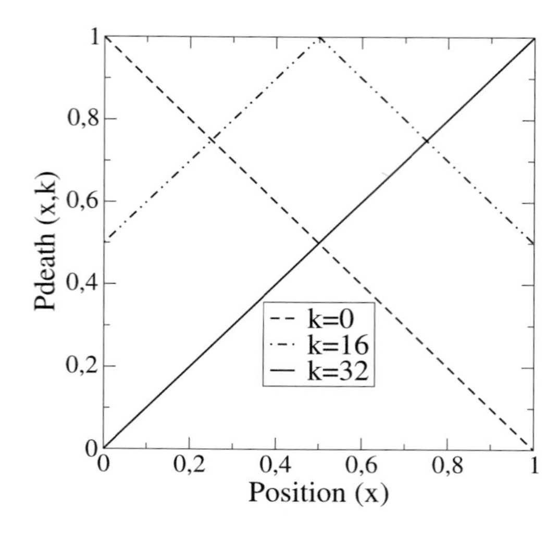

Figure 4.19. $P_{\text{death}}(x, k)$ according to x-position for three different values of the phenotype number, k. Individuals with high or low k survive better on opposite sides of the lattice whereas the ones with intermediate phenotype numbers have a higher death probability everywhere.

integer between zero and $L - 1$. For extreme phenotypes with $k = 0$, the perfect region in which to live corresponds to $x = L - 1$, where $P_{\text{death}}(L - 1, 0) = 0$, while for extreme phenotypes with $k = 32$ the perfect region corresponds to $x = 0$. Individuals with intermediate phenotypes also live better at the extremes of the lattice, but are less fitted than those with extreme phenotypes living in the correct extreme of the lattice. Figure 4.19 illustrates the death probability behaviour for three different values of k.

The results we are going to present were obtained using the following fixed parameters:

- Threshold number of genetic diseases $T = 3$;
- Minimum reproductive age $R = 8$;
- Birth rate $b = 4$;
- Rate of bad mutations in the chronological genome $m = 1$;
- Number of dominant positions in the chronological genome $D = 5$;
- Mutation rate of the phenotypic trait $m_p = 0.15$ or $m_p = 0.2$;
- Number of dominant positions in the phenotypic trait $D_p = 16$.

The simulations started with all genomes randomly filled with zeros and ones, and all individuals randomly distributed on the lattice. During the first 1000 iterations the probability to die given by equation (4.11) is set to zero and the initial distribution of the phenotype numbers is regulated solely by the mutations. It presents the usual Gaussian behaviour shown by the squares in Figure 4.20.

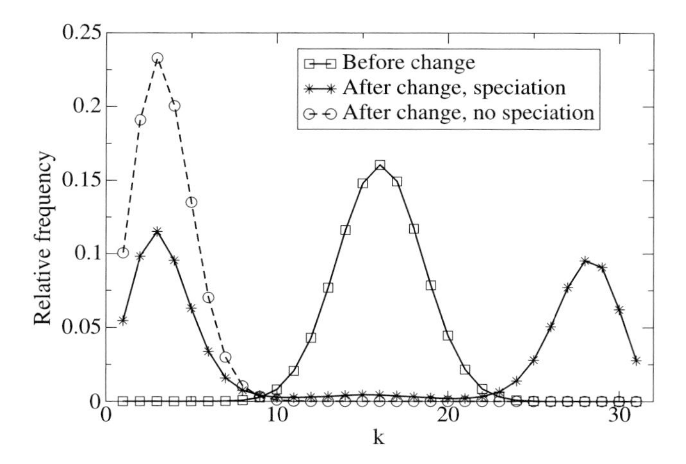

Figure 4.20. Distribution of phenotype numbers. The central curve (squares) corresponds to the initial distribution, before switching on equation (4.11). The circles correspond to the distribution at time-step 2000, in a case where speciation didn't occur after the ecological change. The distribution represented by stars corresponds to a speciation case, also at time-step 2000. In all cases $S = 0.24$ and $m_m = 0.99$.

After these transient steps the ecology is abruptly changed by switching on equation (4.11). Disruptive selection driven by the new ecology leads to a better survival of individuals with high and low phenotype numbers, depending on their current positions on the lattice.

The crucial parameters to obtain speciation in this scenario are the selection pressure S and the movement rate m_m. At low selection pressures, independently of the movement rate, the distribution of phenotype numbers remains unaltered, that is, a Gaussian centred at $k = 16$. The population decreases slightly at intermediate x-positions, but gene flow prevents disruptive selection from dividing the system into two sub-populations.

For intermediate selection pressures and movement rates $m_m \sim 1$, shortly after turning disruptive selection on, equation (4.11), the system reaches an extremely dynamical state where fluctuations may or may not drive the system to divergence. That is, for the same set of parameters, speciation may or may not occur depending on the initial random seed. When it does not occur, the adaptation of the phenotypes on one of the lattice sides is faster and gene flow forces the individuals on the other side to adapt themselves to the opposite phenotype. In this case the phenotypes distribution is unimodal, given by the circles in Figure 4.20, and the lattice is occupied by one of the extreme phenotypes, as shown in Figure 4.21. When speciation occurs, that is, when phenotypic adaptation is balanced, the final distribution is bimodal (represented by stars in Figure 4.20) and there is a clear

Figure 4.21. Final occupation of the lattice in a case where speciation didn't occur. One of the sub-populations with extreme phenotype randomly dominates and finally occupies the whole lattice. The left side remains less populated because in this case it is the worst side for this extreme phenotype to survive (black sites are empty).

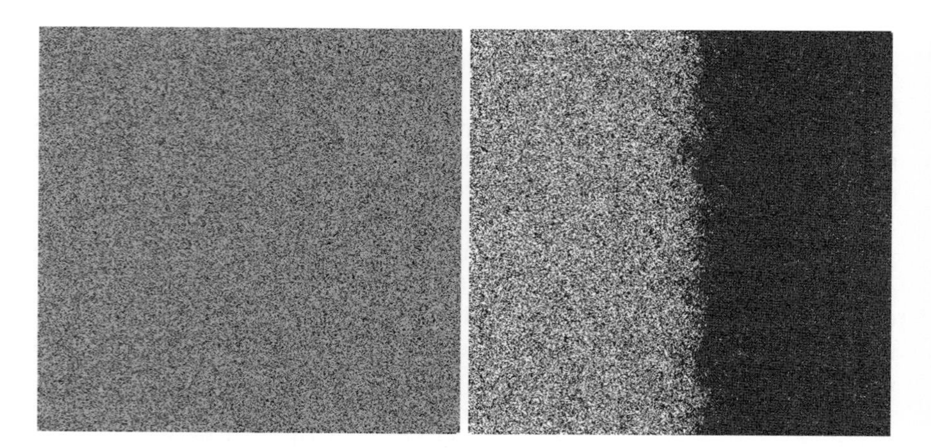

Figure 4.22. Initial (left) and final (right) occupation of the lattice in a case where disruptive selection led to speciation. Black sites are empty; different phenotype numbers are represented by different grey tones, ranging from white to dark grey. Initial phenotypes are randomly distributed between 0 and 32.

division in the lattice occupation between the extreme phenotypes, as shown in Figure 4.22.

From Figure 4.20 we can notice that even when speciation occurs, hybrids (intermediate phenotypes) do not completely disappear, since there is no sexual selection preventing their production. However, Figure 4.23 shows that the

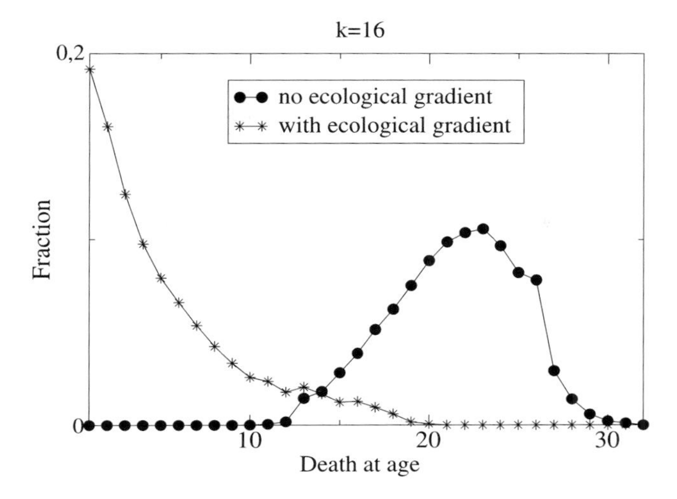

Figure 4.23. Histogram of the final fraction of individuals with $k = 16$ that die at a given age. Circles correspond to a simulation where equation (4.11) was not switched on, presented for comparison.

majority of these hybrids die at low ages and do not generate offspring. Their low viability characterises a speciation process (Porter and Johnson, 2002) in a situation where a small gene flux between different extreme phenotypes persists. Models with small population sizes or mating over large geographical distances need assortative mating in order to obtain speciation (Gavrilets, 1997; Doebeli and Dieckmann, 2003).

A cline is defined as a gradient in a measurable character. Relative to the dispersal rate of a species, the strength of a cline between regions is indicative of the extent to which the inhabitants have differentiated. A steep cline means sharp differentiation while a gentle cline means indistinct divergence between areas (Endler, 1973). In the present case the authors chose the phenotype number k as the measurable character. Figure 4.24 shows the fraction of individuals with $k = 0$ and $k = 32$ at each position x of the lattice, for the case where speciation occurred. A steep cline can be observed for both $k = 0$ and $k = 32$ populations.

In this model low movement rates or very high selection pressures (as also found by Doebeli and Dieckmann (2003)) prevent speciation events. In both cases a great part of the population dies out at the time when the ecological function (4.11) is set. Fluctuations dominate divergent adaptation and the initial Gaussian distribution of phenotypes moves to one of the extremes. For small population sizes fluctuations also seem to always prevent speciation, independently of the movement rate: No speciation events were obtained for lattice sizes smaller than $L = 150$.

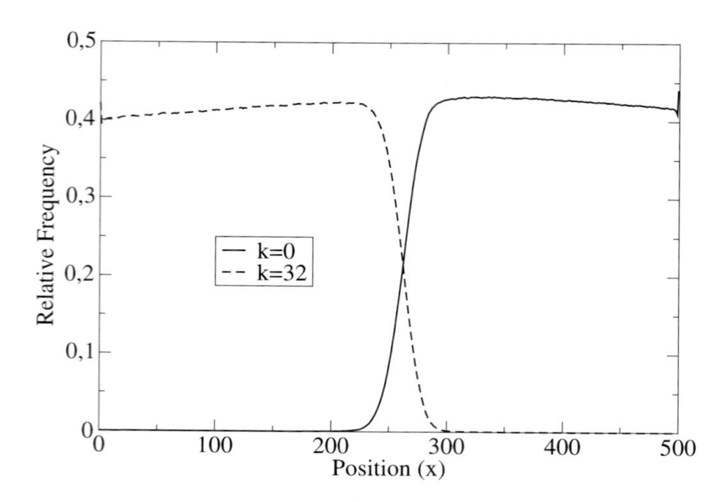

Figure 4.24. Frequency of individuals with phenotype numbers $k = 0$ and $k = 16$, for each position x of the lattice, averaged over the last 10 000 time steps.

It is important to emphasise that the speciation models presented in this chapter allow fluctuations of all quantities, which hinders adaptation and the division of the system into two different phenotypic populations, even for intermediate values of the ecological selection pressure. This could explain the not so frequent occurrence of speciation in Nature, where many environmental factors act on the different population quantities, like the phenotypic distribution, and where fluctuations of these quantities are ubiquitous. Even if the conditions are optimal, speciation remains a statistical event. (The main differences between the mean-field approaches, where fluctuations are neglected, and those presented here have already been explained in detail in Chapter 2.) In the present case the results suggest that parapatric speciation occurs preferably when a large population undergoes a sudden disruptive selection over large geographical distances compared to the range of individuals movements.

4.3. * Many-species models

Biological evolution of species presents some universal behaviour due to its time-and-size scale-free character (see, for instance, Kauffman (1993) and Chapter 2). A parallel between this feature and critical phenomena has already been explored in Sections 2.5 and 4.1.4. Why would the idea of universality apply to evolutionary systems is an interesting and important conceptual question. Some hints towards a possible answer can be seen in Doebeli and Ruxton (1997), Geritz, Kisdi, Meszéna and Mertz (1998) and Parisi (1999). A famous example of a scale-free behaviour is the classification of extinct genera according to their lifetime, a long-term study of fossil data performed by paleontologists John Sepkoski and David Raup (Sepkoski, 1993; Raup, 1986; Raup, 1991). The frequency distribution they found is compatible with a power-law decay with exponent 2. The same exponent was confirmed by at least two distinct theoretical computer models (Newman and Roberts, 1995; Solé and Manrubia, 1996). The computational models we are going to explore in this section are characterised by these scale-free behaviours, related to long-term memory and diversity, as pointed out in Section 2.4.

4.3.1. The Bak–Sneppen model

Slightly modifying an earlier model for surface growth (Sneppen, 1992), Bak and Sneppen (1993) introduced their now-famous model for biological evolution, based on an extreme value dynamics (see also Bak (1997)). In this model there are N species, each one occupying one site of an one-dimensional lattice (a ring). Each species has a random survival fitness $0 \leqslant f_i \leqslant 1$. At the beginning, all fitness are uniformly tossed at random between 0 and 1. The simulation evolves according to the following dynamical rule:

- Search for the smallest f_i corresponding to species i;
- change f_i, f_{i-1} and f_{i+1} into 3 other values randomly chosen between 0 and 1 (mutation or extinction);
- repeat.

One time step consists in iteratively applying this rule N successive times (an N-cycle), where N is the same number which measures the population size. Thus, the time $t = 0, 1, 2, 3 \ldots$ is a discrete variable. $1/N$-fractions of the time unit can also be measured, by considering incomplete N-cycles. Thus, for larger and larger values of N, one obtains more and more the continuous time.

At the start of the simulation the fitness on average grows, although there are fluctuations up and down. However, after a transient period, the system reaches a stationary critical state where the fitness does not grow any further on average: All species have fitness above some threshold (state of "stasis"), which is very close to 2/3 for long enough times (Paczuski, Maslov and Bak, 1995). That is, the band of distributed fitness starts between 0 and 1, but shrinks until all fitness become distributed between the threshold and 1. Figure 4.25 shows the time evolution of the lower bound of this band for a $N = 100\,000$ species system.

Consider a point in time when all species are over the threshold; at the next step the least fit species (right at the threshold) will be selected, eventually starting an avalanche or "punctuation" of mutation events. (Whenever the fitness of a given species changes, it is possible to think that the species has undergone a mutation or that it has become extinct and replaced by a new one.) After a while,

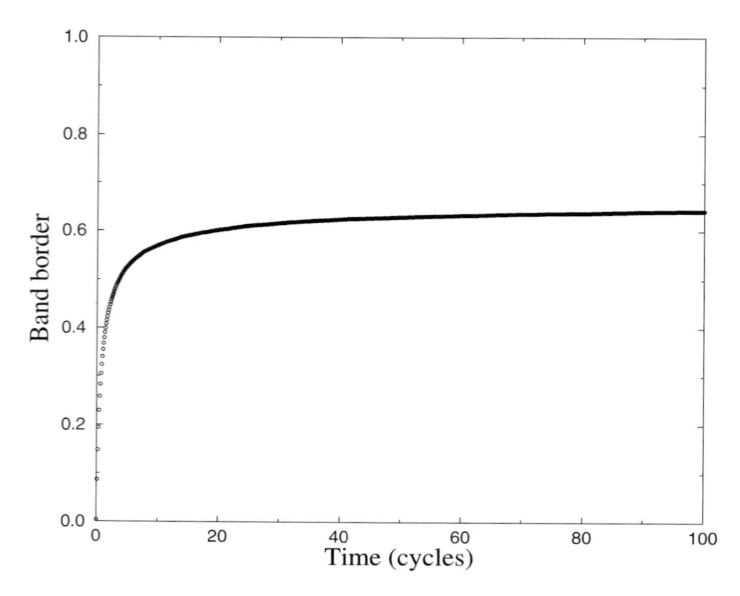

Figure 4.25. Evolution of the lower bound for the band of fitness in the Bak–Sneppen model for a system of 100 000 species.

the avalanche stops, when again all species have fitness above the threshold. The avalanche lifetime corresponds to the number of steps needed to recover the state of stasis and the avalanche size to the number of active species between two consecutive states of stasis. When this process is repeated for large systems (large number of species), one obtains the number $N_a(S)$ of avalanches of size S (involving S species) given by the power law:

$$N_a(S) \propto S^{-\tau}.$$

This distribution means that there is no characteristic size for the avalanches, as would happen if instead of a power law it had an exponential behaviour. The larger the system, the larger the possible maximum size of an avalanche. Small avalanches, in which a few number of species become active, are much more frequent than large ones; however, the probability that a system-sized avalanche occurs, activating all species, is not zero. Such a dynamical behaviour can explain the extinction of the dinosaurs without using any external agent, such as meteorites colliding with Earth (but of course does not exclude such a possibility).

On the other hand, the non-vanishing band-width of this model characterises a diversity which remains forever. In fact, the transient time T one needs to wait in order to reach the threshold $f_c \simeq 2/3$ is of the same order of magnitude as the system size, i.e., $T \sim N$. In the limit of larger and larger sizes, $N \to \infty$, the sys-

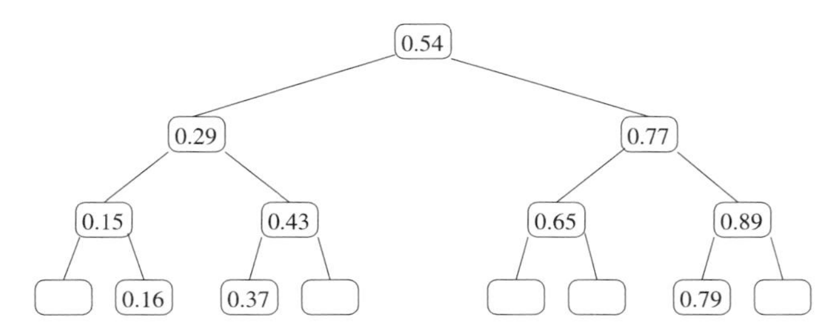

Figure 4.26. Binary tree where the sequence {0.54, 0.29, 0.43, 0.77, 0.15, 0.65, 0.37, 0.16, 0.89, 0.79} is stored in this order.

tem would evolve in an eternal transient. This is the main difference between an evolutionary system and a simple optimisation process. In the latter, one searches for the single best situation among many possibilities, discarding all other options. Evolution, in contrast, preserves many alternative options, not only a single "best", in order to keep the system able to adapt itself to future environmental changes.

In order to implement this model on computers, the program needs to find the minimum fitness among all individuals. The simplest approach is to scan all of them, sequentially, registering at each step the minimum value so far. The corresponding Fortran code can be found in the appendix, Section 9.3. However, this requires N comparisons, which forbids the simulation of very large systems. A good alternative is to construct a *binary tree* with the N fitness. This tree has a root; below it there are two other sites, one on the left and one on the right. Below each new site, another pair, on the left and on the right, and so on, as shown in Figure 4.26.

The sequence of N fitness is stored on this tree as follows. The first entry is located at the root. If the second entry is larger than the first, it is stored on the right side below the root; otherwise, on its left. For each new entry along the sequence, one compares its value with the root, deciding to go downwards to the left or to the right. Then, one repeats the comparison at this new place, deciding again to go downwards to left or right, and so on, until a vacant site is reached. In this way, in order to get the minimum value stored on the tree, one simply goes downwards always to the left, until the last occupied site. That is, one follows only a single branch along the tree, which average length is of the order of $\log_2(N)$, much smaller than N, moreover for large systems. The following C-language routine does the job.

```
unsigned minimum() {
/* finds the minimum on the binary tree */
 unsigned              i,j;
 j = root;
```

```
 do i = j; while(j=left[i]);
 return(i);
}
```

Then, the following routine puts a new entry on the tree.

```
void put(new) unsigned new; {
/* includes a new entry into the binary tree */
 unsigned            i,j,lf,f;
 j = root; f = F[new];
 do {i = j;
  if(f<F[i]) {j = left[i]; lf = 1;}
  else {j = right[i]; lf = 0;}
 } while(j);
 top[new] = i;
 if(lf) left[i] = new; else right[i] = new;
}
```

Finally, below is a third routine designed to remove an entry from the tree. In order to remove entry K, one considers entries L and R below it (left and right, respectively), as well as entry A below L (right). If L is empty, K is replaced by R, otherwise by L. Furthermore, if L, R and A are all three occupied (generating 2 right branches below L, instead of only 1), then A is transferred to the first empty position along the leftmost branch below R.

```
void remove(K) unsigned K; {
/* removes an entry from the binary tree */
 unsigned            t,L,R,A,i,j;
 t = top[K]; L = left[K]; R = right[K];
 top[K] = left[K] = right[K] = 0;
 A = right[L];
 if(t) {
  if(L) {
   if(K==left[t]) left[t] = L; else right[t] = L;
   top[L] = t;
   if(R) {top[R] = L; right[L] = R;
    if(A) { /* re-position of A */
     j = R;
     do i = j; while(j=left[i]);
     top[A] = i; left[i] = A;
    }
   }
  }
  else {
   top[R] = t;
   if(K==left[t]) left[t] = R; else right[t] = R;
  }
```

```
 }
 else {     /* remove root */
  if(L) {root = L; top[root] = 0;
   if(R) {top[R] = root; right[root] = R;
    if(A) {
     j = R;
     do i = j; while(j=left[i]);
     top[A] = i; left[i] = A;
    }
   }
  }
  else {root = R; top[root] = 0;}
 }
}
```

Certainly there are more efficient ways to implement the binary tree, in particular by using recursivity and pointers. However, the resulting code is more complex and so more complicated to understand. A complete C-version of the program using the routines presented here can be found in Martins and de Oliveira (2004).

4.3.2. Lineage branching

Branching processes in general show scale-free behaviour. In this case, an important class, with multiples of $1/4$ as exponents, is ubiquitous. This interesting issue was studied by G.B. West and collaborators (West, 1999; Enquist, Brown and West, 1998; West, Brown and Enquist, 1997). For a recent overview see Savage, Gillooly, Woodruff, West, Allen, Enquist and Brown (2004); see Demetrius (2003) for an alternative. Studying blood transport networks, they proposed a model based on three basic ingredients: A hierarchical branching pattern, where a vessel bifurcates into smaller vessels and so on; a minimum cut-off size for the smallest branches, which makes the branching mechanism a finite process, and a free-energy minimisation constraint. From these three basic hypotheses, they were able to show the emergence of the exponents $1/4$, $1/2$, $3/4$, etc. Of course, not only blood vessel systems follow this general framework, and the same class of exponents, multiples, of $1/4$ were indeed measured within many other contexts. The model that follows, proposed by de Oliveira, Martins, Stauffer and de Oliveira (2004), was developed with the idea that biological speciation could fit very well into the general branching process framework described by West.

The asexual population size is kept constant, with P (typically 10^5 or 10^6) individuals representing a sample of a much larger set. Each individual is characterised only by its genome, represented by an array of g bits (typically 32, 64, 128, ..., 2048) zeros and ones. At the beginning, all bits are zeroed, and all individuals belong to a single lineage. The survival probability of each individual is given by x^{N_i+1}, where N_i is the number of bits 1 along the genome of individual i, that is, the larger the number of bits 1 along the genome, the larger is the individual's death probability. At each time step, a certain fraction f (typically 1% or 2%) of individuals die, each one according to its own death probability, as the outcome of intra-lineage competition.

At each time step, the simulation first obtains the value of x (before the death cycle) by solving, as in Section 2.10, the polynomial equation:

$$\sum_i x^{N_i+1} = P(1-f)$$

where the sum runs over all living individuals. This requirement keeps the population constant. Equivalently, one can solve

$$\sum_N H(N)x^{N+1} = P(1-f)$$

where now the sum runs over N $(0, 1, 2, \ldots)$, and $H(N)$ counts the current number of individuals with precisely N bits set along the genome. After computing the value of x, we scan the whole population ($i = 1, 2, \ldots, P$), tossing a real random number between 0 and 1 for each individual i, in order to compare it with its survival probability: If the random number is larger than x^{N_i+1}, the individual dies.

After each death, another individual is chosen at random to be the parent of a newborn. Its genome is copied, and some random mutations are introduced at a fixed rate per bit (typically 1/32) which does not depend on the genome length. Each mutation flips the current bit state (from 0 to 1 or vice-versa) at a position tossed along the genome. After all mutations are performed, the newborn is included into the population.

If the newborn presents fewer bits 1 than its parent, it receives the label of the potential founder of a new lineage. During the time steps that follow, all its descendents will be monitored: If, at some future time, the number of those descendents still alive reaches or surpasses a minimum threshold s_0 (typically 10), then all descendents of the now confirmed founder, including itself, are considered to belong to a new lineage. On the other hand, extinction occurs when the last individual of a given lineage dies. Although a rare event, a lineage can also become extinct if all its individuals descend from the same potential founder, being altogether transferred to another, new, lineage, by reaching the threshold s_0. A similar model, but without the lineage branching step, was already used before (de Oliveira, de Oliveira and Stauffer, 2003).

Figure 4.27 shows the number of living lineages as a function of time. Each time step corresponds to a scan of the whole population performing deaths and births. The number of living lineages are divided by the constant number of individuals, in order to show that one lineage indeed corresponds to a considerable number of individuals (varying from approximately ten thousand, on average, for

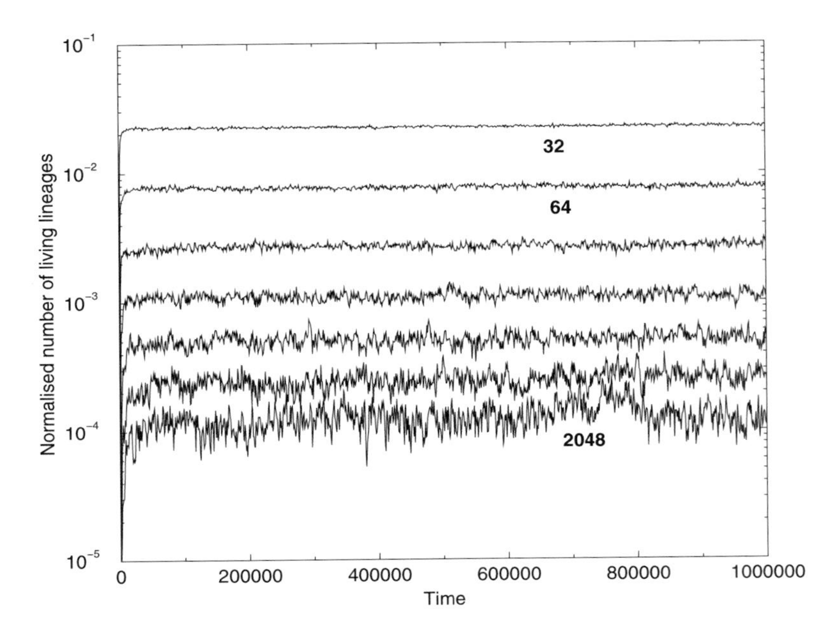

Figure 4.27. Number of living lineages normalised by the population size, as a function of time, for genome lengths 32, 64, 128, 256, 512, 1024 and 2048; $P = 100\,000$, $S_0 = 10$ and $f = 0.02$.

the largest genome length of 2048 bits, down to fifty individuals for the smallest genome length of 32 bits). Figure 4.28 shows the results after averaging over the last 10^5 time steps, in a total of 10^6 time steps. The exponent obtained through a linear fit to the simulation data is -1.24 (error bar within the last digit), which remained the same for other runs with different sets of parameters. Observe that this value is very close to $5/4$, falling into the same family of simple multiples of $1/4$ already mentioned.

Figure 4.29 shows the number of lineages which become extinct each time step, as a function of time. Extinction becomes more difficult for larger genome lengths. Figure 4.30 shows the total number N of extinct lineages, during the whole one-million-time-step history, as a function of the genome length. Again a power-law behaviour is observed, with an exponent very close to 1.

The distribution of lineage lifetimes as a function of genome sizes is illustrated in Figure 4.31. Now the exponent obtained is very close to 2, in agreement with the exponent found by paleontologists John Sepkoski and David Raup (Sepkoski, 1993; Raup, 1986; Raup, 1991) from fossil data.

The authors (de Oliveira, Martins, Stauffer and de Oliveira, 2004) were also able to derive analytically scaling laws and relations between exponents that agree with their simulational results. As they pointed out, their lineage model presents

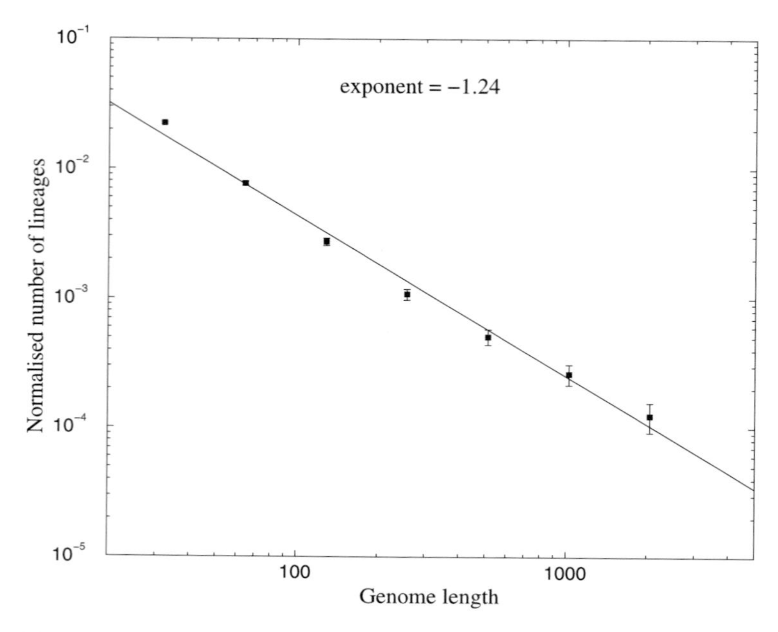

Figure 4.28. Number of living lineages normalised by the population size and averaged over the final 10^5 time steps, as a function of the genome length.

the same three fundamental ingredients that West and collaborators demonstrated to give rise to exponents multiple of $1/4$, namely:

(1) a multiple hierarchical branching: In this case, lineages born from others;
(2) a size-invariant limit for the final branch: In this case, a fixed minimum population s_0 is required in order to have branching;
(3) a free-energy minimisation process: In this case, the growing-entropy tendency provided by the random mutations (in the direction of randomising the bits along the genome as time goes by) is balanced by the selection mechanism (which gives preference to individuals with the smallest possible number of bits 1).

Another interesting model dealing with many species and also presenting scale-free behaviour is now being developed by Schwämmle and Brigatti (2005). The preliminary results we present below were obtained using the age-structured genome of the Penna model plus an integer number k ($0 \leqslant k \leqslant P$) to represent each asexual individual. The integer k accounts for all the phenotypic characteristics that determine the individual's adaptation to a specific ecological niche; individuals with close values of k belong to the same species. Those that succeed in reaching the minimum reproduction age R generate, at each time-step, one off-

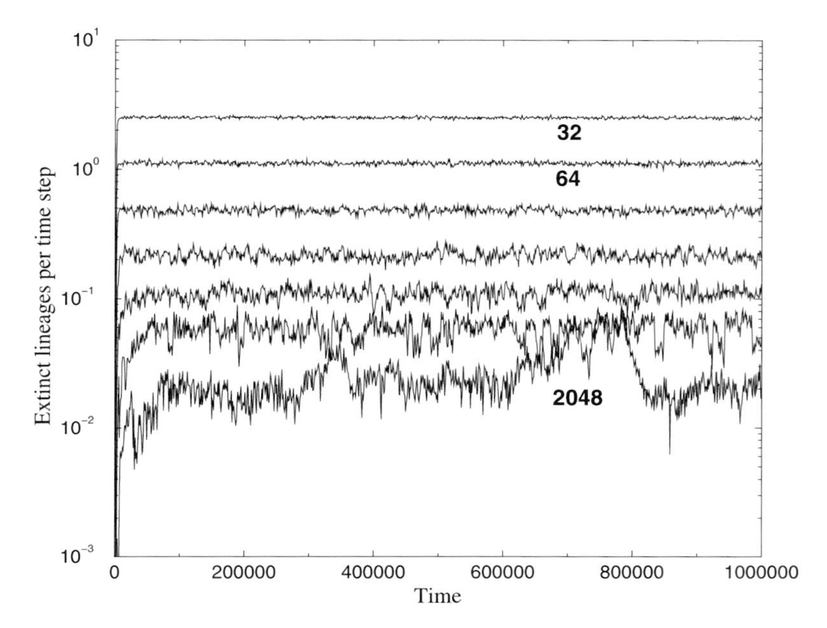

Figure 4.29. Number of lineages which become extinct per time step, averaged over intervals of 10^3 time steps, as a function of time for various genome lengths, as given by the numbers on the curves.

spring. The age-structured part of the offspring genome is inherited according to the asexual Penna model (Section 3.2.1), and its value of k is the same as the parent's one except for an eventual mutation of ± 1, that occurs with a given probability specified at the beginning of the simulation and that is the same for all individuals.

The death probability of any individual with a given k value is given by:

$$D(k) = \frac{1}{N_{\max}} \sum_{l=1}^{P} N_l \exp\left(-\frac{(k-l)^2}{2b^2}\right)$$

where N_l represents the number of individuals with phenotype l, $N_{\max}$ is the carrying capacity and b is a control parameter. Observe that the sum in the above equation spans all the possible configurations in the phenotype space and competition decreases according to phenotypes distance.

Since the model deals with an asexual population, the biological definition of a species as a reproductively isolated population has to be modified: In this case a species is defined as a group of individuals that share most of their phenotypic features but which differ for some traits. According to this definition the used algorithm associates different species to different clusters of individuals that have

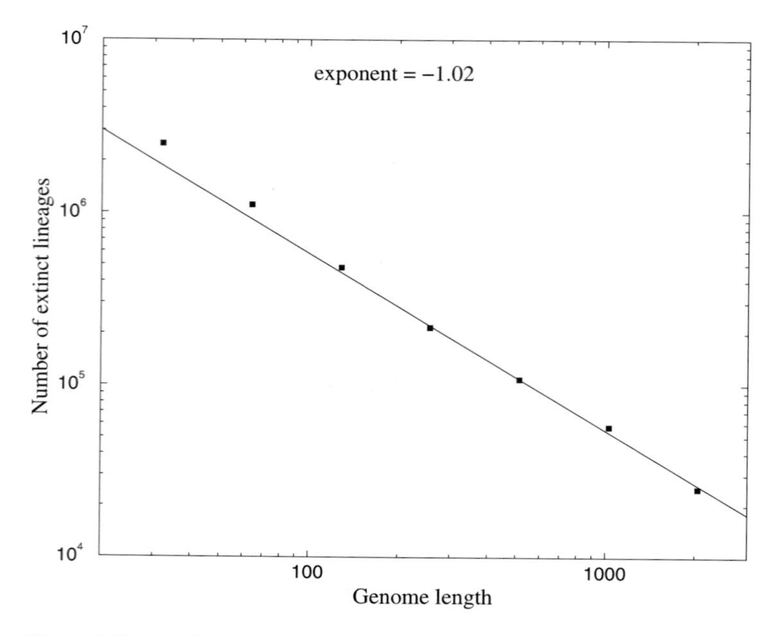

Figure 4.30. Total number of extinct lineages as a function of the genome lengths.

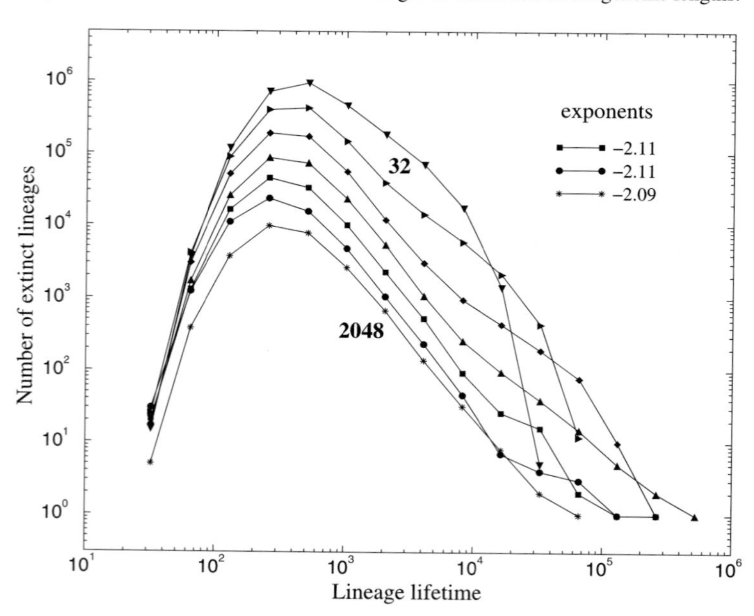

Figure 4.31. Distribution of extinct lineages according to lifetime, for different genome lengths.

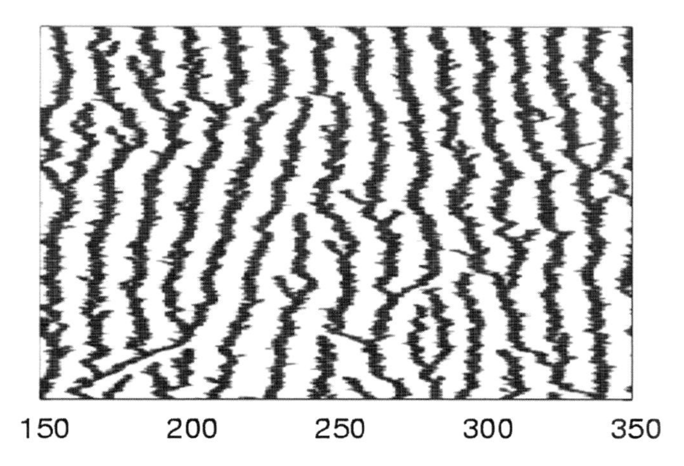

Figure 4.32. Piece of an evolutionary tree generated for $P = 500$. The horizontal axis corresponds to the phenotypic numbers k and the vertical axis to time.

a small (for instance, just one unit) phenotypic distance between them. In this way the space between two clusters cannot be occupied by individuals and the dynamics allows the self-organisation of a varying number of phenotypic clusters, each of them subjected to extinction or branching. An example of a stable and living evolutionary tree generated with standard parameter values can be see in Figure 4.32.

Figure 4.33 shows the number of extinct species according to lifetime. Again the power law behaviour with an exponent close to two is observed, except for very long times (as also obtained by Chowdhury, Stauffer and Kunwar, 2003). A detailed version of this model (without the Penna age-structure) and corresponding results will appear elsewhere.

4.3.3. Ecosystems

Thus far we dealt mostly with the problem how one species can split into two; Schwämmle and Brigatti (2005) had many species but no prey nor predators. Real life contains numerous species, grouped as mentioned in Section 2.8 into genera like homo, then families like hominides, orders like primates, classes like mammalia, phyla like chordata, and kingdoms like eukaryotes of which animals are just one part. Animals kill to eat, and usually the larger animals eat the smaller ones. One may arrange the various prey-predator relations into a food chain (He, Pan and Wang (2005); see also end of Section 4.1.3) where the top animal (e.g., wolves) eats the middle animal (sheep), which eats the lowest species (grass). It is much more realistic to look at food webs, as reviewed by Drossel and McKane,

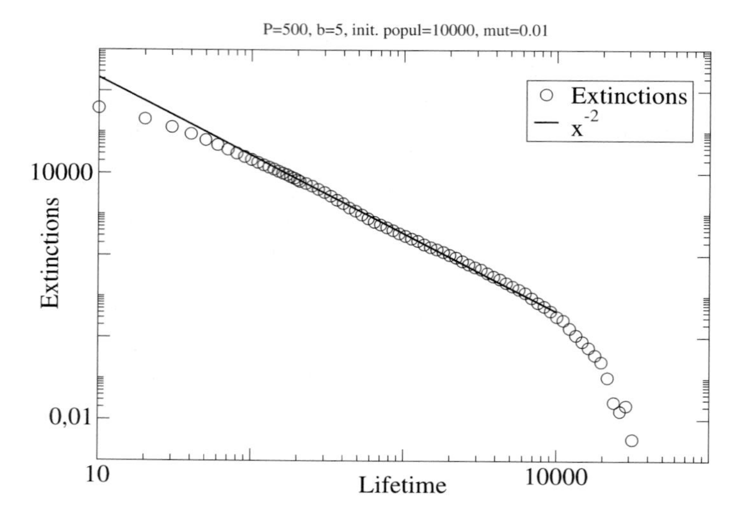

Figure 4.33. Distribution of extinct species as a function of lifetime. Deviations from a power law are observed for very long lifetimes.

p. 218 in Bornholdt and Schuster (2003). There each species may eat several others on a lower food level, and can be eaten by several other species on a higher food level. We neglect parasites which are small animals living in and profiting from a larger species, as well as cannibals or similar cases where animals eat others on the same food level. In agreement with the rest of this book we concentrate on simulations like Droz and Pękalski (2004) which treat each individual separately: birth, maturity, death. The Chowdhury model, starting with the spin-glass approximation of Chowdhury, Stauffer and Kunwar (2003), seems the most complicated one and was recently reviewed in greater detail by Stauffer, Kunwar and Chowdhury (2005) and, in a more general context, by Chowdhury and Stauffer (2005).

On top of the food web in this model is one species in level 1, followed by two species on level 2, four on level 3, and in general $2^{\ell-1}$ on level ℓ. Some species may be extinct; thus this number $2^{\ell-1}$ is the number of possible species, i.e., the number of ecological niches. For each non-extinct species, animals having at least the minimum reproduction age R generate offspring with a probability $(1-V)A$, with a Verhulst death factor $V = N/K$ where N is the current number of living individuals in that species, and K the carrying capacity. Thus the Verhulst deaths occur only at birth (Martins and Cebrat (2000); see Section 3.2.1). The above ageing proportionality factor A decreases linearly with age, from one at age R to zero at the maximum genetic life span taken as $X = 100 \cdot 2^{(1-\ell)/2}$ iterations.

Animals then give birth to M offspring simultaneously. If they are not eaten by others, they die with a Gompertz probability $\exp[0.05(a - X)/M]$ for age $a \geqslant R$.

Animals of species i may eat species k from the lower adjacent food level; then their couplings are $J_{ik} = +1$, $J_{ki} = -1$; otherwise both couplings J are zero. Thus in addition to the above Verhulst and Gompertz deaths, animals may die because they are eaten by predators, or because they do not find enough prey. These deaths are determined by insufficient food, given by a sum over the lower food level of prey, and by predators, given by a sum over the upper food level of predators.

The parameters R, M, and the prey-predator matrix J are self-organised, that means their values emerge via random mutations by ± 1. Also the total number of food levels is self-organised: If the total biomass is below some fixed threshold, then with some probability a new food level is added. Thus Darwinian survival of the fittest brings these values close to their optima but the continuous stream of new mutations also makes them sub-optimal. Therefore these values are not the same for all, but follow some distribution (Stauffer, Kunwar and Chowdhury (2005) and literature cited there).

Speciation may happen when other species became extinct, leaving empty their niches in the ecosystem. Then one species, selected randomly from the highest oc-

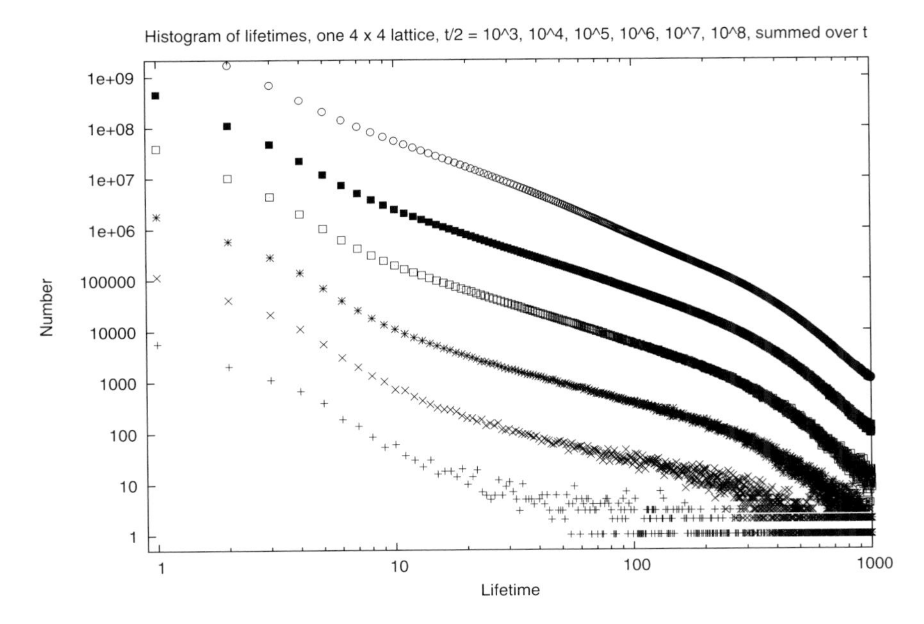

Figure 4.34. Distribution of lifetimes in a 4×4 lattice after small to large numbers of iterations, from bottom to top. (Same simulation as in next figure.)

cupied food level below that of the empty niches, occupies with some probability all the empty niches. There it may from then on have mutations different from those on its original place in the lower food level, and thus in each of the now-filled niches of the upper food level, a new species starts: sympatric speciation.

Finally, one can put each of these ecosystems onto a small square lattice, with different random numbers used for each site. With some probability, a random fraction of the population moves into a neighbour site (with periodic boundary conditions) provided the randomly selected neighbour is empty at the particular niche from which the invading population comes. Thereafter it has new mutations there, thus leading to parapatric speciation.

Figure 4.34 shows that the distribution of species lifetimes roughly but not precisely follows a power law relation with an exponent -2; the tail may be exponential. Different version of this model all gave similar curves. Figure 4.35 shows that the maximum number of food levels, which starts at three, increases about logarithmically with time. The mortality of the top species increases not very strongly with age (Kunwar, 2004) since many die from starvation; if applied to humans the situation thus corresponds to ancient and not to modern times.

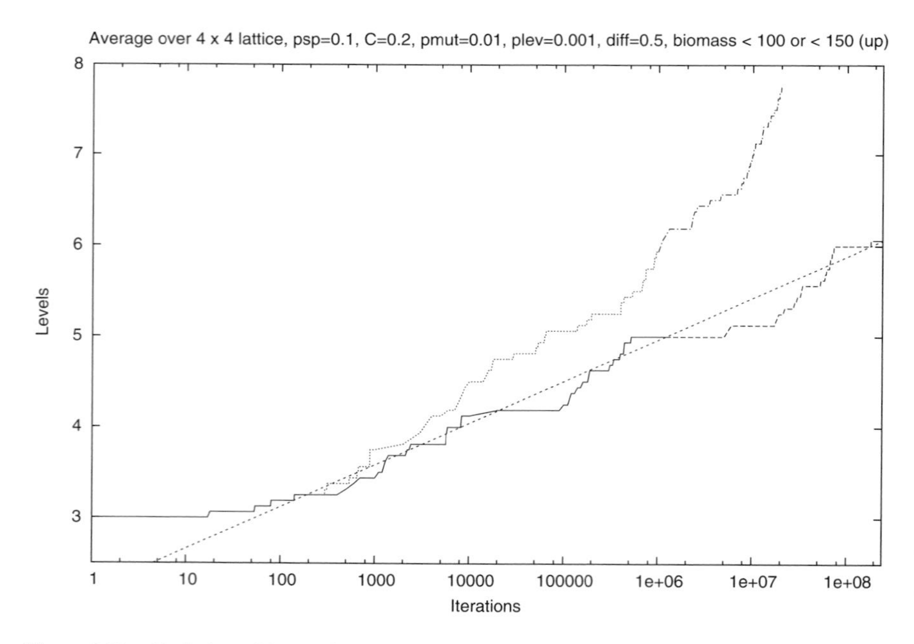

Figure 4.35. Evolution of the maximum number of food levels, averaged over all 16 sites of a 4×4 lattice. Not all levels are necessarily occupied at a give time; a new level may be created only if the total number of animals on that lattice site is below 100 (lower data) or 150 (upper data). Note the logarithmic time axis.

Similar results were found by He and Yu (2006) with a modified model where the prey-predator relations depend on the difference between the food levels.

In contrast to the Rikvold–Zia model (see, e.g., Rikvold (2005)) and the Penna ageing model of Chapter 3 and its application to speciation by Schwämmle and Brigatti (2005), the Chowdhury model does not have a bit-string as an explicit genome. Instead different species are distinguished by different R, M, J_{ik} similar to the Weismann-type ageing model (Stauffer, 2002d; Stauffer and Radomski, 2001). Using these parameters one can build a taxonomy, e.g., by defining a genus as all species having the same R and J but different M. Then one sees (last figure in Stauffer, Kunwar and Chowdhury (2005)) how the probability to belong to the same taxonomic level decreases with time if initially all animals agreed in the relevant parameters: The tree of life grows in this model.

For the simple Lotka–Volterra equations of only two species, prey and predator, the two population numbers oscillate. What are their correlations in the Chowdhury model with many species, often acting as both prey and predator? Figure 4.36 shows that they are slightly anti-correlated: If one level (here: food level 5) has few surviving species, then its prey level (here: level 6) sometimes has few and

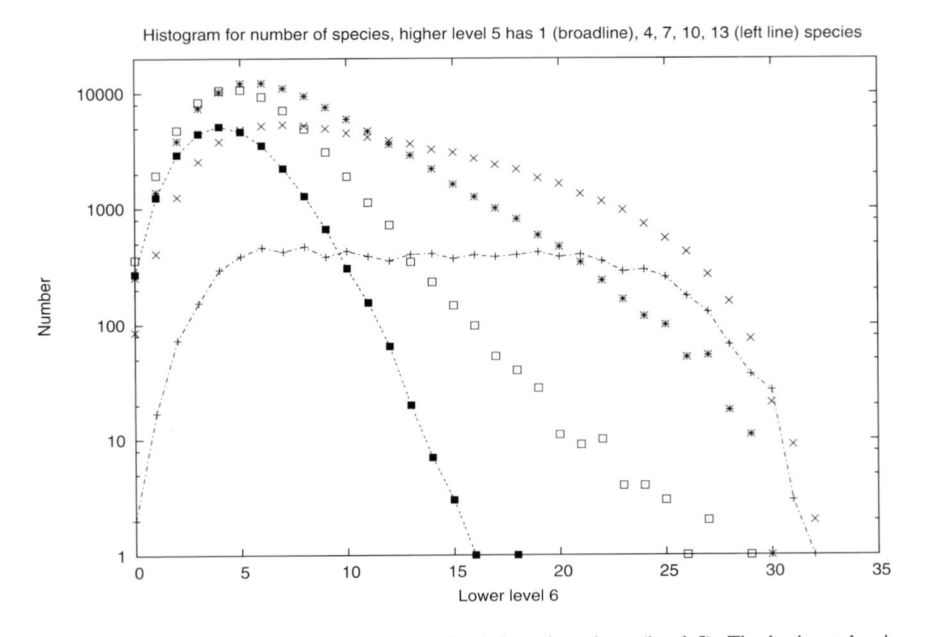

Figure 4.36. Anti-correlations between prey (level 6) and predator (level 5). The horizontal axis shows the number n_2 of species living on level 6 while the vertical axis shows how often for this value of n_2 the level 5 contains n_1 living species. The plus signs (broad curve) correspond to $n_1 = 1$, the full squares to $n_1 = 13$, the other data to intermediate n_1 increasing from right to left. 4×4 lattice.

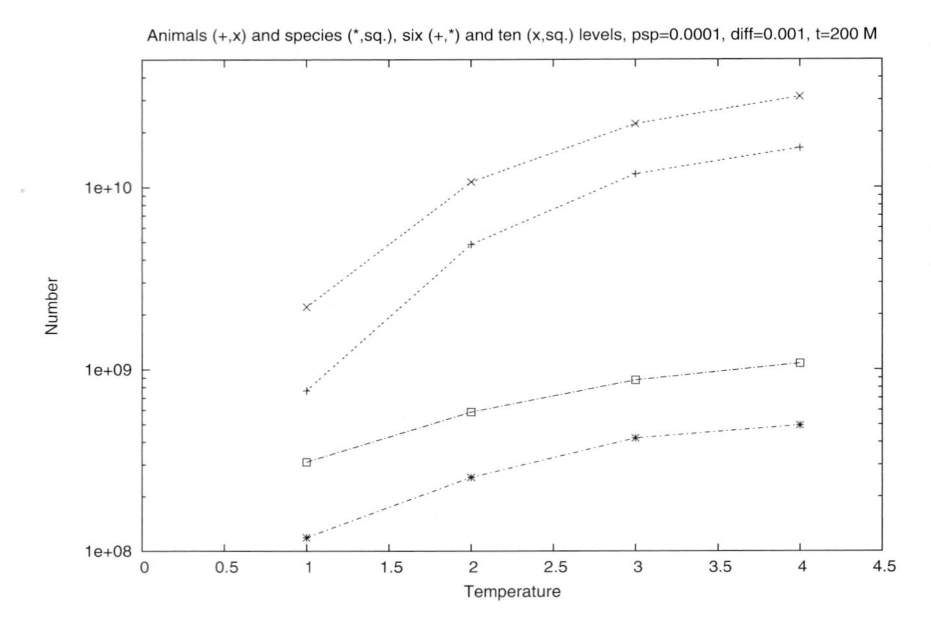

Figure 4.37. Variation of animal numbers (two top curves) and species numbers (two bottom curves) with line number (= geographical latitude characterised by temperature) on a 4×4 lattice, summed over 200 million iterations at a mutation probability of one percent. The number of food levels was fixed at six (smaller numbers) and ten (larger numbers). (See Rohde and Stauffer (2005).)

sometimes has many living species. If, on the other hand, the predator level 5 has many living species, then only few prey species survive on the lower level 6 (black squares in this figure).

Finally, on a 4×4 lattice one may simulate tropical, subtropical, temperate and cold climates, with life getting more difficult for the colder regions (Rohde and Stauffer, 2005). Figure 4.37 shows how the numbers of species (lower data) and animals (upper data) increase with temperature for a fixed number (six or ten) of food layers, a low probability 0.001 of trying to move to a neighbouring lattice site, and a low probability 0.0001 of speciation. This geographical variation was achieved by assuming the rate of successful births to increase proportionally to the line number in the lattice.

Chapter 5

Languages

Whether we human beings deserve to be called *homo sapiens* seems questionable once we read, hear or watch the news about our latest actions. But certainly we talk a lot as *homo loquens*. Nevertheless, also birds have their songs for communication, ants communicate via pheromones (Anderson, 2004), and what we will discuss in this chapter applies to any formalised way of communication, also to human alphabets. The computational methods employed here are very similar to those in biology, and therefore we insert this chapter here between the two biological ones and the following sociophysics Chapter 6.

5.1. Empirical facts

According to the Bible, people started to speak different languages after the tower of Babylon was destroyed. Thus it is somewhat surprising that competition between different languages, similar to survival of the fittest in biology (Cavalli-Sforza, 1997), has only recently been simulated by more than one group at a time.

Children usually learn easily to speak their mother language, old people have much greater difficulties to learn a new language, and scientists have wondered how a language is possible at all. We discuss here not how from the sounds of apes or early humans the first proto-language arose, or how children learn their mother language (Cangelosi and Parisi, 2002; Solé, 2005; Gong and Wang, 2005); instead we concentrate on the competition between different languages of adults, just as our biological discussions in Chapters 3 and 4 dealt with already existing species, not with the origin of life. So, the lack of fossil records for languages is less hindering for us, since we can concentrate more on well-documented history and present reality (Sutherland, 2003). Numerical studies of the 1990s are reviewed by Livingstone (p. 99 in Cangelosi and Parisi (2002)), and thus we concentrate here on later work. A longer review is given by Stauffer and Schulze (2005).

Today thousands of languages are spoken on Earth (Sutherland, 2003), with a particularly high density of languages in New Guinea (Novotny and Drożdż, 2000). Roughly, Figure 5.1, the distribution of language sizes S is log-normal,

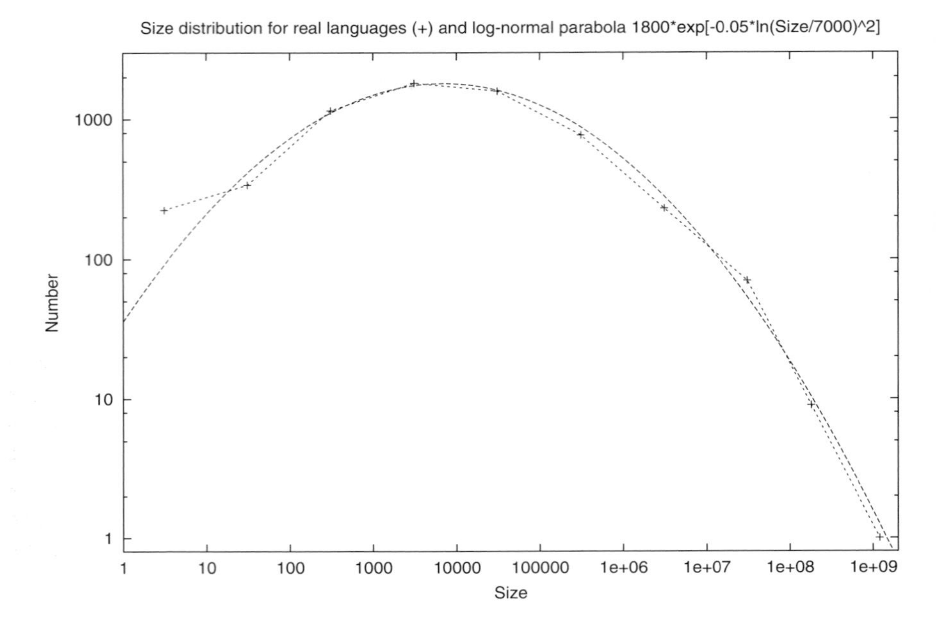

Figure 5.1. Size distribution of human languages (Sutherland, 2003; Science, 2003) with a fitted log-normal function to show the deviation for small sizes. From Stauffer and Schulze (2005). The language sizes are binned by factors of 10 in this log-log plot; e.g., all sizes between 100 and 1000 are put together.

i.e., proportional to $\exp[-\text{const}(\log S)^2]$ where S is the number of people speaking a language as mother tongue. Chinese has the largest size while dozens of languages on their way to extinction are spoken by only one person. The number of languages with less than 10 speakers is larger than what the log-normal distribution would predict. (Gomes, Vasconcelos, Tsang and Tsang (1999) instead fitted two power laws to the right tail of the cumulative language distribution.)

5.2. Differential equations

Let us assume that in one region two languages are spoken, like English and French in Montreal (Canada), with fractions x and $1 - x$ of the total population speaking mainly the first or the second language. Then Abrams and Strogatz (2003) assume

$$\mathrm{d}x/\mathrm{d}t = (1-x)x^a s - x(1-x)^a(1-s) \tag{5.1}$$

with a free exponent $a > 1$ to be fitted on experimental data, and a status variable s with $0 \leqslant s \leqslant 1$ indicating how advantageous it is to use the first of the two lan-

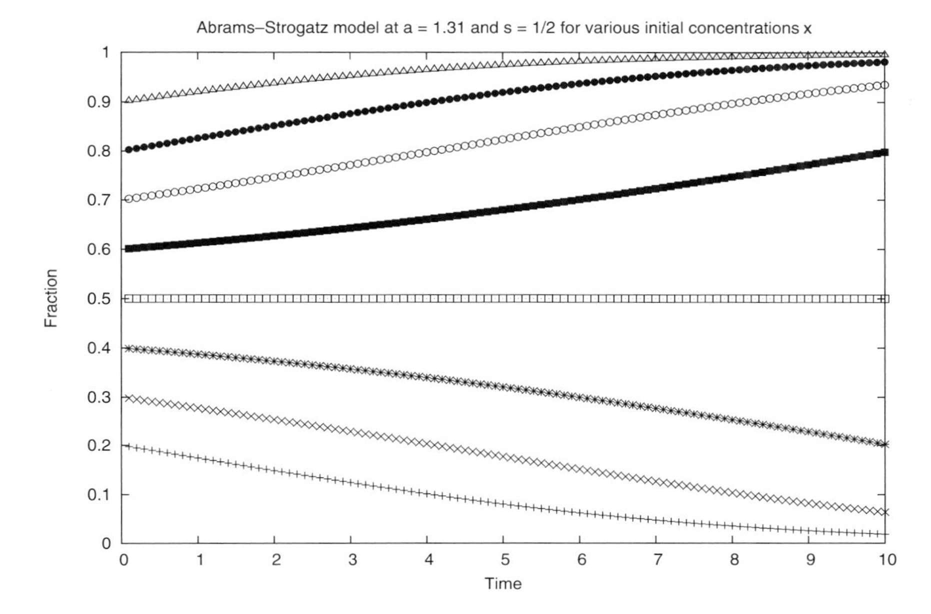

Figure 5.2. Variation of the fraction speaking the first language in the Abrams–Strogatz language model for the neutral case: The initial majority wins.

guages; $s = 1/2$ means neutrality. Figure 5.2 gives the resulting time dependence for various initial fractions x; since the above equation is nonlinear, the results depend on this initial fraction and let the initial minority language die out. For initially equally many speakers of the two languages, Figure 5.3 shows how the language with higher status overwhelms in the course of time the one of lower status. Finally, Figure 5.4 combines the variation of status with an initial fraction of only ten percent for the first language: The higher the status s of the minority language is, the slower is its decay for $s < 0.6$; for $s > 0.7$ the initial minority even wins while the initial majority language approaches extinction.

For the neutral case $s = 1/2$, the free exponent a was taken as 1.31 to fit four examples, like Welsh versus English in Wales. The fitted status s varied slightly between these four examples. However, the empirical data to which a and s were fitted are very poor, a situation happening quite often in sociophysics. Of course, since only two languages are simulated, this model cannot explain the language size distribution of Figure 5.1. Mira and Paredes (2005) generalised equation (5.1) to three populations by adding bilingual people, and found good agreement with the use of languages in Galicia (Spain). The special case $a = 1$ of equation (5.1) leads to Verhulst's logistic equation, applied to languages by Shen already in 1997, as reviewed by Wang, Ke and Minett (2004).

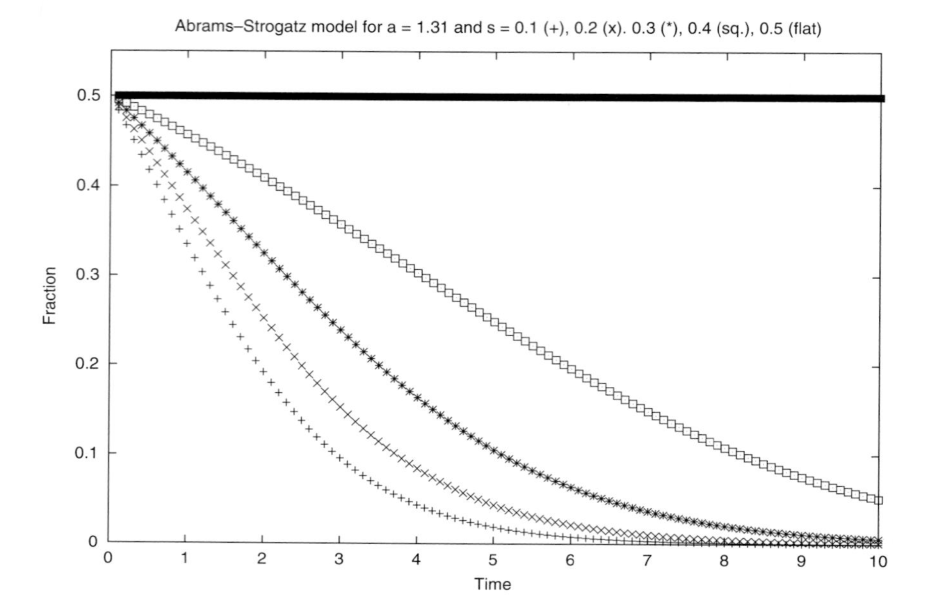

Figure 5.3. Variation of the fraction speaking the first language in the Abrams–Strogatz language model with status: Higher status wins. From Stauffer and Schulze (2005).

This differential equation is a mean-field approximation averaging over all people and ignoring that they are born, mature and die, that they may influence their neighbours, and that their total number is finite. Later we will present different simulations avoiding such unrealistic aspects. Nevertheless the simplicity of this model is very attractive, like the famous Lotka–Volterra equations for prey and predator in biology.

Patriarca and Leppänen (2004) put this model onto a square lattice, in order to study the coexistence of two different languages in neighbouring regions (somewhat similar to parapatric speciation in Chapter 4). To get this coexistence they assume the status s of the first language to be higher in the left part, and of the second language to be higher in the right part. Then the left part speaks the first language, and the right part the second language, with a sharp interface in between. We dislike here that they put in through the status variable what they got out: The first language dominates in the left and the second language in the right part. Their work triggered later simulations to be presented below where this asymmetry built into the parameter s is avoided and where we are thus closer to self-organisation of the interface.

Many languages were studied numerically earlier by Nettle (1999a) using

$$dL/dt = 70/t - L/20 \tag{5.2}$$

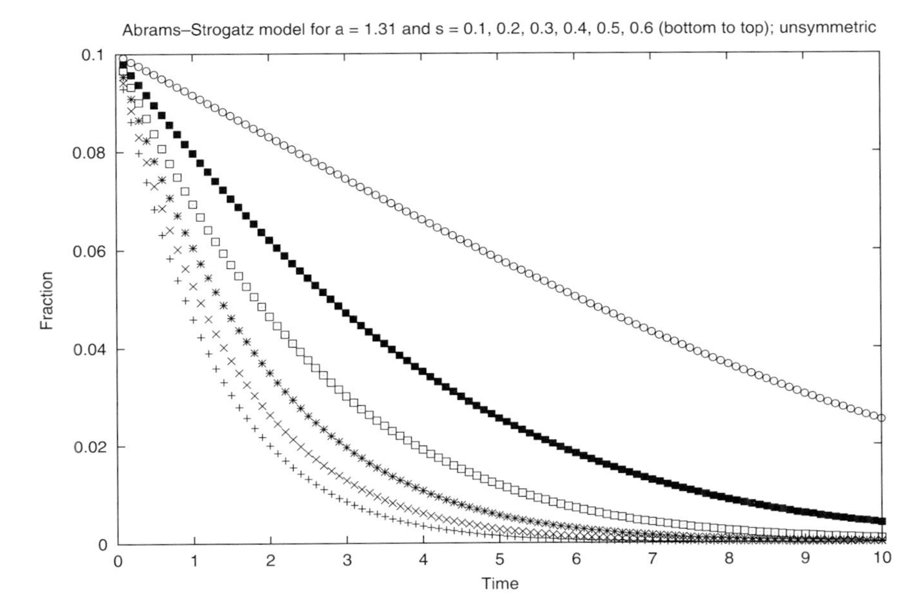

Figure 5.4. Variation of the fraction speaking the first language in the Abrams–Strogatz language model: High status helps initial minority. From Stauffer and Schulze (2005).

for the number L of language groups, e.g., on the American continents. Here the time t is measured in thousands of years. He argues that after the initial settlement by humans the low population density allowed a rapid growth of the number of languages, while later at higher population densities and stronger contact the number of different languages shrinks again. This equation leads to $L(t \to \infty) \to 0$, hardly correct, but only for times longer than the existence of *homo sapiens*.

Nowak, Komarova and Niyogi (2002) reviewed more complicated differential equations to describe a multitude of L languages. They apply them to the learning of languages or grammars, but we think their mathematics can also be used for our problem of competing languages of adults. The fraction x_j of people speaking language j is assumed to be

$$\mathrm{d}x_j/\mathrm{d}t = \left(\sum_i f_i Q_{ij} x_i\right) - \phi x_j \tag{5.3}$$

where $f_i = \sum_k F_{ik} x_k$ is the "fitness" of language i due to the advantage F_{ik} of an i-speaker to be understood by someone who speaks k. Q_{ij} is the probability that children from i-speaking parents will speak j; thus $\sum_j Q_{ij} = 1$ for all i. Finally, $\phi = \sum_i f_i x_i$ is the average fitness and is needed to keep the sum of all fractions constant.

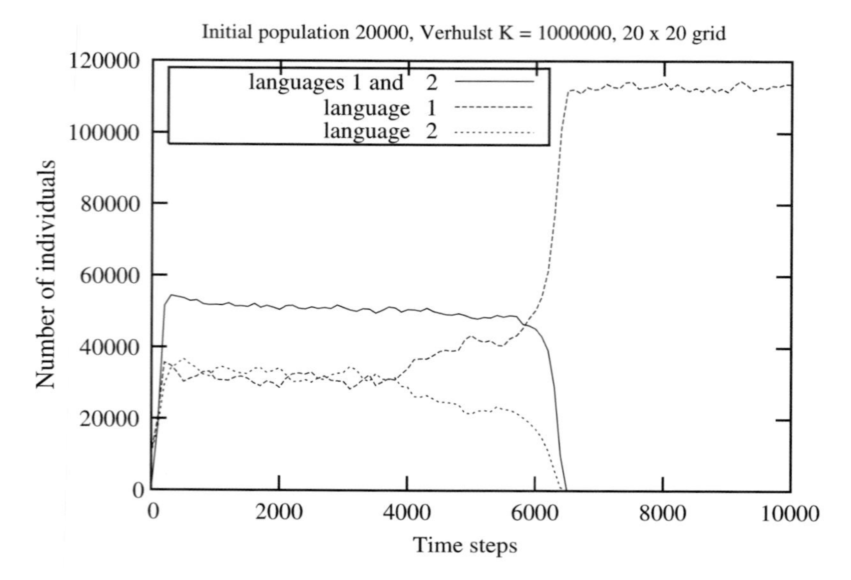

Figure 5.5. Variation with time of the number of people speaking one of the two languages (dashed lines). The solid line gives the number of bilingual people who dominate originally but finally also vanish. From Schwämmle (2005).

The huge number of free parameters in the two $L \times L$ matrices F and Q make general statements difficult. For suitable parameters a phase transition was found where at some "coherence threshold" one switches from one dominating language to a fragmentation into many languages, Section 5.5.

5.3. Agent-based simulations

Agent-based are those simulations where each individual is treated separately, like in Monte Carlo or Molecular Dynamics simulations of physics since half a century. They can give drastically different results than mean-field approximations like the above differential equations. As warned already in Section 2.6, mean field theory predicts for the one-dimensional Ising model a positive phase transition temperature to ferromagnetism, while a proper treatment of the atoms or agents gives no such transition. If, on the other hand, there is no spatial structure and everybody can interact directly with everybody, then for infinitely large populations the agent-based model may agree with mean field theory. These differences and similarities are also relevant for the other chapters, not only for languages. Thus we regard agent-based studies as the appropriate method in general, and for languages that means to treat each person. Again

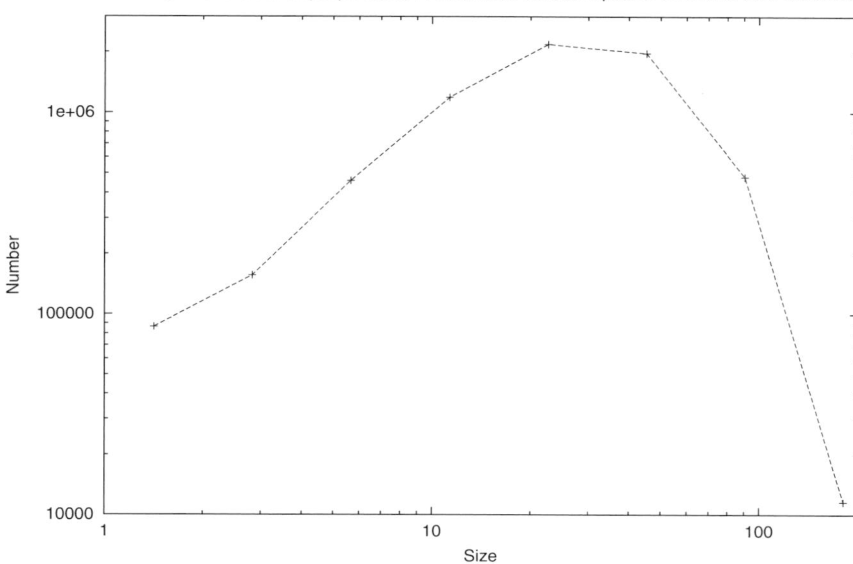

Figure 5.6. Simulated language size distribution, from Schulze and Stauffer (2005); $p = 0.16$, $\ell = 16$, $t \leqslant 1000$. In contrast to binning by a factor 10 in Figure 5.1 we now bin by a factor of 2 since the simulated population of two million is much smaller than the human population on earth.

we can distinguish models for two languages (Kosmidis, Halley and Argyrakis, 2005; Schwämmle, 2005) and for many languages (Schulze and Stauffer, 2005; Teşileanu and Meyer-Ortmanns, 2006; Schwämmle, 2006). In all of them, babies are born, inherit a (changing) language from their parents, mature, produce children, and die. All three models use bit-strings, following a long tradition in biological modelling (Eigen, 1971, Section 2.10); each bit is either zero or one. (Chapter 3 reviewed in detail the Penna bit-string model of ageing, Chapter 4 summarized bit-string models for speciation.) But the interpretation of bits is different in each model.

5.3.1. Two languages

Kosmidis, Halley and Argyrakis (2005) identify each bit with a word of the language. Thus the words of the first language are the first half of the bit-string, and those of the second language are the second half of the bit-string. A bit set to one means the word has been learned by the person while a zero bit means the word has not been learned. Only two languages are possible but people can become bilingual.

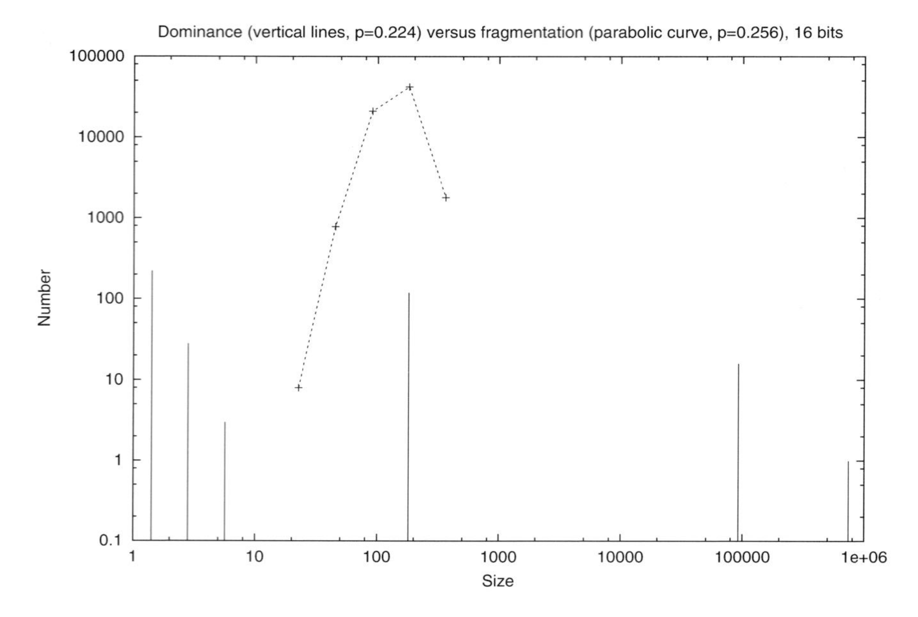

Figure 5.7. Histogram of language sizes for dominance (vertical lines) and fragmentation (connected plus signs), from Stauffer and Schulze (2005).

These agents diffuse on a square lattice, and when they bump into a speaker of the other language (s)he can teach them words from this other language. The more words they learn the higher is their fitness as measured by the chances for reproduction. (Kosmidis, Halley and Argyrakis (2005) fail to cite the movie "Groundhog Days" which is an earlier publication of this principle that learning French helps men in seduction.) In this way their language can contain synonyms, i.e., words meaning the same thing but coming from different original languages. Depending on the details of the model, at the end everybody speaks about half the words from one and half the words from the other language (like the merger of French and German into English after the 1066 Norman conquest), or everyone speaks both languages nearly perfectly (like French and Flemish in Belgium) having nearly all bits set.

Schwämmle (2005) uses the bit-strings to describe the ageing in the Penna model, and also has two languages and a square lattice for motion. This motion is not random as for diffusion but prefers jumps into less occupied neighbour regions. The agents age over 10 to 20 iterations before they die, with the Penna model of Chapter 3. Reproduction is sexual, with the child learning the language of both father and mother and thus possibly becoming bilingual. Also, languages are forgotten if most of the neighbouring people do not use them. Figure 5.5

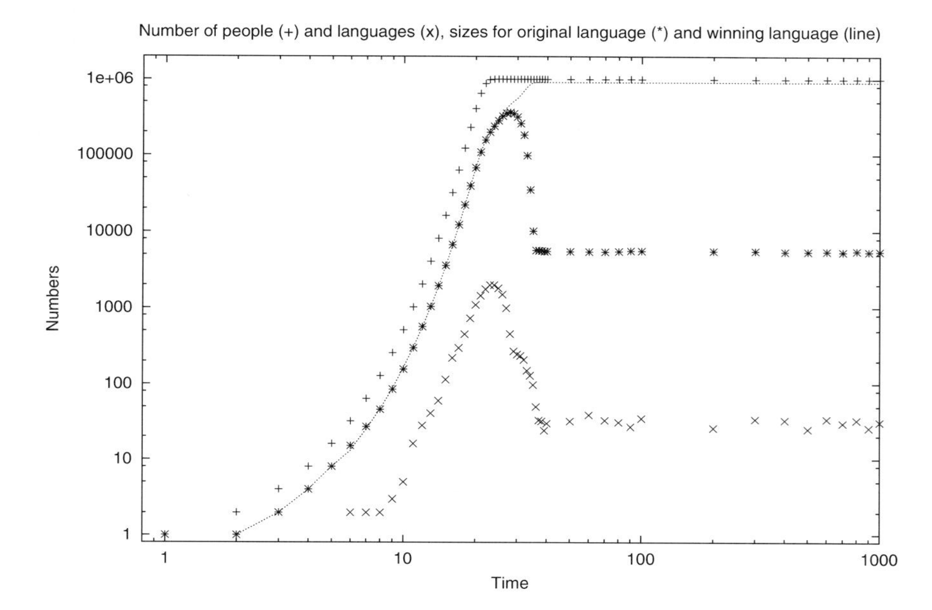

Figure 5.8. Time dependence of population (plus signs) and languages when we start from one person. The number of people speaking the dominating language is shown by a line; accidentally the dominating language is not the same as the original one (stars). The $\times$ signs show the number of languages spoken by at least ten people. From Stauffer and Schulze (2005).

shows that if initially half of the lattice is speaking one and the other half the other language, as in Patriarca and Leppänen (2004), then at the end still one language may dominate and the other language together with the bilinguals dies out. But a stable coexistence, as will be shown below in Figure 5.18 for a different model, is also possible.

Later Schwämmle (2006), after empirically observing the senile author, modified this model and allowed learning of a foreign language only in youth. For small mutation rates, one language dominates, for larger ones fragmentation is observed. This phase transition seems to be of first order for many possible languages and of second order for only two languages. The threshold mutation rate increases if the age limit for learning a foreign language is increased.

5.3.2. Many languages: Homogeneous systems

Schulze and Stauffer (2005), followed by Teşileanu and Meyer-Ortmanns (2006), interpret each different bit-string of length ℓ as a different language and thus allow for 2^ℓ languages, i.e., 256 for 8 bits and 65536 for 16 bits. Each bit may be a grammatical structure like the ordering subject-verb-object or subject-object-verb

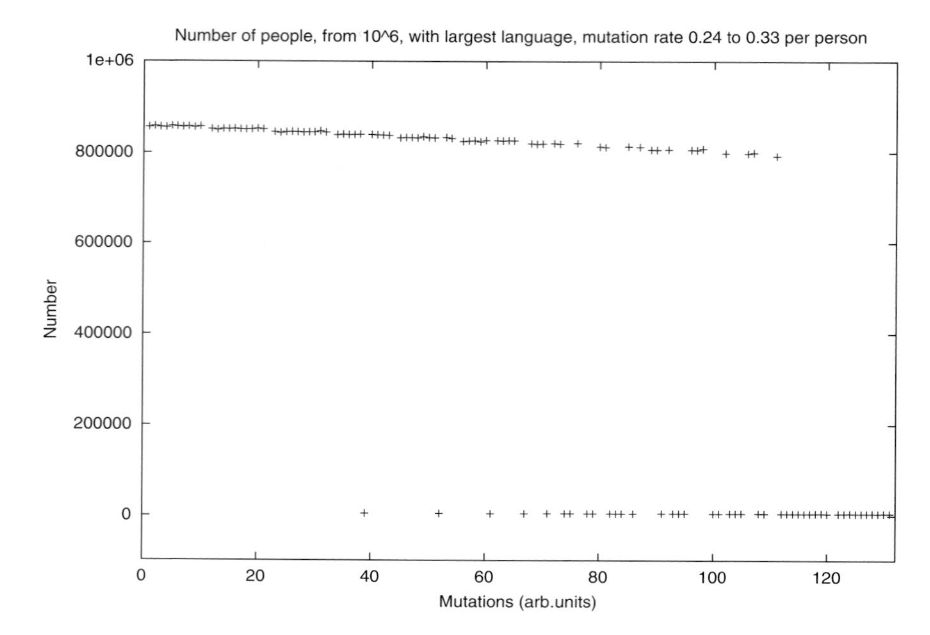

Figure 5.9. Evidence of first-order transition with hysteresis; the mutation rate increases from 0.24 on the left margin to 0.33 on the right margin of the figure. Since 8 bits were used, fragmentation means that a language is spoken by typically $N/256$ of the $N = 10^6$ simulated people. From Stauffer and Schulze (2005).

(Science, 2003). A set of 30 independent binary grammatical parameters, i.e., $\ell = 30$, was regarded as reasonable by Briscoe (2000), and $8 \leqslant \ell \leqslant 64$ was simulated (Stauffer and Schulze, 2005).

A new child gets the bit-string from the mother (asexual reproduction) except that with probability p one randomly selected bit is toggled (changed from 0 to 1 or from 1 to 0). (Up to here the simulation is similar to one mentioned for biological speciation: de Oliveira, Martins, Stauffer and de Oliveira, 2004.) Also, at each iteration t, a person with language i adopts the language of another randomly selected person with probability

$$r = \left[N(t)/N(t \to \infty)\right] \left(1 - x_i^2\right) \tag{5.4}$$

where x_i is the current fraction of people speaking language i in the whole population $N(t)$, and $N(t \to \infty)$ is the equilibrium population established by a Verhulst death probability $\propto N(t)$ applied to everybody; as in Chapter 2, $\propto$ denotes proportionality. (The first factor [...] can be omitted if one starts already with the equilibrium population size; the second factor was replaced by $(1 - x_i)^2$ in some simulations.) The first factor in this equation means, as was argued before by

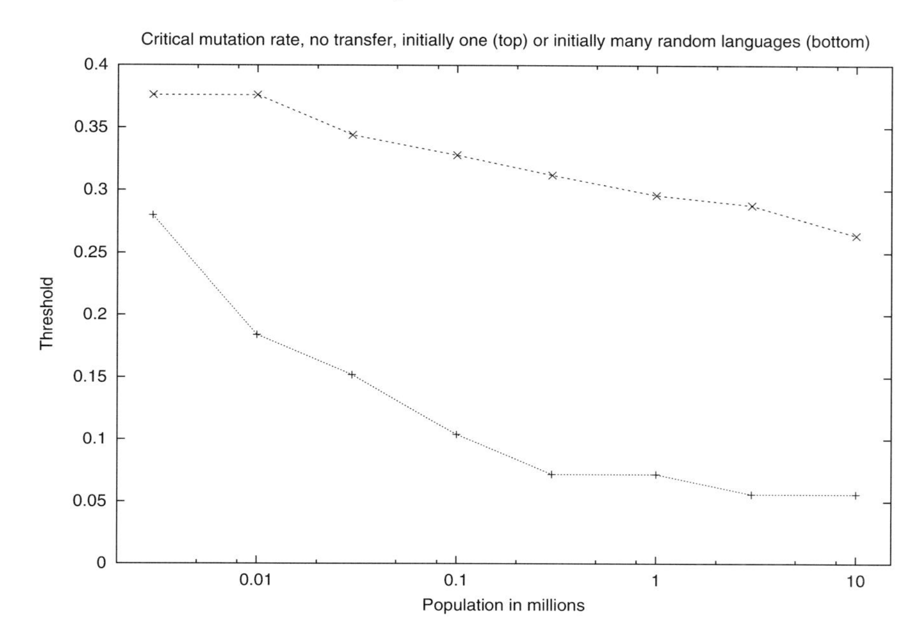

Figure 5.10. Limits of stability if we start from fragmentation (lower curve) or one language (upper curve); we show the values of the mutation rate p for the border between fragmentation (above the curves) and dominance (below the curves); at these values the initial state starts to switch to the opposite behaviour. Note the strong relative variation of the lower curve with the population size. From Stauffer and Schulze (2005).

Nettle (1999a), that for low population densities languages barely compete with each other. For high densities when the first factor is near unity, the second factor induces speakers of small languages to switch to a more widespread language. (If instead of selecting randomly another person one selects randomly another language then large languages are not favoured enough and the dominance to be mentioned below may become impossible.) Section 9.4 in the appendix lists a program.

Figure 5.6 from these simulations has some similarity with Figure 5.1 from reality: roughly the distribution of language sizes is log-normal with a higher value for the smallest sizes.

If we vary p we see a first-order phase transition, the position of which depends on the length of the bit-string and the size of the population: For small p one language comprises about four fifths or more of the population, and most of the remaining people speak a language differing from this dominating language by only one bit. For larger p we get a fragmentation into many languages of roughly equal size; the relative width of the size distribution shrinks towards zero if the

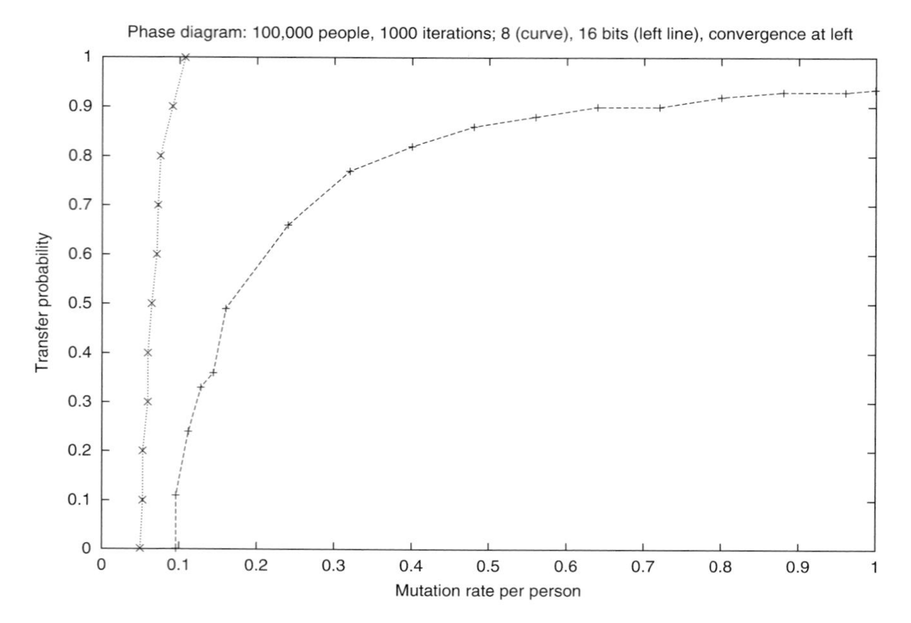

Figure 5.11. Phase diagram for dominance (upper left part) versus fragmentation (lower right part), when we start with the equilibrium population of 10^5 distributed randomly over all possible languages. The curve corresponds to 8 bits, the nearly straight line to 16 bits. From Stauffer and Schulze (2005).

population size increases. This fragmentation-dominance transition seems similar to the coherence transition of Nowak, Komarova and Niyogi (2002). The simulations do not exclude that the creation of dominance out of an initially fragmented population is a fluctuation effect which would vanish for infinite populations. Figure 5.7 compares the distribution of language sizes for dominance (vertical bars, $p = 0.224$ for 16 bits) and fragmentation (parabolic curve, $p = 0.256$). The dominating language is accompanied by 16 languages differing from the dominating one by only one of the 16 bits, and by many much smaller languages; in the other case of fragmentation, all language sizes are between 20 and 500, roughly log-normally distributed.

When we start with one "Eve", the population first grows, and so does the language diversity, Figure 5.8. When the population reaches its saturation value, then language competition becomes fierce, as already stated by Nettle (1999b), most languages die out, and the surviving languages are the dominating one and those which differ from it by one of the 16 bits. Due to accidental mutations during the first time steps, not the original language but one of its mutants becomes dominating.

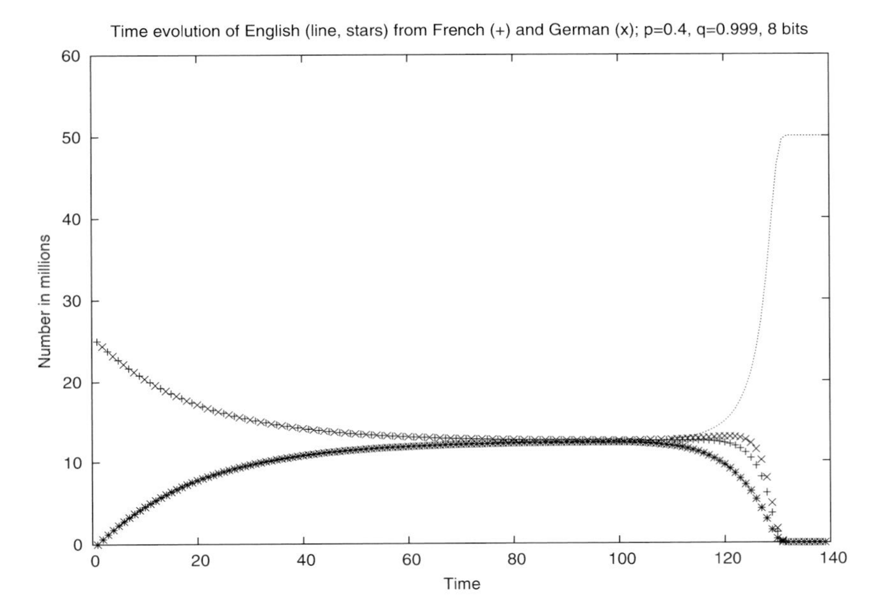

Figure 5.12. Emergence of a mixture language 2 (English, line) out of languages 0 (French, +) and 3 (German, ×). Initially French and German are spoken equally often, thus their symbols are shown for even and odd times, respectively, in the left upper curve. The lower stars correspond to language 1, which dies out, and overlaps with the line in the left part. The flight from small languages, equation (5.4), is switched on after 100 iterations.

The transition from dominance at low mutation rate to fragmentation at high p is of first order, Figure 5.9, that means the size of the largest language jumps by several orders of magnitude at some p value which differs from sample to sample. We start with one person and thus initially have dominance. The figure thus shows when this dominance is stable for 1000 iterations (giving a population of nearly one million speaking this language) and when it decays into fragmentation when the largest language is spoken by only about 5000 people. 10 samples are shown with p increasing from 0.24 to 0.33.

The hysteresis as well as strong finite-size effects are shown in Figure 5.10 where we start with the equilibrium population and either have them initially all speak the same language (upper curve) or distribute them initially over all 256 possible languages (lower curve). For mutation rates p above the curve we end up with fragmentation, while below the curve we get dominance. Thus the final result strongly depends on the initial distribution, just as at low temperatures a ferromagnet stays with the magnetisation direction we started with, even if later a small external magnetic field wants to turn the magnetisation into the opposite direction.

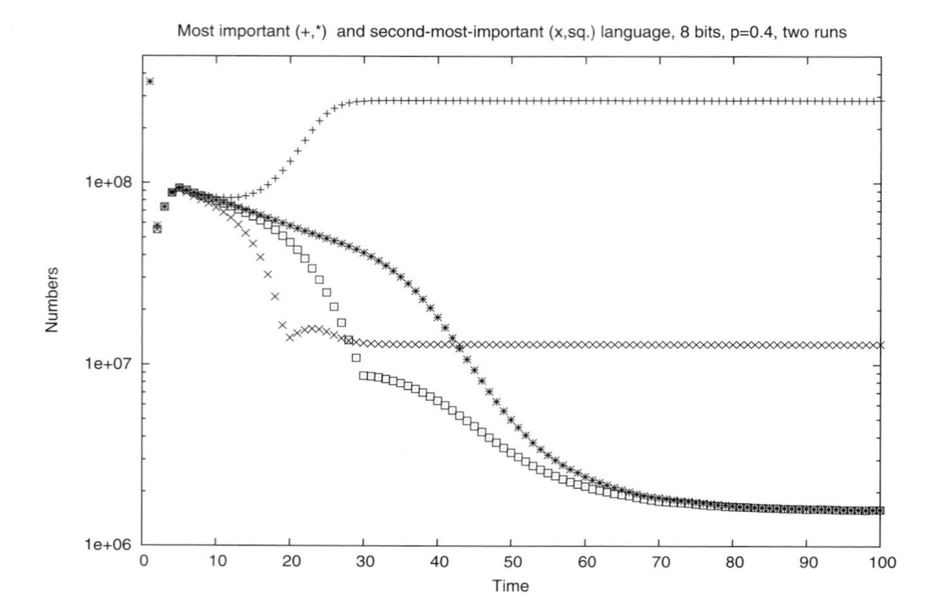

Figure 5.13. Competition of languages 00000000 and 11111111 at intermediate mutation rate $p = 0.4$. We show the sizes of the most important (+, stars) and the second-most important ($\times$, squares) language versus time. The two pairs of curves differ only by the random number seeds. As seen already in Figure 5.9, sometimes one gets dominance (+, $\times$) and sometimes fragmentation (stars, squares).

Figure 5.10 also shows strong trends of the transition lines when we increase the population size from 3000 to 10 million; it thus seems possible that in an infinite population no dominance would arise out of a fragmented population. Thus a mean-field approach to infinite populations by deterministic differential equations for equivalent languages, as in Section 5.2 might not get this transition to dominance. More simulations regarding the nucleation of dominance are given by Stauffer and Schulze (2005).

We may generalise the model by introducing a transfer probability q in addition to the mutation probability p. Then, at each mutation, the bit is not flipped randomly but with probability q assumes the value which the corresponding bit has in a randomly selected other agent. Thus with probability q we learn from other people, who may speak a different language. The higher this transfer probability is, the easier it is to get dominance. Figure 5.11 shows this effect more quantitatively for 8 and 16 bits. In the lower right part of this p-q-diagram we have fragmentation, in the upper left part we have dominance. We see here an unusually strong difference between 8 bits (curve) and 16 bits (nearly vertical line).

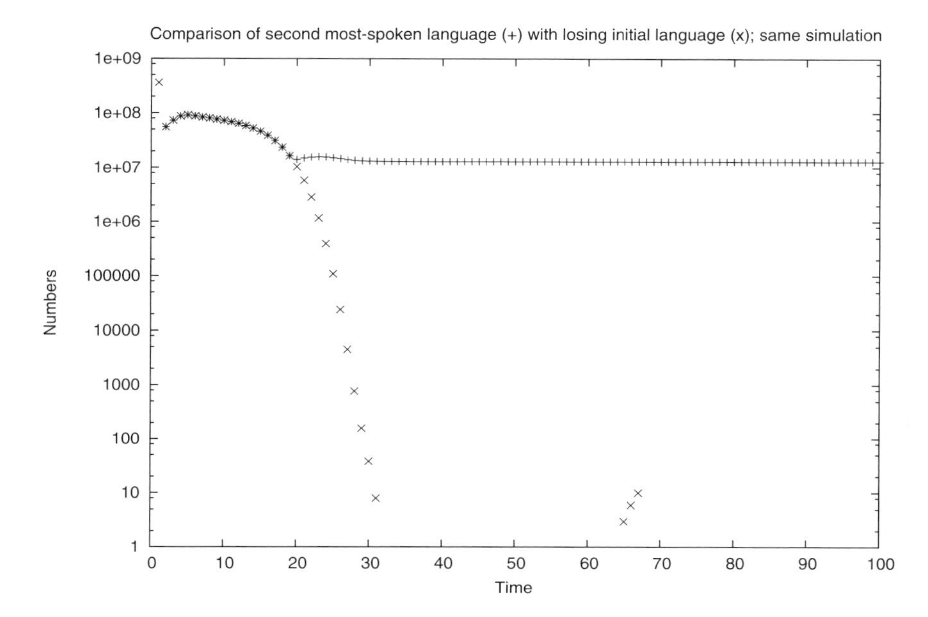

Figure 5.14. Semilogarithmic plot of the dominance case of Figure 5.13: The + signs symbolise the second-most important language while the × signs now represent the losing initial language which only for short times is the second-most important language, then dies out, and later accidentally is spoken again for a few iterations by a few people.

If we use 32 and 64 bits the transition point for $q = 0$ remains about the same as for 16 bits (not shown).

Teşileanu and Meyer-Ortmanns (2006) consider the case where the replacement of a language by another one is determined by their mutual Hamming distance, see Figure 2.19. Again a transition between dominance and fragmentation is found, which is also reflected in the Hamming distance between the two languages with the largest and second to largest number of speakers. They also consider the case where the population is localised on a square lattice and the interaction of individuals is restricted to a certain distance.

We mentioned at the end of Section 5.3.1 that the model of Schwämmle (2006) was simulated for both two and several languages.

5.3.3. *Many languages: Mixing, nucleation, interface*

The English language is a mixture of German words, spoken by the Anglo-Saxons, and French words, spoken by the Norman invaders of 1066. Kosmidis, Halley and Argyrakis (2005) obtained such a mixture quite easily under conditions where the 20 bits are more or less random: Then on average 5 of the first

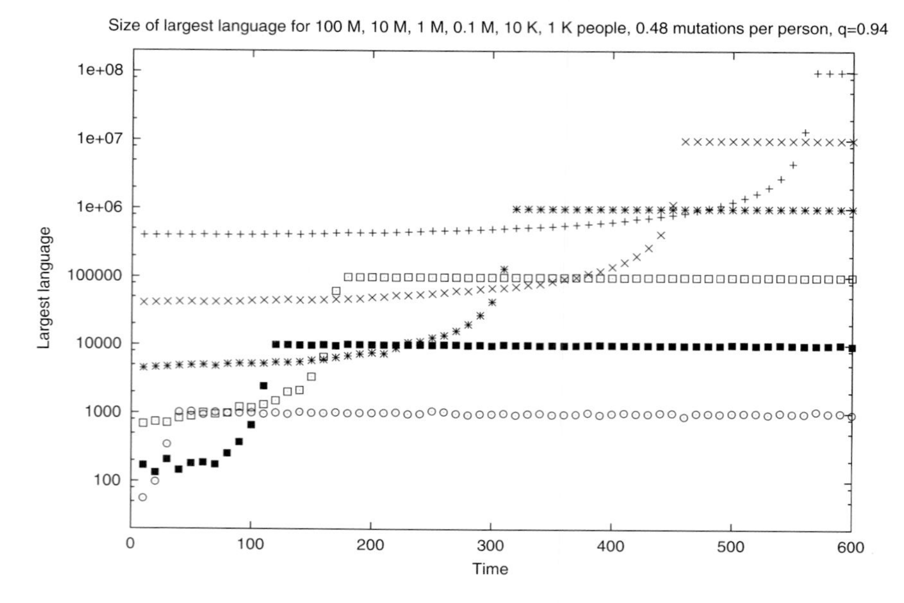

Figure 5.15. Self-organisation of dominance out of fragmentation, for population size varying in powers of 10 from 1000 (circles) to 100 million (+). One sample only, 8 bits. From Stauffer and Schulze (2005).

10 bits (meaning French words) are set, and so are 5 of the last 10 bits (meaning German words). The average person then knows 10 words of which half are French and half are German. Also in the Schulze and Stauffer (2005) model one may find roughly random bit-strings, and this was called fragmentation. But there the interpretation is different, since different bit-strings mean different languages. Thus the mixing into English was simulated differently by Schulze and Stauffer: For eight bits, they used 00000000 for French, 00000011 for German, and at the beginning do not apply the flight of equation (5.4) from small to large languages whenever the small language starts with six zeroes. Starting with half the population speaking French and the other half German, under conditions where fragmentation is avoided, after some time nearly one quarter still speaks French, nearly one quarter still speaks German, but nearly one half speak 00000001 or 00000010 which is interpreted as English. If then the flight via equation (5.4) is switched on, meaning closer social interaction between the ethnic groups, one of the four languages $0 = 00000000$, $1 = 00000001$, $2 = 00000010$, $3 = 00000011$ dominates; this happens to be language 00000010 = English in Figure 5.12. The other English possibility 00000001 dies out together with German and French.

For a better comparison with the several models mentioned above having only two languages, let us start with two equally strong languages in the standard

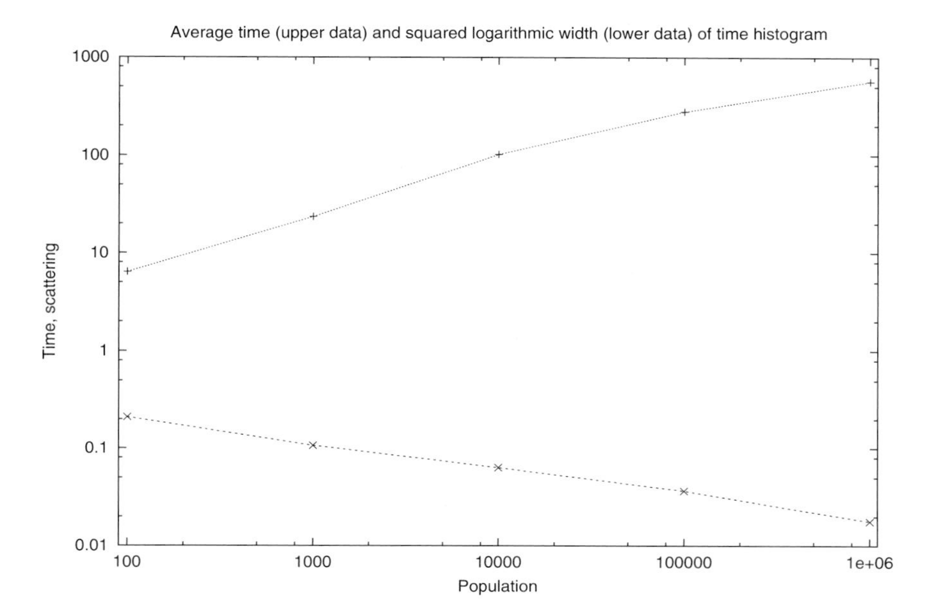

Figure 5.16. Average (top) and mean square fluctuation (bottom) of the (logarithm of the) time needed to get dominance nucleated out of a fragmented initial language distribution. From Stauffer and Schulze (2005).

Schulze–Stauffer model without this trick to avoid flight from languages starting with six zeroes. Figure 5.13 shows how depending on the random numbers, for the same parameters we may either get one of the languages dominating such that it is spoken by hundreds of million people, while the other language loses out; or we get fragmentation where both initial languages are spoken by about $N/256$ of the N people (8 bits). For dominance, the losing language dies out completely, as Figure 5.14 shows; some 1-bit mutant of the dominating language becomes the second-most important language.

If in the Schulze–Stauffer model we start with fragmentation and check for the development of dominance, then Figure 5.15 shows that we have to wait the longer the larger the population N is; here N varies between 10^3 and 10^8. With 1000 samples instead of only one, Figure 5.16 shows how the average time $\langle t \rangle$ to get dominance from fragmentation increases with increasing N, while its scattering σ decreases; the distributions for $N \geqslant 10^4$ are roughly log-normal (not shown here; see Stauffer and Schulze (2005)). Here we define the average t as $\exp(\langle \ln t \rangle)$ and $\sigma^2 = \langle (\ln t)^2 \rangle - \langle \ln t \rangle^2$. Thus for huge populations the distribution of the times needed to nucleate dominance seems to become a delta function in log(time), while the times themselves diverge. Some explanations of this size effect were attempted by Stauffer and Schulze (2005).

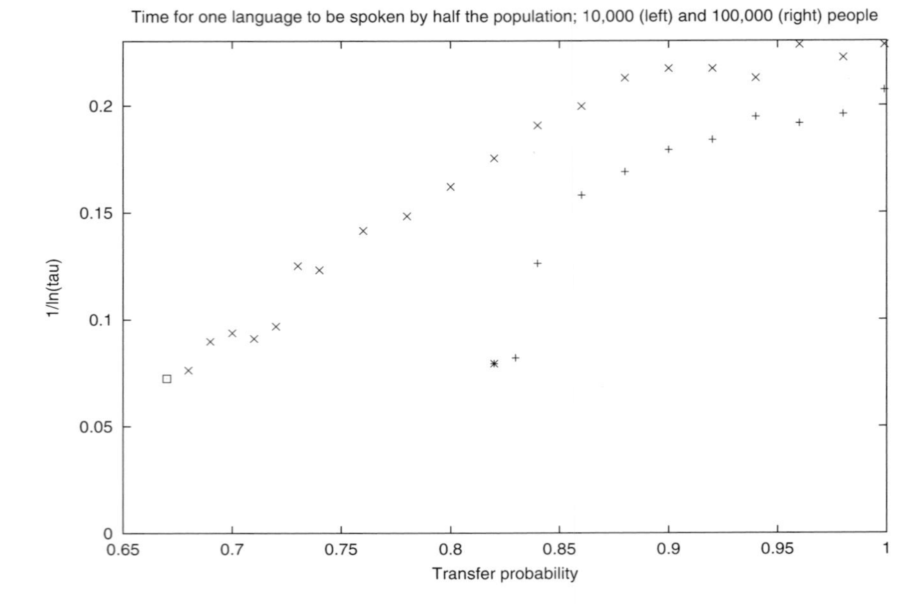

Figure 5.17. Reciprocal logarithm of the median time to reach dominance. The special symbols at the lower left ends of the data sets indicate that no median was found and that the median must be beyond the time indicted by the star and square. From Stauffer and Schulze (2005).

With these slowly increasing times one may question whether the phase diagram in Figure 5.11 is reasonably equilibrated. Thus for two different population sizes 10^4 and 10^5 we determined from 9 samples the median time τ after which dominance emerged from fragmentation. Does this time just become exponentially large like $\exp(\text{const}/q)$ instead of diverging at some finite q_c? Figure 5.17 says no: the quantity $1/\ln\tau$ seems to vanish at some q_c near 0.8, and not at $q_c = 0$, for $N = 10^5$; for $N = 10^4$ the data are more difficult to interpret. Thus for a large but finite population we seem to have a phase transition, not a gradual freezing in.

Similarly to Patriarca and Leppänen (2004), one may also put this model onto a square lattice, Figures 5.18 and 5.19, with thousands of people per lattice site and a small probability 0.01 to move to a neighbour site. With probability q they learn a bit from a speaker on the same site or on one of its four neighbour sites, and with probability r they select the whole language of another speaker on the same site. In the left half of the 20×20 lattice initially everybody speaks 00000000, in the right part everybody speaks 11111111; otherwise no status difference was assumed. Then a smooth interface develops where the probability of a zero speaker to survive in the right half decays exponentially with the distance from the interface center, similar to electrons in quantum mechanics penetrating into regions

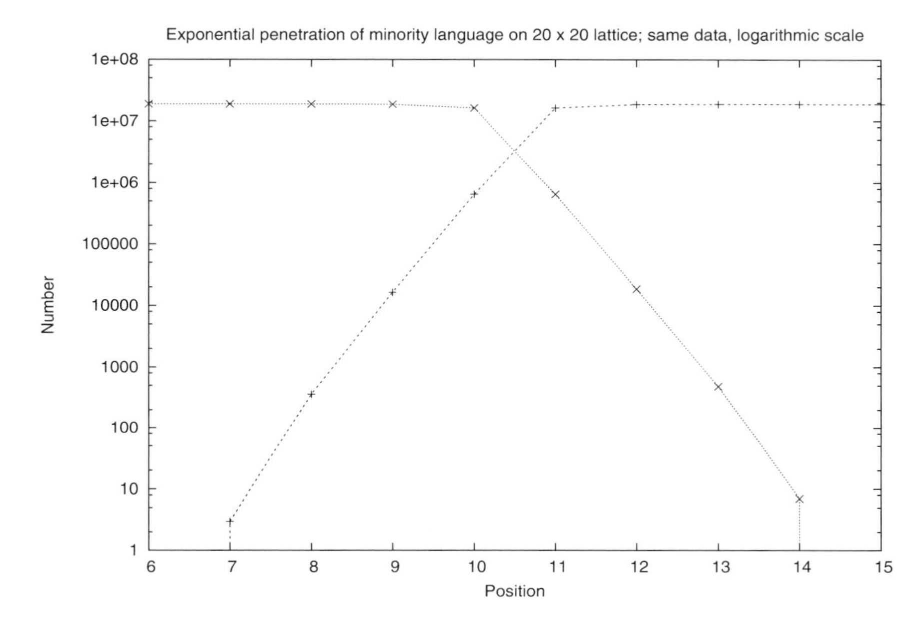

Figure 5.18. Interface profile of the two dominating languages in a square lattice, plotted semi-logarithmically. From Stauffer and Schulze (2005).

forbidden to them without quantum tunneling. We did not succeed in getting such phase separation by starting with a fragmented population. Results similar to Figure 5.18 were obtained in the Schwämmle model (2005). Finally, on a lattice the quantitative agreement with real language-size distributions can be improved; Figure 5.20 could apply to a region with many small languages like New Guinea (Novotny and Drożdż, 2000).

Penna (2005) used this lattice model to check if a language spoken in a geographically compact region, can win over a language spoken by as many people but geographically scattered over a larger region, where also many other languages exist. Usually, the scattered language ends up as a majority but the compact language does not die out, for large populations.

A different lattice model was simulated by de Oliveira, Gomes and Tsang (2006), with a fitness of a language proportional to the number of people speaking it, and with a mutation probability inversely proportional to this fitness. The number of languages spoken in an area A of the lattice varied as $A^{0.4}$ as in reality; strong interactions between the populations reduced this exponent for large $A > 10^5$. This variation with area does not come out as nicely in the model of Schulze and Stauffer (2005), Figure 5.21; as we see there the results depend only weakly on the transfer probability q.

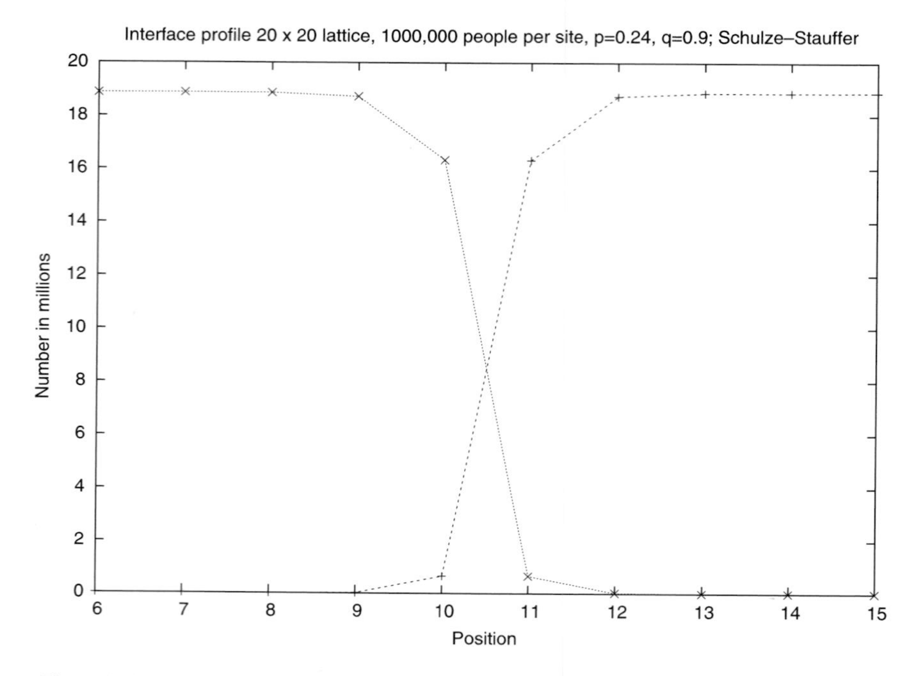

Figure 5.19. Same data as in Figure 5.18 but plotted linearly. From Stauffer and Schulze (2005).

5.4. * Wang–Minett model

The Monte Carlo simulations of Wang and Minett (2005b) come from a combination of linguistic and mathematical experience but are somewhat similar to the above physics models.

Each language is a string of integers 0, 1, 2, ... First, a small language tree is constructed by random bifurcations; thus some early languages split into many present languages, and some only into two. Then each node on this tree is filled with a language; the proto-language at the root of the tree is a string of zeroes. For all present languages the time since the start with the proto-language is the same, and thus not given by the number of branchings in the tree. As in Figure 5.11 above, two competing processes change the languages: mutations and transfer. The mutation probability during time t is $1-r^t$ where r is the retention probability per unit time; a mutation gives the integer in the string a unique new value, like 2 if it was 1 before. The transfer takes the integer value of another language with which the considered language is in contact. The retention and transfer rates are not always the same but may fluctuate within some interval. (It is useful to get the

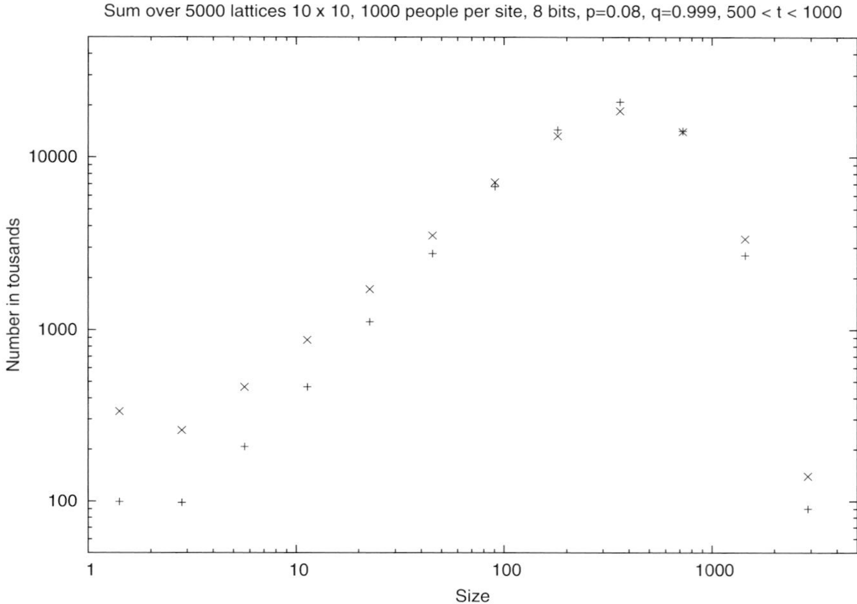

Figure 5.20. Simulated size distribution on small lattices with diffusivity 0 and 0.01; 8 bits. As in Figure 5.6 the language sizes are binned by factors of 2, e.g., sizes 9 to 16 are put together.

additional material of these authors from www.philsoc.org.uk/transactions.asp in order to understand their simulations.)

The model deals with interacting languages, not with interacting speakers of languages. It thus does not include the flight of speakers from small to large languages. The intention of the model was not the competition between languages and the possible dominance of one of them, but to help linguists analyze the historical language tree when transfer between languages in contact makes this analysis difficult.

Finally we remark that both the "vertical" transmission of languages from one generation to the next, and the "horizontal" transfer of language elements through personal contact, in this model as well as in that of Schulze and Stauffer (2005) of the previous section, are analogous to biology, e.g., for bacteria (de Oliveira, de Oliveira and Stauffer, 2003).

5.5. * Additional remarks

The differential equations (5.3) were intended for the *learning* of a language (Nowak, Komarova and Niyogi, 2002; Komarova, 2004) but we see no reason why

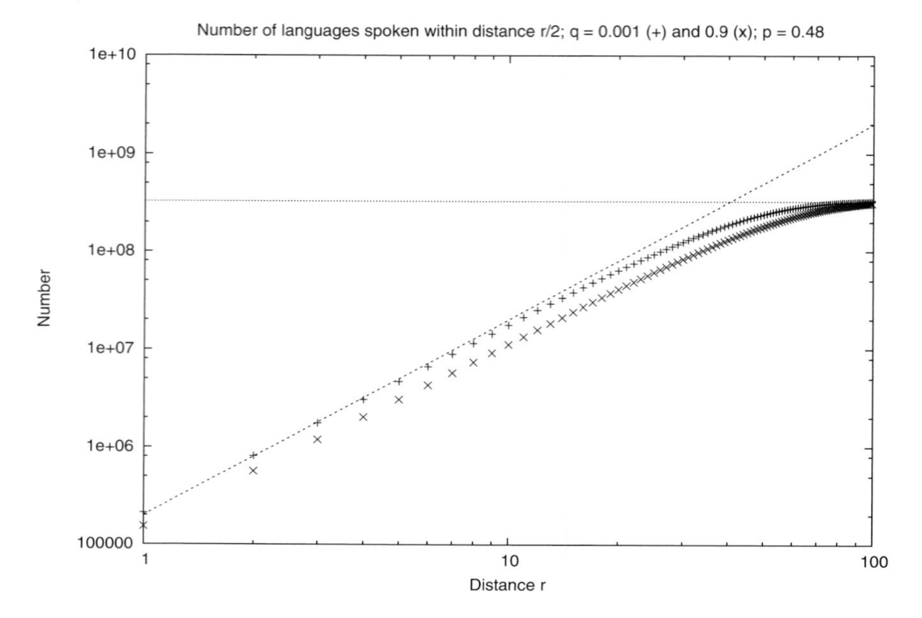

Figure 5.21. Fractal language geometry? Number of different languages spoken in a $r \times r$ square within a larger 99×99 lattice of the model of Section 5.3.2 with $p = 0.48$, $\ell = 16$, $q = 0.001$ and 0.9, 100 people per site and varying transfer probabilities q. The two straight lines correspond to the trivial limits of 40 new languages per site (slope 2) and all 65536 languages present (plateau). Sum over 100 samples.

the same mathematics cannot be applied for the *competition* between languages of adults, emphasised in this chapter. The fundamental problems with differential equations were already discussed in Chapter 2, but since in most of this present Chapter 5 we reported on agent-based simulations, we can compare the results.

For a large number L of such languages, the then quite huge $L \times L$ matrices F_{ij} and Q_{ij} contain numerous free parameters. Thus first we follow Komarova (2004) and assume $F_{ij} = a$ except $F_{ii} = 1$ and $Q_{ij} = p/(L-1)$ except $Q_{ii} = 1 - p$ where a and p are free parameters between 0 and 1; $1 - p$ is called "learning accuracy" by Komarova (who denotes p by q), and thus p is called here the mutation rate, analogous to our previous sections. (This a has nothing to do with the exponent in equation (5.1).) We also follow her in setting $a = 0.3$ and start the simulation either fragmented (all languages equally strong) or dominated (one language spoken by everybody). Figure 5.22 for $L = 30$ languages confirms Figure 1 of Komarova: Already a small mutation rate p prevents a fragmented state to change into a dominated one, but a much larger p is needed to destroy an initial dominance. (When we start fragmented, the fraction x_1 for the first language is enlarged by 0.001 to allow an instability to develop.)

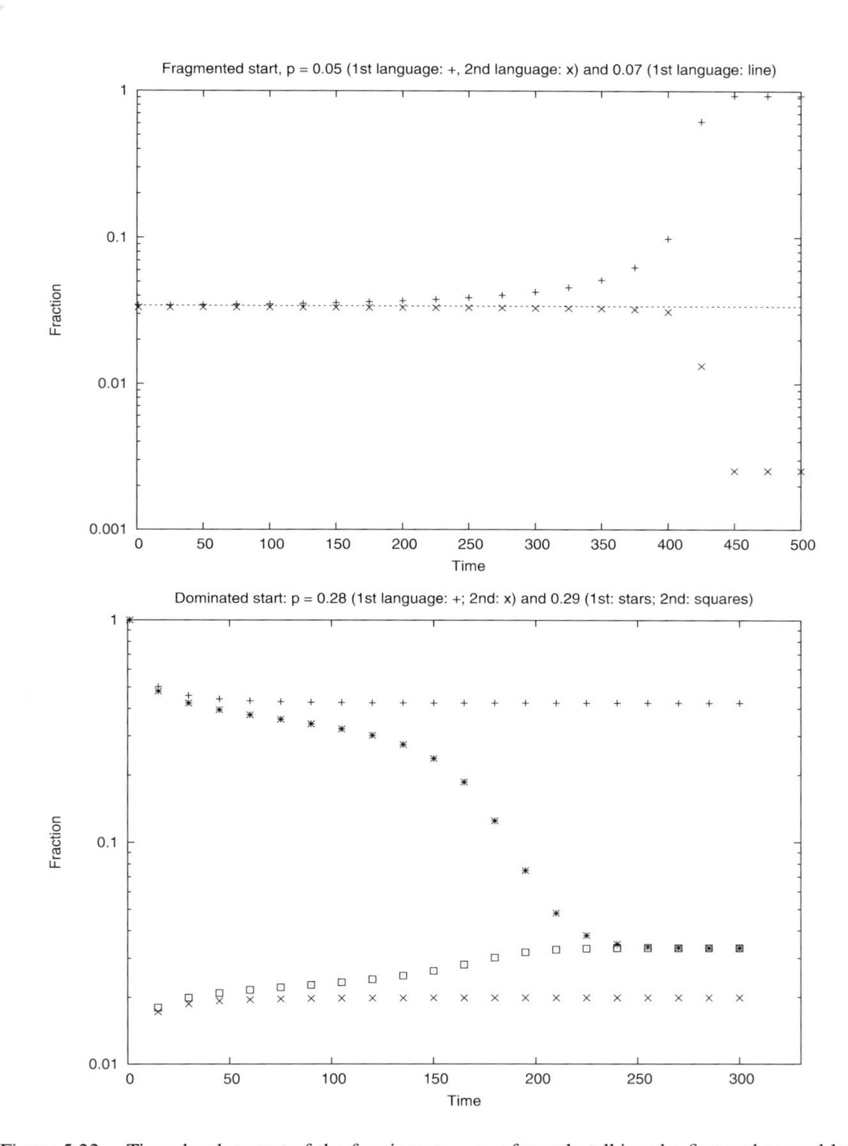

Figure 5.22. Time development of the fractions x_1, x_2 of people talking the first and second language. Top: fragmented start; transition near mutation rate 0.06; bottom: initially everybody speaks first language; transition near mutation rate 0.28. This mutation rate is the probability for errors in learning.

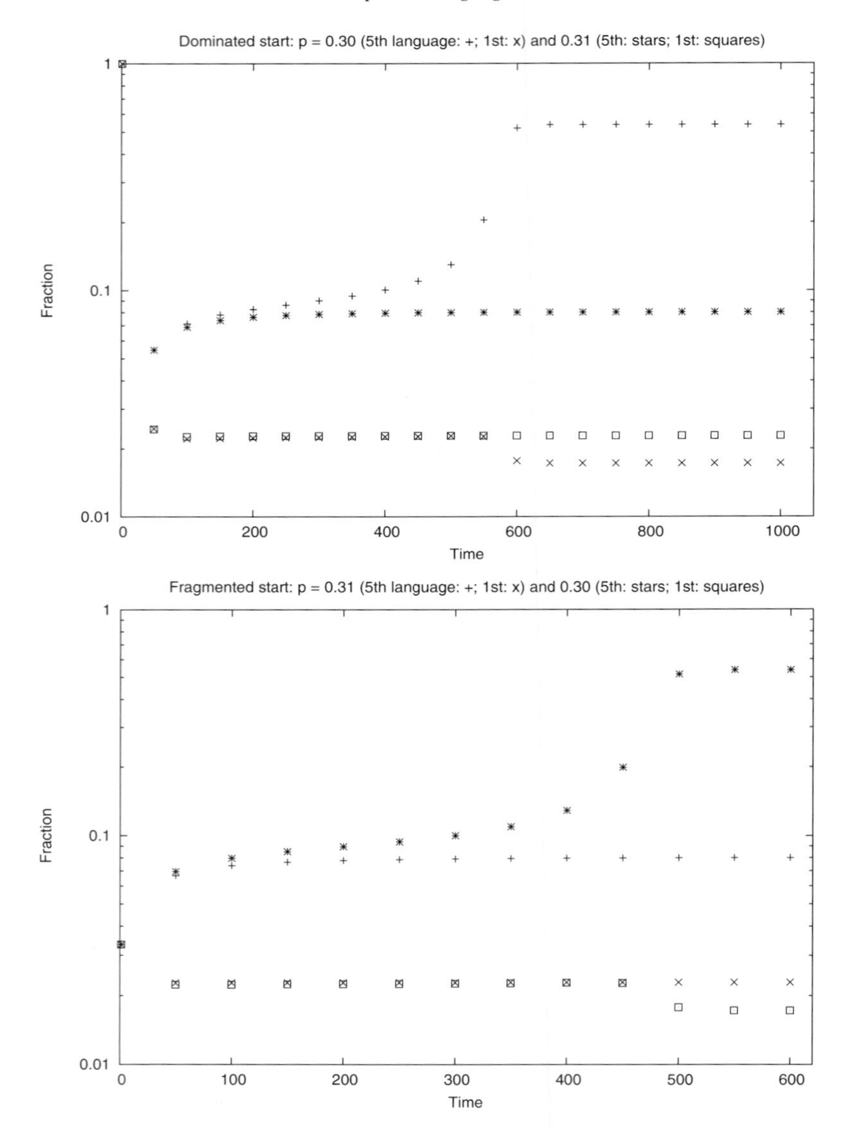

Figure 5.23. As previous figure but with random matrices F_{ij} and Q_{ij} and showing the first and the fifth of 30 languages.

But why should all Q_{ij} be equal except for the diagonal elements? The same question arises for the F_{ij} matrix. In the tradition of statistical physics and some previous models in this chapter, we therefore assume all off-diagonal matrix el-

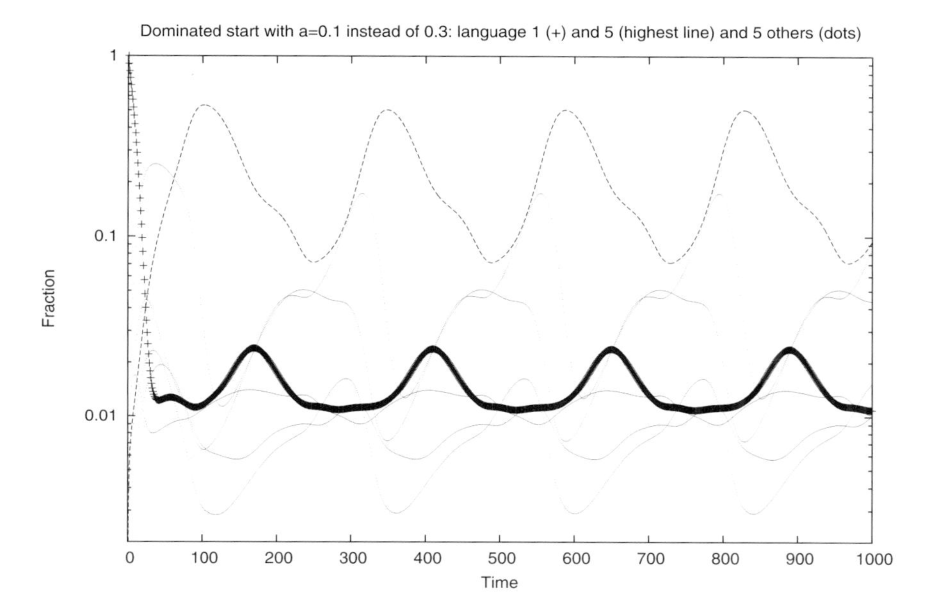

Figure 5.24. As in Figure 5.3 bottom, but with a only 0.1 instead of 0.3. Besides languages 1 (+, starting language) and 5 (most successful language) also several other of the 30 languages are shown.

ements to be random between zero and twice the above value. Then Figure 5.23 again shows the transitions of the preceding figure, except that the decay of dominance now also happens near 0.3. Note that in the bottom part of this figure everybody speaks the first language at the beginning, but it is the fifth language which may dominate. For larger number L of languages and observation time 1000, the maximum mutation rate p allowing nucleation of dominance out of fragmentation gets much smaller, between 0.02 and 0.03 for 4000 languages, while the minimum p to destroy dominance is roughly independent of L, near 0.275 for $L = 4000$.

Thus the results are quite similar to those of Schulze and Stauffer (2005): There is a transition between dominance and fragmentation. However, the crazy behaviour in the next Figure 5.24, with $a = 0.1$ instead of 0.3, shows that not everything is about the same; we used the same random number sequence as in Figure 5.23. Also, the histogram of language sizes is rather narrow in the case of fragmentation, though we cannot exclude that with suitable values for the numerous free parameters it gives what we want: something like Figure 5.1. If for the case of dominance we ignore the largest language, the histogram is nicer: Figure 5.25. In this last figure up to about as many languages are simulated as there exist now for mankind in reality. In the traditional version of this

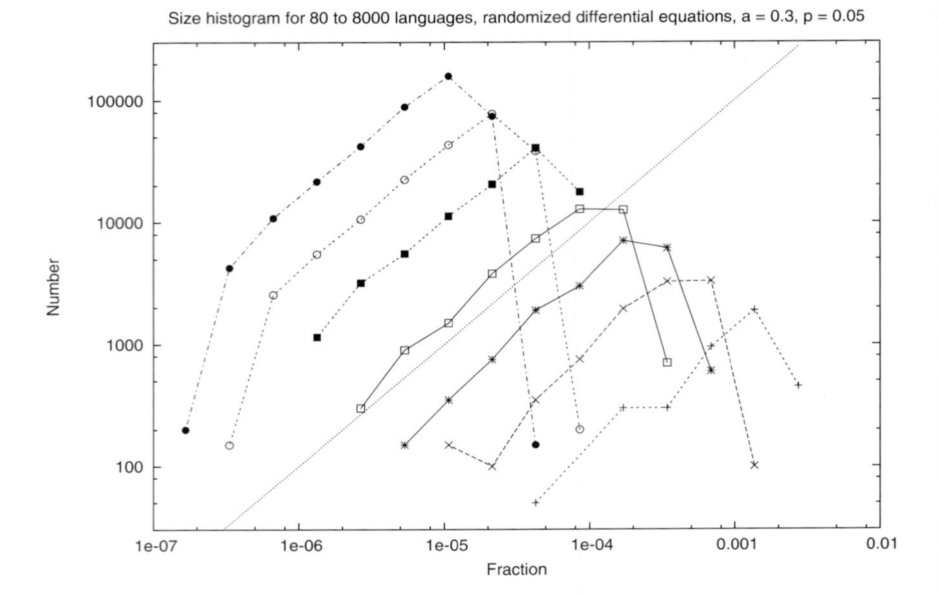

Figure 5.25. Histogram of language sizes, to be compared with Figure 5.1, $a = 0.3$, $p = 0.05$, after 100 iterations for 80, 200, 400, 800, 2000, 4000 and 8000 languages, from right to left. The straight line has slope 1.

model (Nowak, Komarova and Niyogi, 2002; Komarova, 2004), all languages except the possibly dominating one have the same size; it is the randomness of the coefficients, not the differential equation, which now produces this better result of Figure 5.25. To the left side of the maximum near $0.1/L$ in this log-log plot, the curves increase roughly with unit slope, which means (due to our exponential binning) that the number $\mathrm{d}L$ of languages with a size within a small interval $\mathrm{d}x$ is roughly constant for small x and then decays rapidly for large x.

One may reduce drastically the number of free parameters and the computer storage requirements for the matrices Q_{ij} and F_{ij} by assuming that they can be factorised: $Q_{ij} = Q_i Q_j$ and $F_{ij} = F_i F_j$. Now only $2L$ coefficients F_i and Q_i are needed and selected randomly, instead of $2L^2$ for L languages; Q_1 and F_1 are taken as large. But now no transition between dominance and fragmentation was found in tests; the first language was always stronger than the others.

Finally we mention that Kinouchi, Martinez, Lima, Laurenço and Risau-Gusman (2001) made computer simulations not of linguistic graphs but on them: A walker looks for "neighbours" of words in a large table of synonyms, and in this way finds out the network structure.

5.6. Conclusions

Agent-based simulations of language competition seem to start only now, thanks to the simple model of Abrams and Strogatz (2003), while emergence and learning of languages has a longer history (Nowak, Komarova and Niyogi, 2002; Cangelosi and Parisi, 2002; Wang, Ke and Minett, 2004; Wang and Minett, 2005a; Cavalli-Sforza, 1996; Cavalli-Sforza, 1997). The size distribution of real languages is qualitatively recovered, phase transitions are found, languages can merge like French and German into English or into a bilingual population, and stable interfaces may exist separating regions of different languages. The ageing model of Schwämmle (2005) could serve as a bridge between the language competition discussed here and the language learning reviewed by Nowak, Komarova and Niyogi (2002). It would be nice to have a more quantitative agreement between the language size distributions in reality, Figure 5.1, and in simulations, Figure 5.6. The field would profit from further constructive criticism from linguistics, as made by Wang and Minett (2005a) about the above models of Abrams and Strogatz, Patriarca and Leppänen, and the Nowak group.

Chapter 6

Social Sciences

In biology, people are accustomed to think that from simple animals up to dinosaurs, all species originated by Darwinian evolution and selection of the fittest. Once the principle is applied to human beings, some people dislike it and rely instead on creationism. Similarly for sociophysics, not much emotion is aroused if ants are simulated by mathematically defined probabilities. But to apply the same type of modelling to humans is disliked by some: We are not just atoms. Of course we are not; neither is the planet Earth a point mass. Nevertheless, for Kepler's laws of how Earth rotates around our sun, a point mass is a good approximation. And with respect to humans, already more than two thousand years ago Empedokles observed that some groups of people are like wine and water, mixing well, while others are like oil and water, mixing badly (according to J. Mimkes). More recently, Edmond Halley (famous through his comet) three centuries ago tried to estimate the survival probabilities of people; Gompertz in 1825 was more successful in that task; and life insurances, pension plans, and similar tools of modern society are well established since decades. If they go bankrupt it's not because people's death cannot be predicted by probabilities but only because changes in these probabilities were wrongly estimated. All these well known methods rely on the idea that the individual is difficult to predict, but averages over thousands and millions can be estimated quite accurately. Similarly, mass psychology is different from individual psychology. In this well-established spirit we review here some aspects of human action and thinking, as recently simulated on computers. Before any of the authors were born, the theoretical physicist Ettore Majorana (1942) suggested to apply quantum statistical physics to social sciences; Weidlich (2000) studied such questions since 1971, the same year in which non-physicist and 2005 economics Nobelist Schelling (1971) published his Ising-like agent-based simulations for ghetto formation; Galam (2004) gave a personal testimony of sociophysics going back to his 1982 publication. Besides this Weidlich book, other books were written by Schweitzer (2003) and Arnopoulos (2005). As was wisely remarked by W. Selke, if already the Ising model is so difficult to understand (e.g., Sumour and Shabat, 2005), why should human relations be simpler?

6.1. Retirement demography

As Chapter 3 discussed, ageing is part of life. Who supports us in retirement? We may have beautiful rights written down in present laws, supported by accumulations of our payments into pension funds or by our savings in various forms. But one must not forget that also by the year 2030 retired people will be consuming mostly only those goods and services which are produced at about that time by working people. If more money is available than things one can buy, inflation eats up the excess money. And if you want to play it safe and buy gold, first ask one of us who did that in 1980. In short, retirement problems are related primarily to the number of people within and outside working age; money is a secondary effect. Here we review the socio-econo-biophysics papers (Stauffer, 2002a; Bomsdorf, 2004; Martins and Stauffer, 2004; Zekri and Stauffer, 2005) where the ratio of retired people to working-age people is extrapolated for the next decades. Both the increased life expectancy and the decreased birth rate will make retirement support more difficult in the future than it was in the past.

6.1.1. Mortality and birth rates

As discussed in Chapter 3, the mortality function μ at middle age increases exponentially with age a:

$$\mu(a) \propto \exp(ba). \tag{6.1a}$$

Roughly this is the fraction of people who are alive at age a and die within one year; more precisely, $\mu = -d \ln S/da$ where $S(a)$ is the number of people still alive a years after their birth. Moreover, for different calendar years and different countries the Gompertz slope parameter b differs, but other parameters are the same (Strehler and Mildvan, 1960; Gavrilov and Gavrilova, 1991; Azbel', 1996, 2005), if the mortality function (which has the dimension 1/time since a has the dimension time) is written in a dimensionless form:

$$\mu(a)/b = A \exp\big[b(a - X)\big] \tag{6.1b}$$

where $A \simeq 10$ and $X \simeq 100$ years for all humans.

This Gompertz law is not valid for ages below 30 years, and perhaps also not for centenarians, Figure 6.1. However, in rich and peaceful countries the number of people dying before the Gompertz law starts to be valid, is of the order of one percent, less than the accuracy of our extrapolations. And the "mortality deceleration" claimed, e.g., by the Vaupel group (1998) for the oldest old will be questioned below. So we work with equation (6.1a).

Perhaps the universality of equation (6.1b) is not valid since about 1970 in Western Europe: Yashin, Begun, Boiko and Ukraintseva (2001) found that the

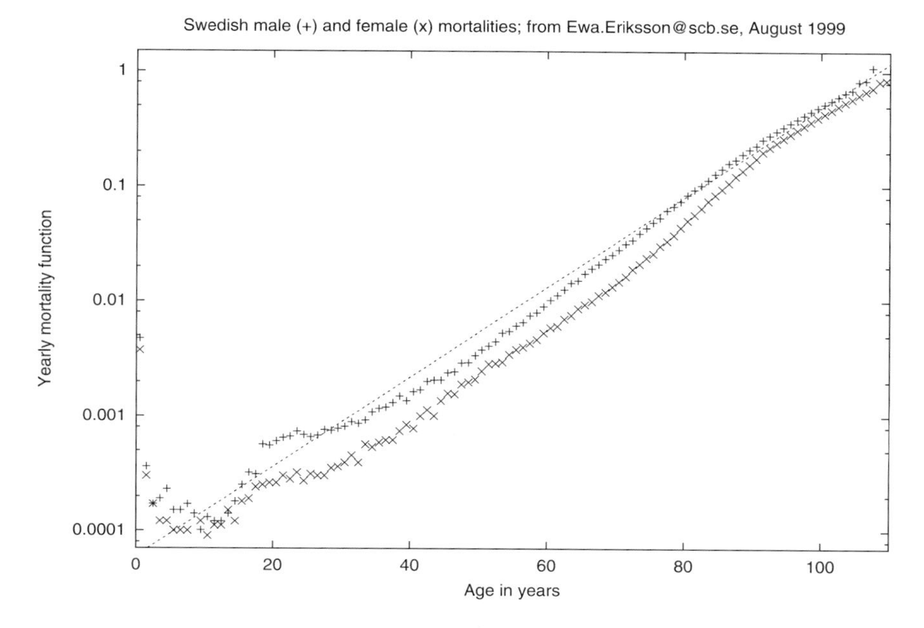

Figure 6.1. Mortality functions for Swedish men (top) and women (bottom), 1993–1997. From Stauffer (2002d).

survival curves since then are shifted parallel to higher ages, instead of getting more and more rectangular; see also Figure 4 in Tuljapurkar (2005). (We could not confirm this for West Germany.) Wilmoth, Deegan, Lundström and Horiuchi (2000) similarly give a change around 1970, in the age of the oldest Swede dying in each year. If correct, this means that before 1970 the Gompertz slope b increased at constant characteristic age X, while afterwards X increased at constant b. We thus assume b to increase linearly with calendar year from 0.070 in 1821 to 0.093 in 1971 and to stay constant thereafter, while X was constant at 103 years before 1971, and increased by 1.8 months every year thereafter, to give a rising life expectancy. (The remaining life expectancy at the median age, emphasised by Sanderson and Scherbov (2005), then remains roughly constant after 1970.) This possible change of behaviour around 1970 is not yet established very well, and so Bomsdorf (2004), in contrast to us, ignores it and extrapolates the logarithm of the mortality linearly in time.

This rising life expectancy, whether due to growing b or to growing X and shown already in Chapter 3, Figure 3.1, is complemented by the fall in birth rates after the use of contraceptive pills and the improvement of living conditions. For West Germany, the average number of children born to a woman (the fecundity) is approximated as

$$2.2 - 0.4 * \left[1 + \tanh\left((t - 1971)/3\right)\right] \tag{6.1c}$$

decreasing from about 2.2 to about 1.4 within a few years. Actually, East and West Germany, then separated enemies, were around 1970 the first industrialised countries where the birth rate sunk below the replacement rate of slightly more than two children per woman. At that time the Club of Rome had warned about the limits of growth and the dangers of overpopulation, and only a quarter of a century before, World War II had ended which was started by Hitler's Germany demanding more living space for its master race. Thus at the time a reduction of births was seen as favourable. Italy and Spain followed later but sunk to a lower level with 1.2 children per woman, while in France the reduction was weaker, and in the last decade the number of children per woman even increased in France from 1.7 to 1.9 (www.ined.fr). Thus Germany (now united enemies) is about typical for continental West-Central Europe; retirement in countries like Bulgaria with emigration of young people will be even worse. Michard and Bouchaud (2005) fitted numerous European fecundities to an error function (integral over Gaussian function instead of the tanh in equation (6.1c)) and found scaling.

6.1.2. Extrapolation

With the above assumptions, i.e., with a linear increase of life expectancy of 1.8 months per year and a constant number of 1.4 children per woman one can extrapolate into the future. The farther away this future is, the less reliable is the extrapolation. The numerical evaluation starts centuries ago to give a good "equilibrium" today and changes the population in each yearly age cohort by the above mortalities and birth rates. First we assume everybody to work from age 20 to 62 years and prevent net immigration. Children below the age of 20 are added to the people above retirement age since both groups need support and do not earn money.

Then the top curve in Figure 6.2 shows the extrapolation of the status quo: An enormous increase of the ratio of people needing support to the people of working age, and no end in sight. Also (not shown), the total population shrinks. A much better picture is seen in the middle curve where a net immigration (= immigration minus emigration) of 0.38 percent per year of the total population stabilises the total population; the immigrants are assumed to be 6 to 40 years old. (According to the Census Bureau of the USA, 22 percent of the whole population there were at least 55 years old in 2004, while for the fastest growing group, the "Hispanics", this percentage was only 11; Files (2005).) In addition, retirement age after 2010 is assumed (similarly Sanderson and Scherbov (2005)) to increase from 62 years by about half the increase of the life expectancy (just as a similar fraction of our salary's increases goes into taxes etc.). Then one can stabilise the support

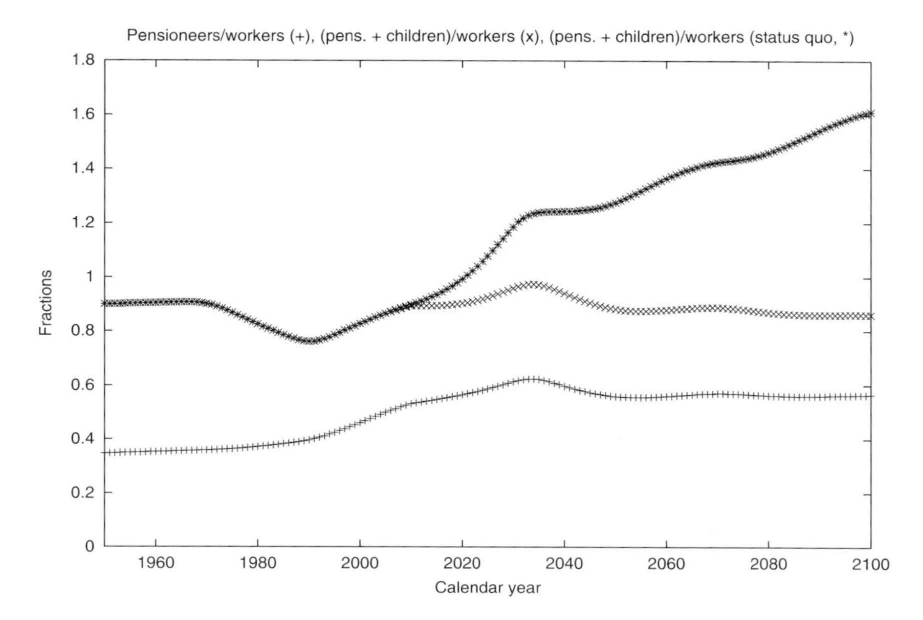

Figure 6.2. Extrapolation of the status quo (top curve) for the ratio of old plus young people to working-age people. The middle and lower curves assume an increase of the retirement age after 2010 and a controlled immigration of less than 0.4 percent per year. In the top and middle curves the children are added to the old people, while they are ignored in the bottom curve. From Zekri and Stauffer (2005).

ratio after the year 2040, as shown in the middle curve of Figure 6.2. Omitting the children in this ratio one recovers the lower curve as published by Martins and Stauffer (2004). The years around 2030 are the problem years: Then the 70-year old people are expected to be the strongest age cohort in Germany, while at present the majority of Germans want to retire at age 60 or earlier. Also with respect to immigration, Germans at present seem more worried about illegal immigration from Ukraine than about the need of immigration once unemployment has been reduced appreciably. Also experts who should know better have proclaimed the nonsense that immigration does not help since also immigrants get older. Bomsdorf pointed out that one could similarly claim that births do not help since also babies get older; one does not need our computer program (available as file "rente16.f" from stauffer@thp.uni-koeln.de; simpler version in our appendix, Section 9.5) to understand that births are needed for a stable population.

In developing countries (Berquó and Cavenaghi, 2005) like Algeria (Zekri and Stauffer, 2005) the extrapolations are much more optimistic if the present number of births per woman is assumed to be constant. Then the status quo is quite stable. They should learn from the European errors.

As pointed out above, extrapolations are dangerous. We assumed for the future a linear increase of life expectancy by 0.15 years every year. Watched over 250 calendar years t, the life expectancy does not increase linearly and can for Swedish women be approximated by $60 + 26 * \tanh[(t - 1910) * 0.013]$ as if there is a maximal life expectancy of 86 years, Figure 3.1 in our ageing Chapter 3. Indeed, from 1998 to 2002 this life expectancy increased only from 81.94 to 82.11; but then it jumped to 82.43 in 2003 and 82.68 in 2004 (from www.scb.se), killing the idea that a plateau was nearly reached. Only a comparison of different extrapolations by different reasonable assumptions using many past years can give a plausible prophecy.

A more "microscopic" simulation of demography, based on individuals who are born, age and die (analogous to physics simulations based on individual atoms instead of our mean-field averages which could become wrong after many generations, Section 2.8) was given by the geneticists Cebrat and Łaszkiewicz (2005); that journal issue also contains other ageing reviews. These and our simulations gave general trends only; specific studies for particular countries have to take into account, e.g., the changes in the percentage of people still working at ages 55 to 64 years, which in 2004 was 69.1 in Sweden, 30.5 in Italy, and 40.5 (36.6 in 2000) for the whole European Union.

6.1.3. Mortality deceleration?

For flies, the Gompertz law is violated for the oldest ages, where the mortality reaches roughly a plateau after 99.9 percent of the flies have died. Chapter 3 mentioned computer simulations reproducing this effect. But do humans die like flies (Stauffer, 2002d)? On pp. 18, 47 and 122 of the collection of reviews edited by Wachter and Finch (1997) we find: "mortality decelerates at older ages. ... the rate of increase slows down"; "mortality continues to rise throughout adult life, but at a decreasing rate after the age of 75 or 80"; "beyond 85 years, the mortality rate stops increasing exponentially and becomes constant, or actually decreases". Earlier such claims were recently reviewed by Gavrilov and Gavrilova (2005). It is difficult to reconcile all these claims with the Swedish data of Figure 6.1. Older data from the USA showed mortality maxima above the age of 100, and also numerous people living beyond 124 years (Klement and Doubal, 1997); presumably this effect was due to incorrect statistics and is no longer seen in more recent US life tables. A downward deviation from the Gompertz law, as seen in Figure 6.1, does not necessarily mean a deceleration (negative second time derivative): if at the start of a Formula 1 car race, Michael Schumacher (Cologne district) overtakes Rubens Barichello (Brazil), it does not mean that Barichello used the brakes to decelerate; he just accelerated less strongly than Schumacher.

The downward deviation seen for Sweden in Figure 6.1 is weaker than that found in England etc. (Thatcher, Kannisto and Vaupel, 1998) and much weaker than in the old data from the USA. The USA control their citizens less than Europe (no national identity card) and thus their demographic statistics are less reliable. Sweden avoided war since nearly two centuries and has a tradition of caring for its citizens; therefore Sweden may have the best statistics for natural development. Thus the better the data, the smaller are the deviations from the Gompertz law; perhaps with no errors in the birth dates and no war, the deviations would vanish completely.

Also, as found out in 2005 by the president of Harvard University, women are different from men. They are usually less law abiding then men, as can be seen in Figure 6.1: While men there obey the Gompertz law up to about 90 years, the lawless women after about 75 years try to catch up with men, and their mortality becomes *higher* than the Gompertz extrapolation from middle age. At even older ages, the two mortality curves in this figure become nearly parallel and show slight downward curvature. Thus women should not be trusted to test deviations from the Gompertz law at old age, since they disobey it anyhow. Males usually follow Gompertz up to a higher age and thus test better for downward deviations of the oldest old. But less men survive up to 100 years than women, making the statistics worse.

Thatcher, Kannisto and Vaupel (1998) found excellent fits for mortality functions above 80 years of age by the Kannisto expression:

$$\mu(a) = \exp(ba)/\big[\text{const} + \exp(ba)\big] \tag{6.2}$$

which gives for old age a mortality plateau $\mu(a \to \infty) = 1$. But here a dimensionless exponential function is mixed with the mortality which has the dimension 1/time. In other words this fit works only if we look at the yearly mortality function, not at the 12 times smaller monthly mortality. It seems questionable that human life should be determined that much by the yearly seasons.

The Vaupel group (1998) claims a mortality deceleration for the oldest old for humans, flies and several other animals. The paper does not make clear whether it plots the mortality function $\mu(a)$ (also called the "force of mortality" or "hazard factor"), or the fraction $q(a)$ of individuals living at age a which die within the next time interval; obviously this q cannot become larger than one and gives a mortality plateau or maximum as a mathematical triviality.

One of the best single papers on the statistics of the oldest people, besides Gavrilova and Gavrilov (2005) for the USA, is Robine and Vaupel (2001). They name more than 100 Europeans aged 110 years and above and find a mortality plateau. They distinguish between medium and high quality data. With both sets combined the data are not a smooth continuation of the Swedish data of Figure 6.1 if we plot survivors versus age; with only the high-quality data included the slopes are continuous but the curvature changes at an age of about 110. Again, the higher the quality the weaker are the deviations from Gompertz. They do not cite Suematsu and Kohno (1999) who found without quality check a plateau for

Japanese mortalities, but already after 100 years. Moreover, Robine and Vaupel plot the age of the oldest person dying in one calendar year. Including the data of medium quality they get a break such that after 1970 the data increase much faster than before, as in Wilmoth, Deegan, Lundström and Horiuchi (2000); with only the high-quality data included, this effect nearly vanishes.

Similarly, Gavrilova and Gavrilov (2005) show in their Figures 3 to 6 for the USA: The better the data, the smaller the deviation from Gompertz. For their Americans born in 1891, the monthly mortality function increases exponentially up to 105 years, and then scatters in both directions from the extrapolated exponential. Thus they "expect that cohorts born after 1891 would demonstrate even better fit by the Gompertz model than the older ones because of improved quality of reporting".

Thus there are still important systematic errors due to overstated ages; elimination of these errors may eliminate the deviations from the Gompertz law for the oldest old (Stauffer (2002d), as warned also by Gavrilova and Gavrilov (2005)), but may also invalidate our above assumptions in Section 6.1.2 for the breaks in b and X around 1970, in favour of the linear extrapolations of Bomsdorf (2004). Careful checks for deviations from Gompertz law should rely on male mortalities, using all ages above 30 years for the fit, and should watch out for and possibly reduce systematic errors. And since the human mortality plateau of Robine and Vaupel applies to one person in a million, while that of flies helped one out of a thousand, a repetition of the fly experiments (Curtsinger, Fukui, Townsend and Vaupel, 1992; Carey, Liedo, Orozco and Vaupel, 1992) would be useful.

6.1.4. Conclusions

The literature needed for the ageing simulations of Chapter 3 gives a worrying future for retirement in some parts of the world, and mankind's hope to get closer to eternal youth via mortality deceleration and mortality plateau may not be justified yet. People in the rich countries with low birth rates may have to retire later; role models are bank robbers in Germany in their seventies, or the American woman who at age 78 shot her "boy" friend, aged 85, when he became attached to another woman.

6.2. Self-organisation of hierarchies

Societies, not only of humans, often develop hierarchies; some leaders are on top and others follow them. Why? How come one guy is a full professor, living luxuriously surrounded by several women, while others are badly paid associate professors, close to starvation. The idea that kings are in that position by the grace of God is less widespread now. Instead we can explain things by higher

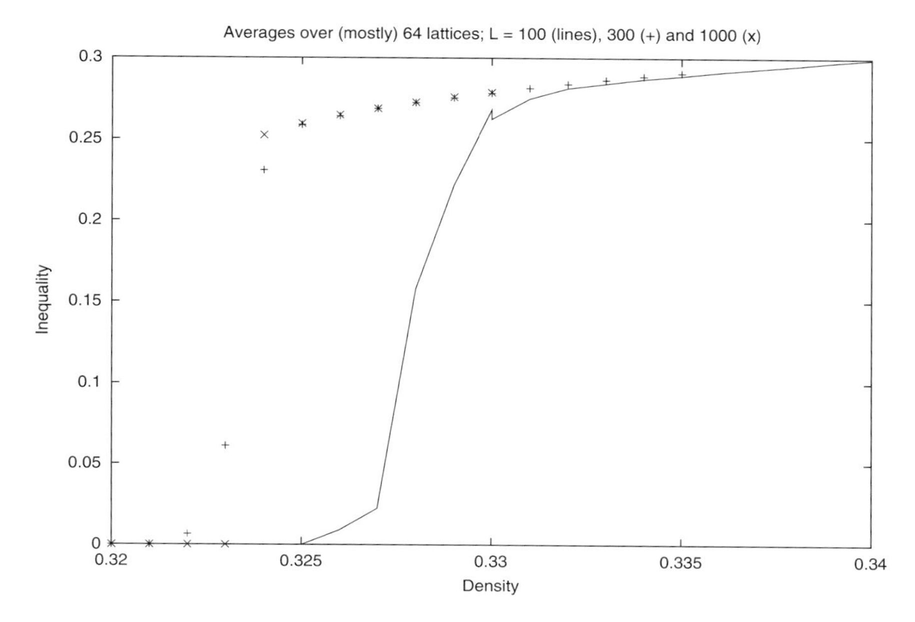

Figure 6.3. First-order phase transition from egalitarian society at low population densities to hierarchies at high densities on $L \times L$ square lattice. From Stauffer (2003).

intelligence, racial quotas, etc. On the other hand, if statistical physics is applied to this question, the basic assumption is obvious: Everybody is originally equal, and then due to random events some people get ahead of others. This indeed is what Bonabeau, Theraulaz and Deneubourg (1995) did (in a paper which lay dormant for several years before it was simulated again (Sousa and Stauffer, 2000; Stauffer, 2003); a program and more details are published in Stauffer (2005)).

In this Bonabeau model, people diffuse on a square lattice filled with density p. Whenever a person wants to move onto a site already occupied by someone else, a fight erupts which is won by the invader with probability q and lost with probability $1 - q$. If the invader wins, the winner moves into the contested site whereas the loser moves into the site left free by the winner; otherwise nobody moves. Each visitor adds $+1$ to a history variable h, and each loss adds -1 to h. At each iteration, the current h is diminished by ten percent, so that roughly only the last ten time steps are kept in memory h. The probability q for fighter i to win against fighter k is a Fermi function:

$$q = 1/\big[1 + \exp\big((h_k - h_i)\sigma\big)\big] \tag{6.3}$$

where σ with $\sigma^2 = \langle q^2 \rangle - \langle q \rangle^2$ is the standard deviation in these probabilities. Initially everybody starts with $h = 0$; then $q = 1/2$ for all fights. After some

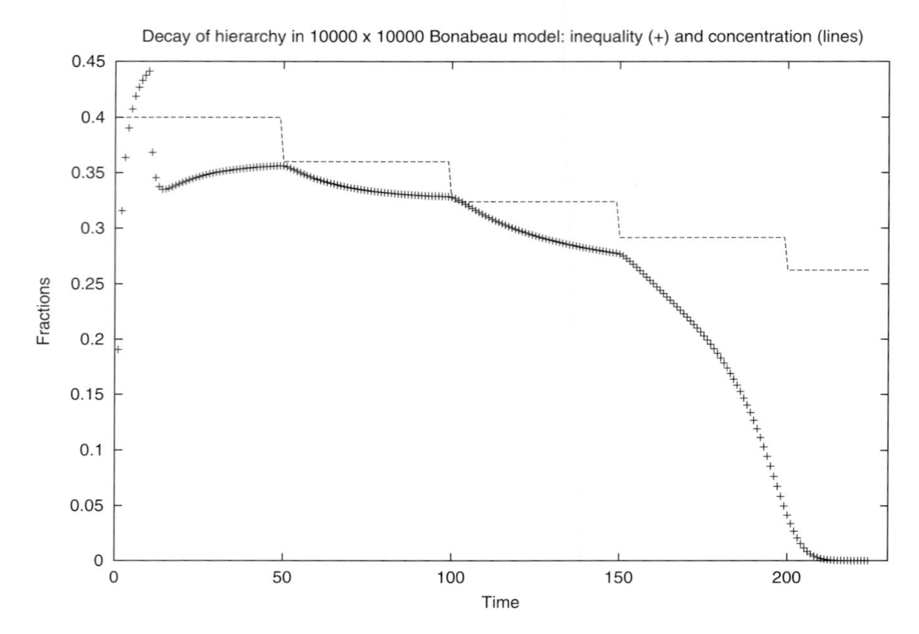

Figure 6.4. Revolution in 10000 × 10000 Bonabeau lattice. The concentration p (horizontal lines) starts at 0.4 and then is diminished every 50 time steps by ten percent of its current value. The + symbols give the inequality σ. Basically, the horizontal lines separated by steps are the input which causes the downward curve or revolution.

time, history h accumulates in memory, q differs from $1/2$, $\sigma(t)$ becomes positive, measures the amount of inequalities in society at that time step t and is obtained by averaging over all fights occurring during this iteration t. We thus have a feedback: σ enters into the calculation of the q and afterwards is calculated from these q values for the next iteration.

A phase transition is observed if the concentration p increases above a threshold near 0.32, Figure 6.3: The social inequality σ jumps from zero to a nonzero value, which in the history of humanity may correspond to the transition from the more egalitarian nomadic society to agricultural life with property of land, cities, and nobility. Wealth can develop only if there is a surplus of food etc. (Angle, 1986).

If a hierarchy has developed at high concentration, it can be destroyed again if the concentration is lowered, Figure 6.4. We can interpret this lowering as an increase of the number of people who refuse to follow the rules and thus do not participate in the power game. This destruction of hierarchies can be interpreted as a glorious revolution where we get rid of the politicians, or as universities descending into chaos and no longer regarding us professors as infallible.

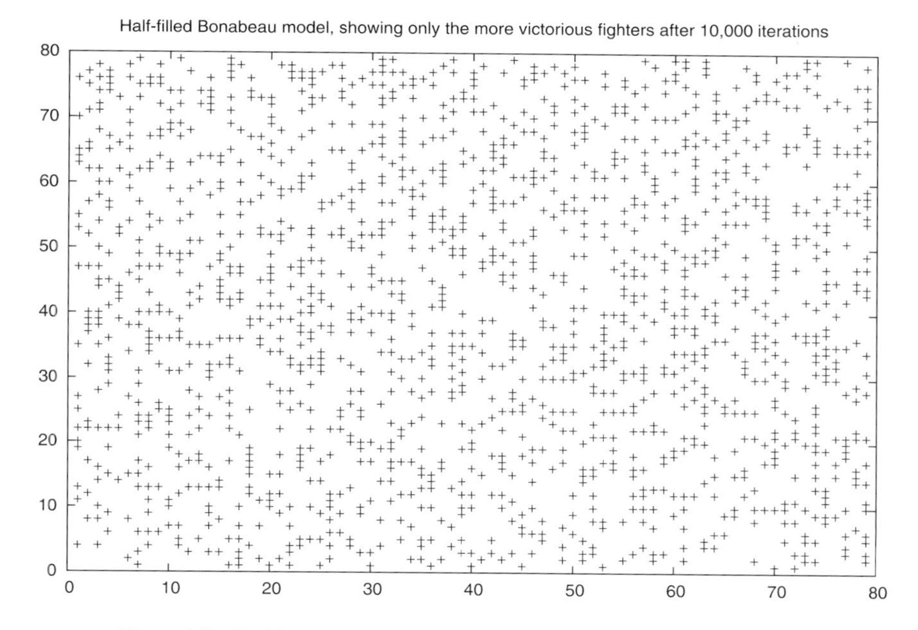

Figure 6.5. Positions of Bonabeau winners, having more victories than losses.

One might think that winners repel each other since in any collision between them one will be the loser. Their stupidity, however, prevents any obvious spatial ordering, as Figure 6.5 indicates.

In the above model, people fight for the purpose of fighting; for example, an invader may win a fight and thus move to the right, and later may again win a fight and move to the left, forcing the inhabitant of the left site to move back to the same right site from which this loser was expelled before. Typical male? Naumis, del Castellino-Mussot, Pérez and Vázquez (2006) behaved more reasonably and distinguished attractive from less attractive sites; then the concentration threshold is decreased.

In this simple version, as well as when different people have different fighting abilities distributed symmetrically (Schulze and Stauffer, 2004), there are as many leaders as are followers. An asymmetric version with less leaders was given by Stauffer and Martins (2003) who also looked at people on a scale-free network instead of a square lattice. An asymmetry was also found in a modified mean-field version by Ben-Naim and Redner (2005). These and similar networks were investigated more thoroughly by Malarz, Stauffer and Kułakowski (2005), Gallos (2005), and Sousa (2005) and may correspond to fights over important positions in society instead of over territory on the square lattice.

In conclusion, being on top is an accident; if you are unhappy with your life, blame the random numbers.

6.3. Opinion dynamics

How do people convince each other to reach a consensus, or fail to do so? In line with the rest of this book we do not look at specific people and specific arguments, but at the general principles. Mostly we will review work starting in 2000, but first we look at some older models, some of which were reviewed by us already in our earlier book (de Oliveira, de Oliveira and Stauffer, 1999). While also some animals must find a consensus on where to move (Couzin, Krause, Franks and Levin, 2005), we concentrate here on humans.

6.3.1. Before 2000

Ising-type models

The Ising model was already applied decades ago to explain how a school of fish aligns into one direction for swimming (Callen and Shapero, 1974) or how workers decide whether or not to go on strike (Galam, Gefen and Shapir, 1982); no new simulations were needed then. The social impact model of Latané (1981), modified by Kohring (1996), was also applied to languages (Nettle, 1999b) and gave a phase transition (Bordogna and Albano, 2006). In this Latané model the Ising spins $S_i = \pm 1$ are updated, $S_i(t+1) = S_i(t)\,\mathrm{sign}\,h_i(t)$, according to the sign of their local field

$$h_i = \sum_j \left[J^s_{ij}(1 + S_i S_j) - J^p_{ij}(1 - S_i S_j)\right] \tag{6.4}$$

coming from the other people j. Here J^s is a supporting force encouraging not to change opinion, and J^p a persuasive force trying to change it. The model may give a consensus, a fragmentation into many different opinions, or a leadership effect when a few people change the opinion of lots of others. A thorough review was given by Hołyst, Kacperski and Schweitzer (2001). A random-field Ising model was suggested early by Galam in a rare collaboration with a psycho-sociologist (Galam and Moscovici, 1991). They were able to ground it in the field of opinion formation on the basis on real experiments from experimental psychology. Random-field Ising models were also applied, e.g., by Michard and Bouchaud (2005). Later Galam focused on linking the zero temperature properties to rational decision making (Galam, 1997) and several limited agent-based numerical experiments were subsequently performed (Galam and Zucker, 2000). The influence of contrarians who are always against the majority was studied in several papers since 2004, as cited by de La Lama, López and Wio (2005) and Caiafa and Proto (2006).

To some extent the voter model of Liggett (1985) is an Ising-type model: Opinions follow the majority of the neighbourhood, similar to Schelling (1971); see Dornic, Chaté, Chavé and Hinrichsen (2001), Suchecki, Eguíluz and San Miguel (2005) and Castellano, Loreto, Barratt, Cecconi and Parisi (2005) for recent applications, and San Miguel, Eguíluz, Toral and Klemm (2005) for a nice introduction. Efros and Désesquelles (2005) showed how a few zealots can produce a phase transition to bad behaviour in a community of individually good people; the transition from Bach and Beethoven to Auschwitz comes to our mind (see also Michard and Bouchaud (2005)).

Also related are models for the formation of ghettos where within one large city one sees large districts with ethnically rather pure populations coexisting with each other. In the simplest case this is approximated as black and white, simulated with two Ising spin orientations. Then, of course, if the temperature is above the Curie temperature of the Ising models, black and white mix apart from some short-range correlations, whereas at lower temperature one has "infinitely" large domains of parallel spins (e.g., mostly black) with equally large domains of opposite spin direction (e.g., mostly white). Schelling (1971) in the first issue of Journal of Mathematical Sociology simulated this with conserved "magnetisation" ("Kawasaki" dynamics) on a dilute lattice. Thus one person can migrate to a neighbouring empty site with a probability depending on the number of neighbours of the same or the opposite colour. Later work (Meyer-Ortmanns, 2003), with up to seven different groups (Schulze, 2005), studied the effect of temperature increasing with time while a minority group immigrates into a country. This social temperature measured the tolerance towards people of other ethnic groups. If this temperature increases fast enough, no large domains are formed and ghetto formation is avoided.

Schelling's paper does not mention the Ising model of 1925 or related physics work, and presumably in 1971 no physicist had yet published simulations of Ising models with Kawasaki dynamics and annealed dilution. This excuse does not apply to much later work (Zhang, 2004). The pioneering work of Schelling shows that the same type of simple models made by physicists are also made in the fields to which these simulations are applied. Physicists should not claim they were first, and sociologists should not put down physics models.

Galam conservatism

Galam (1990) suggested an analytically solvable model on how majority decisions are arrived at in complex societies. A group of four people may elect one representative, who together with the elected representatives of other groups selects a super-representative, and so on further up in the hierarchy until the highest group of four people makes the final decision based on their four opinions. Only two choices, yes and no, are open for this decision. If on any level there are two votes for yes and two votes for no, then this means a no (or status quo). For a random distribution of yes and no votes on the lowest level, one needs a high percentage of yes votes, about three quarters, to arrive after many hierarchies at the top with a yes decision. This model could explain how a dictatorship, once established (corresponding to no or status quo) can stay in power even if some not overwhelming majority of the people want change (yes). The model was extended to any group size and also mixture of sizes. When the dynamics is applied to opinion dynamics instead of voting it could explain why any reforms are so difficult. Even the leading French journal Le Monde reported about it on Feb. 26, 2005, in connection with the referendum on the European constitution. Also some computer simulations were made (Galam, Chopard, Masselot and Droz, 1998; Galam and Wonczak, 2000; Stauffer, 2002b; Tessone, Toral, Amengual, Wio and San Miguel, 2004; Schneider, 2004; Stauffer and Martins, 2004). More literature on Galam models is cited in Sousa, Malarz and Galam (2005).

Axelrod multiculturality

Axelrod (1997) simulated how people from different cultural backgrounds can interact with each other and still keep different cultural identity. This model triggered lots of follow-up, see, e.g., Klemm, Eguíluz, Toral and San Miguel (2003) or the very nice review of San Miguel, Eguíluz, Toral and Klemm (2005). Each agent's culture is represented by S variables (e.g., binary: zero or one), and when two agents meet one may take over one of the variables of the other. Thus in contrast to most opinion models, people form an opinion not only on one question, but on S different questions. This pioneering paper was also the foundation for the multi-opinion papers, Section 6.3.3, on missionaries (Sznajd-Weron and Sznajd, 2005), negotiators (Jacobmeier, 2005) and opportunists (Fortunato, Latora, Pluchino and Rapisarda, 2005) of the next subsection.

6.3.2. Three recent models

This subsection deals with the Deffuant negotiators (Deffuant, Amblard, Weisbuch and Faure, 2002), the opportunists of Hegselmann and Krause (2002) and the missionaries of Sznajd-Weron and Sznajd (2000), all three invented apparently independently from each other around 2000. All three are different in their rules of opinion change but quite similar in their results: They lead to a final status with, depending on parameters, one opinion ("consensus"), two opinions ("polarisation") or more than two opinions ("fragmentation"). These names, of course, are arbitrary and should be changed for different applications where, e.g., consensus means dictatorship. They were studied with continuous opinions, or with integer opinions including the binary case (yes or no) as the simplest choice. People were put onto lattices, networks, or simply allowed to interact with everybody. A somewhat longer review is given by Stauffer (2005), and programs by Stauffer (2002c).

In all three models, the concept of bounded confidence was used. It means that we authors from Niterói might be willing to discuss and to agree with people from Rio de Janeiro, but certainly not with Argentinians (particularly if born near Neanderthal). Thus if we have five opinions, corresponding, e.g., to five parties represented in parliament and ordered politically from left to right, then party 4 might make a compromise with parties 3 and 5, but hardly with party 1. Thus only people whose opinions differ by not more than a confidence bound (equal to one in this example) discuss with each other. Since the extremists (1 and 5) have opinion neighbours on only one side, a consensus usually is based on a centrist opinion like 3. In the simplest case, when we have only two possible opinions, this confidence bound makes no sense and is omitted.

Negotiators

At present the Deffuant negotiators seem the most realistic agents simulated by more than only one group of authors. Two people do not agree immediately but get closer in their opinions after their discussion. At every iteration, each agent selects randomly one other agent for discussion. If their opinions differ by more than the confidence bound no opinion changes; otherwise each opinion moves closer to the other, by an amount proportional to the difference between the opinions. (If for integer opinions the two agents differ by only one unit, then one of them, randomly selected, accepts the opinion of the other.) For continuous opinions between 0 and 1, up to 450 million people were simulated, corresponding to the opinions on the constitution draft for the European Union. Figure 6.6 shows the similarity between these negotiators and the opportunists reviewed below, at a confidence bound of 0.15 for opinions between 0 and 1. (Only for the presentation of this figure we rescaled opinions from 0 to 5 and binned them into unit intervals.) For confidence bound above $1/2$, consensus is found (Fortunato, 2004b); if the confidence bound decreases below this threshold, first polarisation happens and then fragmentation, with the number of surviving opinions inversely proportional to the confidence bound. For this case of everybody having a chance to talk with everybody, Ben-Naim, Krapivsky and Redner (2003) made an accurate mean-field approximation. (Typically, if two opinions move towards each other, they move by 30 or 31.6 percent of their difference in the simulations presented here.)

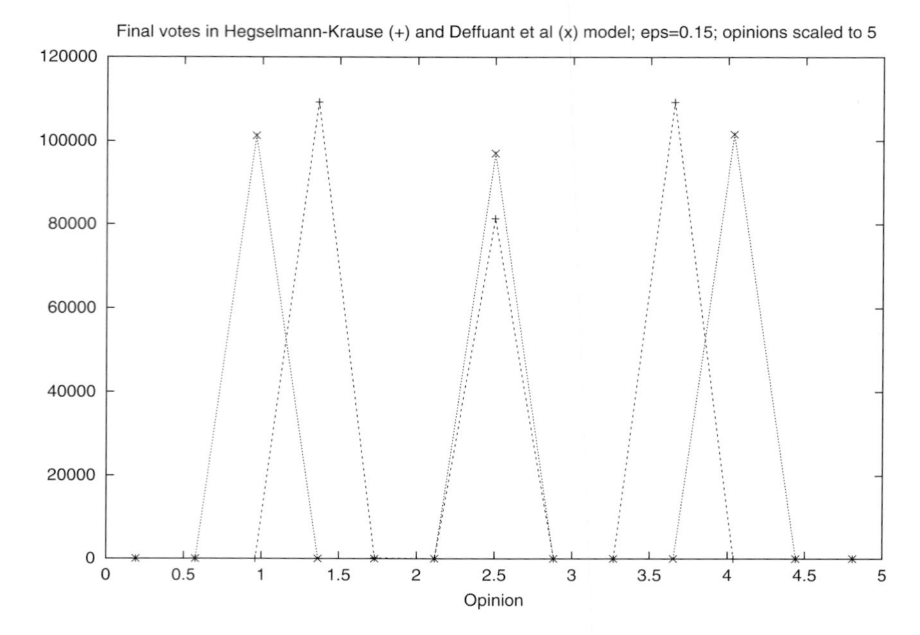

Figure 6.6. Comparison of the opinion distribution for 300 000 negotiators (×) and opportunists (+) at the end of their discussions at a confidence range of 15 percent, allowing the survival of three major opinions. From Stauffer (2005).

Opportunists

In the Krause–Hegselmann model (Hegselmann and Krause, 2002) each opportunist also can talk to everyone, but now each agent talks to all agents simultaneously and assumes their (weighted) average opinion. This arithmetic average (other averages were also used: Hegselmann and Krause (2005a)) ignores all agents outside the confidence bound. Again, for large enough confidence bounds (above 0.2) only one opinion survives, for smaller ones we have two or more opinions. (Krause prefers the characterisation by compromise instead of opportunism.)

Missionaries

The Sznajd model (Sznajd-Weron and Sznajd (2000); see Sznajd-Weron (2005) for a recent review) of missionaries assumes that two lattice neighbours sharing the same opinion force this opinion onto all their neighbours. Thus in contrast to the opportunist, voter and Ising models, information now flows from inside out instead of from outside in. It is the most popular of the three recent models and

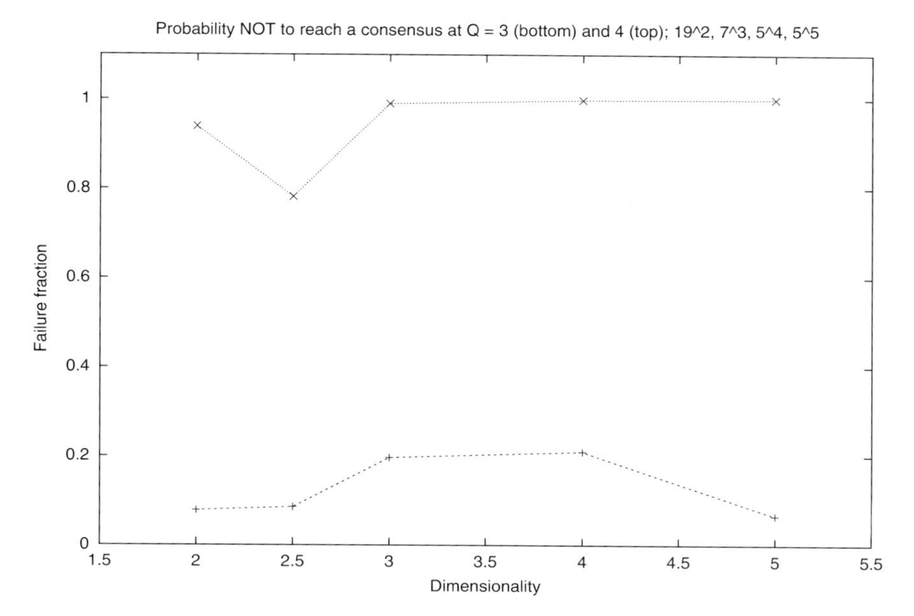

Figure 6.7. Sznajd missionaries: Comparison of three and four possible opinions, at unit confidence bound, for square, triangular, simple cubic, 4D hypercubic and 5D hypercubic lattices. We plot the probability not to find a consensus which is high for four and low for three possible opinions. From Fortunato and Stauffer (2005).

mostly simulated with integer opinions, in particular only two. For unit confidence bound, in general one gets a consensus for $Q = 2$ or 3 possible opinions, but not for $Q \geqslant 4$. Figure 6.7 shows this for various lattices; of course, the technology of the 21st century will be based mainly on five dimensions, the last point in this figure.

Missionaries with two possible opinions have a phase transition in two and more dimensions but not in one: If initially half the people have opinion A and the other half opinion B, randomly distributed on the square lattice, then at the end everybody has the same opinion: A in half of the samples and B in the other half. If, however, initially one opinion occurs slightly more often than the other, then at the end in large square lattices everybody has the initial majority opinion. The larger the lattice is the sharper is this transition (Sousa and Stauffer, 2000); for a modified model, a thousand $L \times L$ lattices were simulated (Schulze, 2004) to show that the width of the transition, as measured in the size of the initial majority, vanishes as $1/L$ for $10 < L \leqslant 10\,000$.

Applications

All three models were applied to the question of how a single event influences the whole development. How would the 20th century look like if Adolf Hitler would have been killed during the first World War? Thus one simulates the same system twice, with the same random numbers and the same initial distribution. The only difference is that in the lattice centre we change one opinion. Physicists call this type of study "damage spreading". The influence of this initial damage on the later development depends on the parameters and was recently reviewed by Fortunato and Stauffer (2005); see Weisbuch, Deffuant and Amblard (2005), and Sahimi and Stauffer (2005) in a different model, for the influence of a few extremists. Chapter 2 mentioned already the genetic analog: How would mankind's genes look like if Ghengis-Khan or Emperor Pedro I would have had no children.

The Penna ageing model of Chapter 3 was combined with the missionaries by Sun, Luo, Mao and He (2005) such that young children follow the parents while older people follow the opinion of similar age groups; the influence of the family then hinders the spread of new opinions.

Only Sznajd missionaries were thus far found to agree with election results in Brazil and India, perhaps because the other models were not yet tested on these accurate social data. Empirically, in elections of many candidates the probability that one candidate gets v votes decays as $1/v$ for intermediate v while for both very large and very small v the probabilities are smaller: No candidate can get half a vote, or more votes than were cast. This power law plus the deviations from it were well reproduced by missionaries on a Barabási–Albert scale-free network at intermediate times before a complete consensus was reached (Bernardes, Stauffer and Kertész, 2002; González, Sousa and Herrmann, 2004): Figure 6.8.

Physicists like scaling laws: Having measured the magnetisation as a function of magnetic field at one temperature near the Curie point, scaling laws predict the magnetisation-field curve also for a temperature only half as far away from the critical temperature. Mathematically, scaling means that a function $F(x, y)$ is a scaled function of only one variable $z = y/x^a$ if a function f can be found such that

$$F(x, y)/x^b = f\big(y/x^a\big) = f(z) \tag{6.5a}$$

holds for small x and y, where a and b are two critical exponents. Thus the quantity (magnetisation) F, scaled by some power of the variable (temperature difference) x, is a x-independent function of the variable y, provided that variable is scaled by x^a. A more physical presentation of scaling, using the renormalisation group, was already given in Chapter 2, tested for a biological model as shown in Figure 2.29, and Section 3.3.3 applied it to the shape of the mortality function. Scaling thus means more than just power laws; it means the similarity of whole functions $F(x, y)$. (If x and/or y goes to infinity instead of zero in the region of

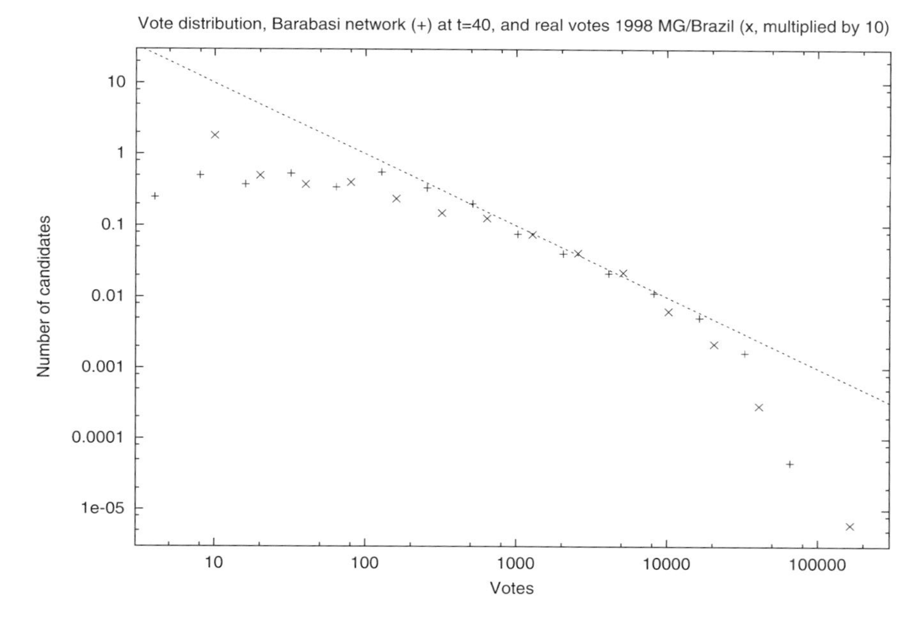

Figure 6.8. Sznajd missionaries: Comparison of election results in Brazil with computer simulations of missionaries on Barabási–Albert networks. The number of candidates getting a certain number of votes is plotted double-logarithmically versus the number of votes. Only the shapes should be compared; the number of votes in the simulations was multiplied with fitted scale factors. From Bernardes, Stauffer and Kertész (2002).

interest, then the above scaling can be tried in the reciprocal variable as we use it below.) Actually, it is quite difficult to invent a function $F(x, y)$ of two variables which does not obey this scaling asymptotically.

For opinion dynamics, scaling is particularly simple in all three dynamics (negotiators, opportunists, missionaries) if we make the opinions discrete and assume that people discuss only with others sharing the same opinion or differing by only one unit from it (Stauffer, Sousa and Schulze, 2004; Fortunato, 2004a; Rodrigues and Costa, 2005). The number of surviving opinions, corresponding to $1/F$ in the above formula, depends on the number Q of possible opinions, corresponding to $1/x$, and the number of people, corresponding to $1/y$. But the exponents a and b, which general scaling theory does not predict, are trivially equal to one here. This can be seen by defining $z = y/x =$ number of possible opinions per person, and by looking at the limits $z \gg 1$ (i.e., number of people much smaller than the number of possible opinions) and $z \ll 1$ (much more people than possible opinions). In the first case, everybody can stick to his/her original opinion meaning $F = y$; in the second case, there are so many people that each opinion can find a follower, and therefore $F = x$. Thus the scaling result is

$$F(x, y)/x = f(z), z = y/x \tag{6.5b}$$

where $f(z \gg 1) = z$ (i.e., $F = y$) and $f(z \ll 1) = 1$ (i.e., $F = x$). The numerical results (summarised by Stauffer (2005)) confirm this scaling law for large numbers of opinions and people, with deviations clearly visible in all three models for small numbers of possible opinions: Scaling is valid only asymptotically. In this sense, opinion dynamics offers a simple introduction into the scaling laws near critical points in physics. (Things become more complicated if the confidence interval is not just ± 1 as assumed above, but becomes a third independent variable in addition to x and y.)

6.3.3. * *Additional remarks*

Cluster sizes

If at the end of an opinion simulation, everybody shares the same opinion, one may say that all people form one cluster. If, on the other hand, all people keep their original opinion which is not shared by anybody else, then each person can be said to form a separate cluster of size one. Thus in general, people can be grouped into clusters such that within one cluster everybody has the same opinion, and people in different clusters have different opinions. Following the tradition in percolation theory (Section 6.6 below) we denote the number of people in a cluster by the cluster size s, and the number of clusters of size s by n_s. The above Figure 6.8 is, in this sense, a cluster size distribution. Numerically, it is much easier to check whether or not two opinions agree if the opinions are discrete integers between 1 and Q; for continuous variables one needs a small threshold $\sim 10^{-6}$ such that opinions count as identical if they did not differ by more than this threshold.

Figure 6.9 shows for continuous Deffuant negotiators the cluster size distribution on directed scale-free networks, and the following figure illustrates how they are formed as a function of time. (In normal or symmetric networks, if node A is neighbour to node B then B is also neighbour to A; in directed networks, if A influences B then B does not influence A.) It is important to vary the size N of the whole system. Then one sees for large N that there are two types of clusters. Small ones have an n_s/N roughly independent of N, i.e., the number n_s of isolated sites $s = 1$, pairs $s = 2$ and triplets $s = 3$ increases proportional to N. Large clusters, on the other hand, have a size s proportional to N and exist perhaps only about once in a network. Percolation experts know that such behaviour is expected for concentrations slightly above the percolation threshold, see Section 6.6 below.

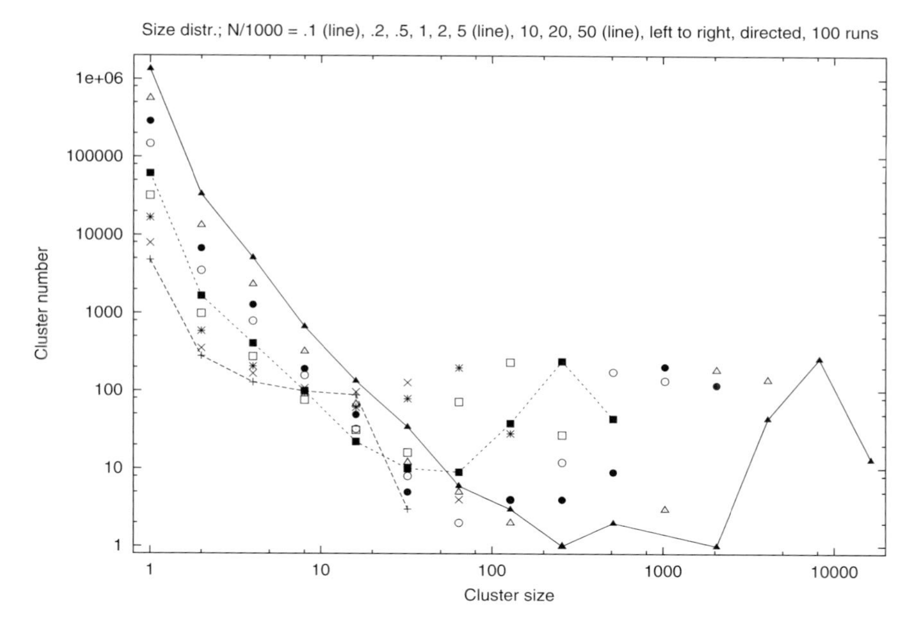

Figure 6.9. Deffuant negotiators: Cluster size distribution n_s for $N = 100$ to 50 000 on directed BA networks (see Section 6.5.2). Note the distinction between small and large clusters when N becomes large enough. From Stauffer, Sousa and Schulze (2004).

Ising comparison

Physicists like Ising models where usually spins are up and down, and thus opinion dynamics with only two choices can be compared with Ising models. (We mentioned already the 1971 Schelling model of sociology which is an Ising variant.) For negotiators and opportunists, only two opinions make little sense, but missionaries were invented for and mostly simulated with two choices only, and sometimes the literature even denotes them as Ising models. Figure 6.11 shows that the similarity is limited: The white regions (one opinion) are very white, black regions (other opinion) are very black, and there are a only few isolated black points in white regions, and only a few single white points in black regions. In contrast, for an Ising model at finite temperatures one always will see some isolated overturned spins or small clusters, while at zero temperature the standard Ising model with nearest neighbours on the square lattice does not order well: The domain growth is blocked for Ising, while the missionaries finally convince the whole lattice of one opinion.

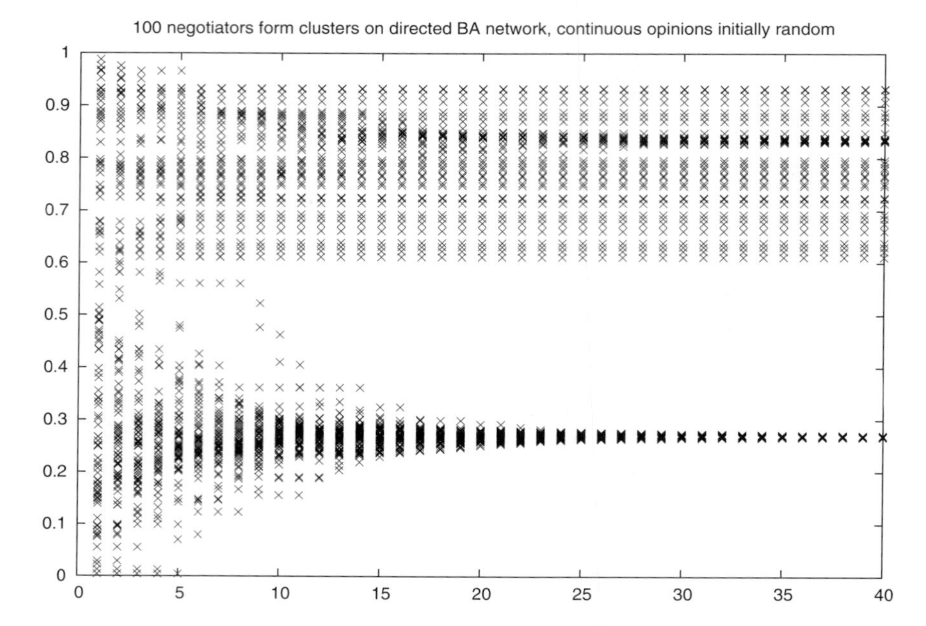

Figure 6.10. Deffuant negotiators: Formation of opinion clusters; as in the previous figure but with only one sample of 100 nodes at confidence bound 0.3 for clarity. We end up with 28 isolated opinions and two larger clusters. Similar pictures without an underlying network were published by Deffuant, Amblard, Weisbuch and Faure (2002).

Several themes

The Axelrod model of 1997 already introduced the possibility that people have opinions on S different subjects; on each question the opinion is either an integer between 1 and Q, or a real number between 0 and 1. For example, people have one opinion about politics and another about football. If these opinions are treated completely separately, then nothing new comes out, one just has S separate systems of opinion dynamics. Thus it is more interesting to introduce some coupling through the bounded confidence. For the above three recent models in Section 6.3.2 two people did not even discuss with each other when their opinions were too far away from each other. Thus now one calculates the opinion distance between two people by summing up, e.g., the absolute values of their differences in each of the S fields. Then either this sum must be smaller than some threshold to allow for discussion (Sznajd-Weron and Sznajd, 2005; Jacobmeier, 2005; Fortunato, Latora, Pluchino and Rapisarda, 2005), or the probability for discussions to happen diminishes with increasing sum of differences (San Miguel, Eguíluz, Toral and Klemm, 2005). This last paper applied these methods to the Axelrod model where one person takes over the opinion of another,

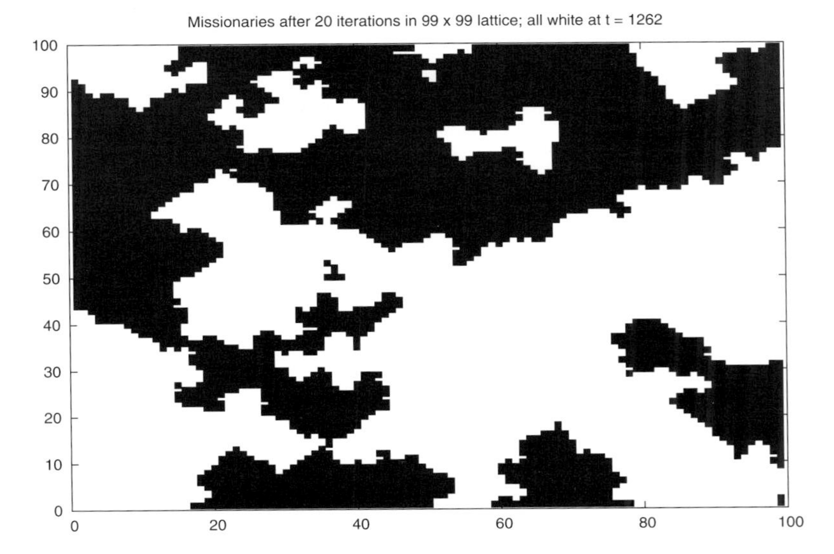

Figure 6.11. Sznajd missionaries: Formation of opinion domains for two opinions at intermediate times. For longer simulations the whole lattice would become all of one colour.

while Sznajd-Weron and Sznajd (2005) used the missionary model, Jacobmeier (2005) negotiators with discrete opinions, and Fortunato, Latora, Pluchino and Rapisarda (2005) opportunists.

For example, a consensus in the Axelrod model on scale-free networks (see Section 6.5.2 below) was possible only if the number Q of possible opinions was lower than a threshold Q_c (see also Figure 6.7 above), which increased towards infinity if the network size grew (San Miguel, Eguíluz, Toral and Klemm, 2005).

Negotiators for $S = 10$ themes on such networks at the end either agreed in most themes, or in none or only few, because of the coupling; very rarely they agreed in half of the themes as shown in Figure 6.12 (Jacobmeier, 2005). These negotiators discussed only if the sum of the absolute differences in their opinions was at most 10 (+), 20 ($\times$), 30 (starts) and 40 (squares). This figure summed up 5000 samples with 1000 people (upper part) or 100 samples with 5000 people (lower part), surrounding a core of three people, with each surrounding node having three neighbours. We see that size effects seem to be weak.

For the opportunists with opinions between 0 and 1, the threshold for consensus was about the same for $S = 2$ as for $S = 1$ (Fortunato, Latora, Pluchino and Rapisarda, 2005). Also for the missionaries, $S = 2$ was chosen representing political and economic opinions (Sznajd-Weron and Sznajd, 2005). But these two themes were treated with different rules; as a result, initial opinion differences on economics could be removed more easily than those on politics.

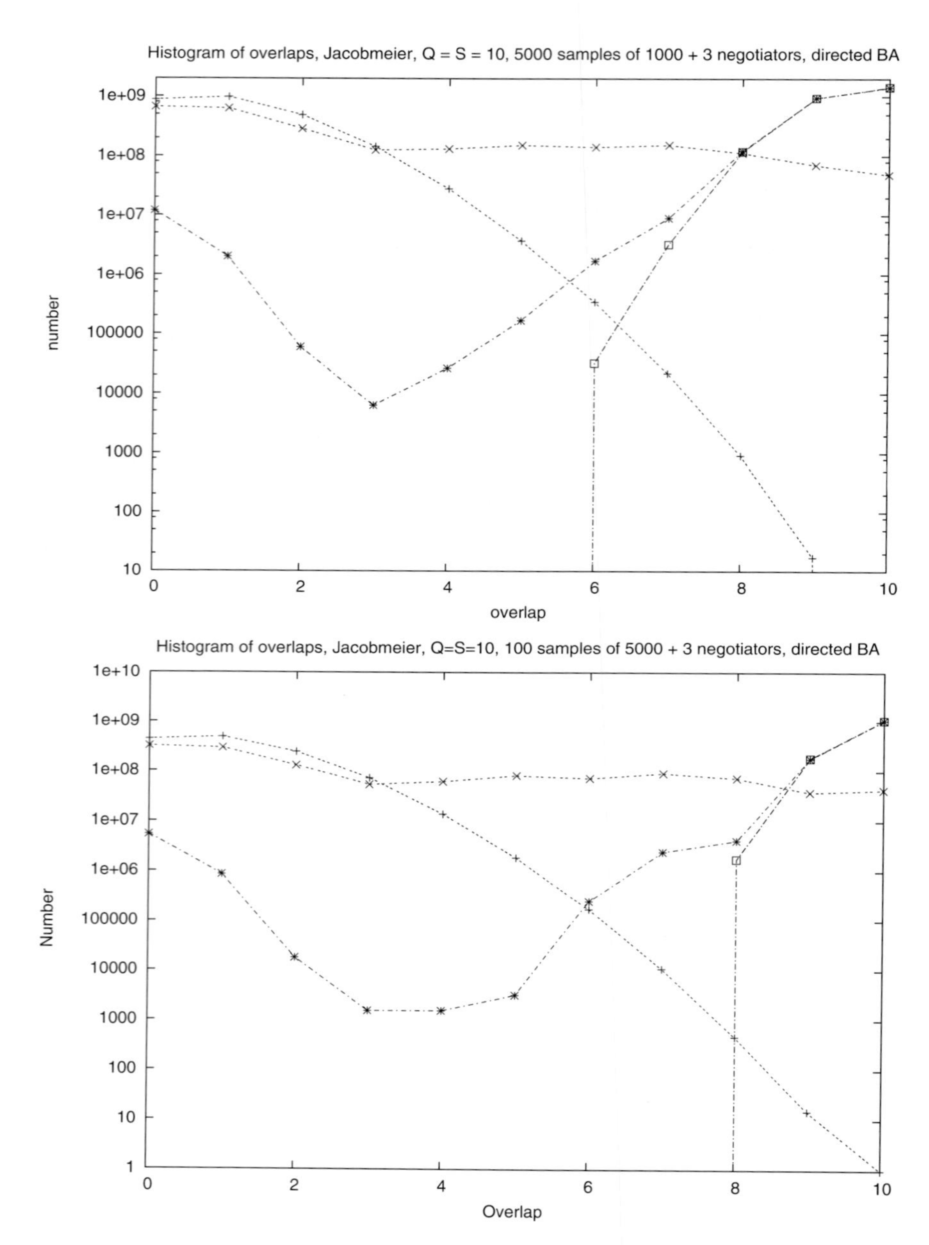

Figure 6.12. 1000 (top) and 5000 (bottom) negotiators for ten discrete opinions on each of ten different themes, on a directed scale-free network, as in Jacobmeier (2005).

Extreme events

Reactions of human beings on extreme events can be simulated by damage spreading, as reviewed by Fortunato and Stauffer (2005) for single themes, $S = 1$: One person changes opinion due to the event, and then the question is how the opinions of the other people evolve compared to the case when the initial opinion change did not happen. For Jacobmeier's (2005) version of $S = 10$ themes with $Q = 10$ possible opinions on each (negotiators on a directed scale-free network), Figure 6.13 shows that for a bounded-confidence parameter $\varepsilon = 0.1$ the damage is very small (four plus signs in lower left corner); for 0.2 and 0.3 it affects a large part of the population, and for 0.4, 0.5 and 0.6 it dies out after some time. (The simulations stop when the damage has gone to zero or when no opinion changed anymore. The initial damage for opinions distributed randomly between 1 and Q changes opinion O to opinion $Q + 1 - O$.) This result is quite different from Figure 4 in Fortunato and Stauffer (2005) with single-theme simulation of *continuous* opinions, where the damage remained large for all $\varepsilon > 0.05$. (ε here is the confidence bound: For Q opinions and S subjects, the sum of the absolute opinion

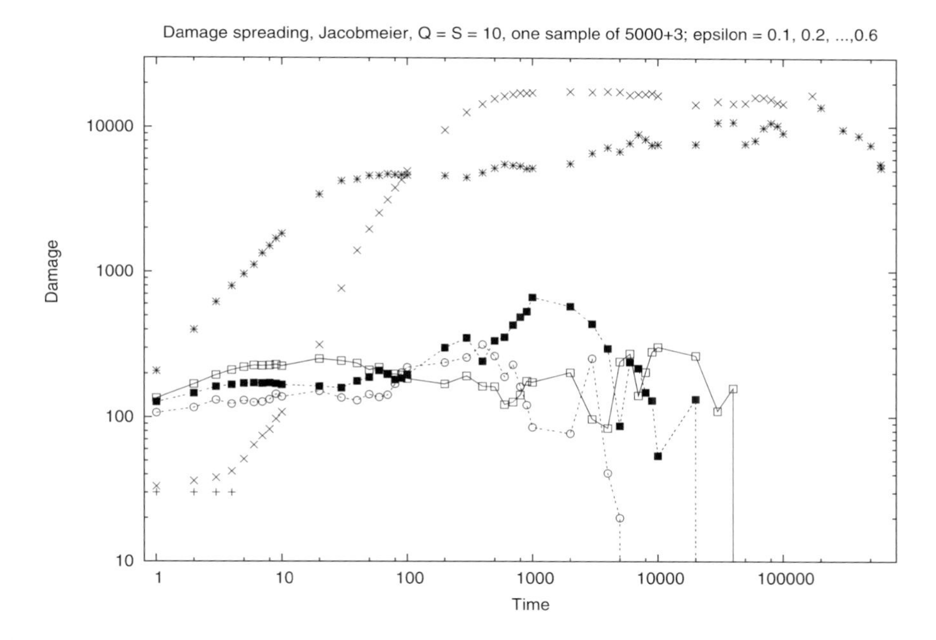

Figure 6.13. Multi-theme damage spreading: Influence of an initial opinion change for the network core on the rest of the population, as a function of time. The confidence-bound parameter is 0.1 (+), 0.2 (×), 0.3 (stars), 0.4, 0.5, and 0.6 (all with lines). Only for intermediate values of this parameter can this "damage" spread. 5000 + 3 people were simulated once.

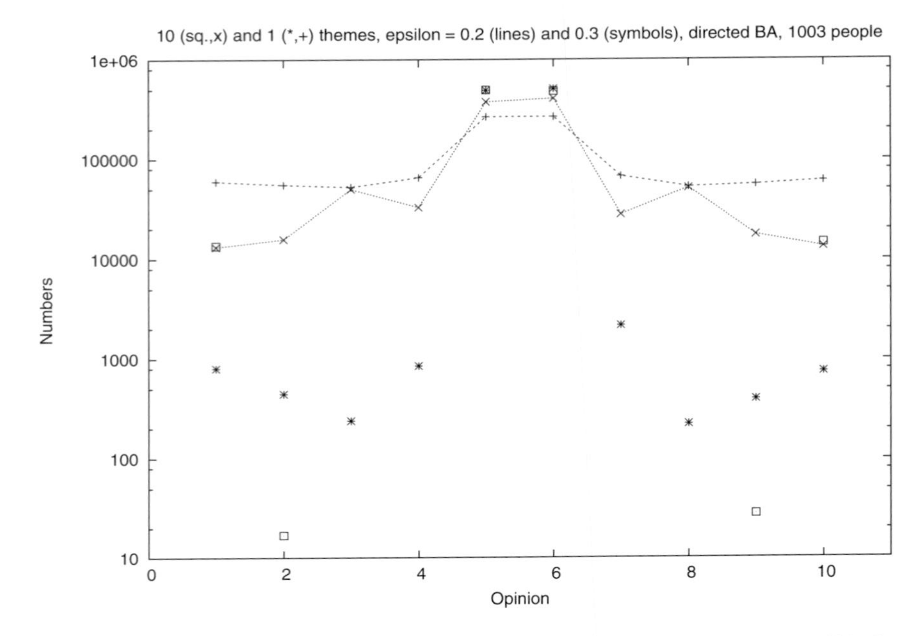

Figure 6.14. Opinion histogram: Final distribution of opinions when ten opinions were possible, for $\varepsilon = 0.2$ (broad distribution) and 0.3 (narrow distribution) for ten themes (100 samples) and one theme (1000 samples).

differences has to be not larger than εQS to allow negotiations. The damage is the sum of the themes at which one person changes opinion due to the extreme initial event, summed over the whole population.)

100 samples, each with 1000 people surrounding the core of 3 people who switch opinion, and $Q = S = 10$, confirm the picture of Figure 6.13 more quantitatively: For $\varepsilon = 0.1$ the damage mostly stays at its small initial value 30; at 0.2 it remains large in all and at 0.3 in nearly all cases; and at 0.4, 0.5 and 0.6 it survives in only a fifth of the cases. This last fraction decreases with population size increasing from 200 to 5000. For symmetric instead of directed networks, the situation is similar. Figure 6.14 shows in the directed case the opinion spectrum: A consensus is impossible, but we see in this figure a transition from a broad opinion spectrum to a near-consensus on the two centrist opinions 5 and 6. For 9 instead of 10 possible opinions, only the centrist opinion 5 becomes heavily adopted. In the directed case with $Q = S = 10$, with increasing population size the equilibration time increases and may develop a singularity at a phase transition near $\varepsilon \simeq 0.25$, Figure 6.15. This time is the number of iterations needed until either the damage died out or both simulated samples simultaneously no longer changed at nonzero damage. If ε increases from below to above this transition

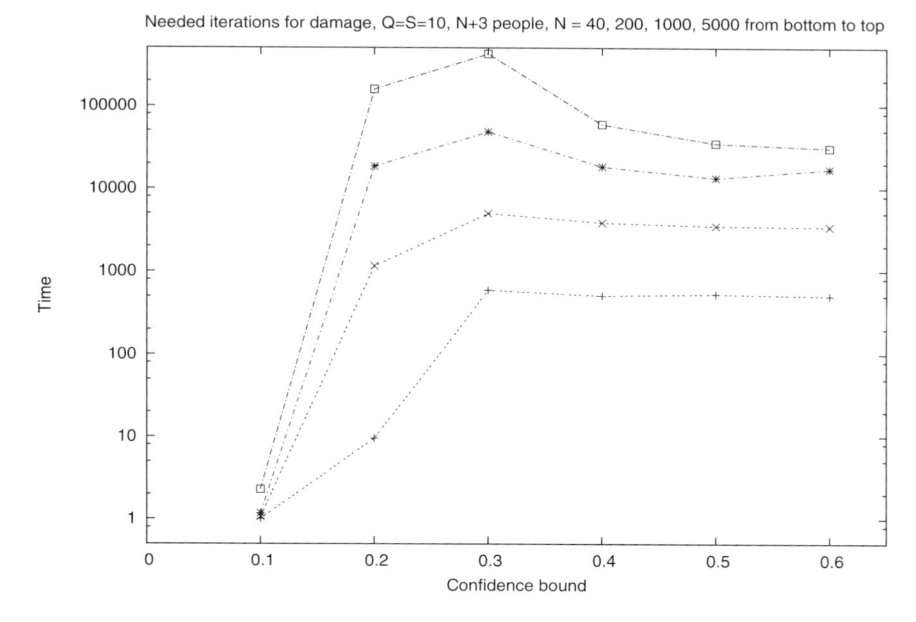

Figure 6.15. Onset of critical slowing down, Section 2.5? Average equilibration times versus ε for a population of 40 (+), 200 ($\times$), 1000 (stars) and 5000 (squares) surrounding the initially damaged core of three. The maximum can be seen only for the larger populations, as is customary near critical points (Chapter 2).

value, most people agree on a centrist opinion, while at a higher $\varepsilon \simeq 0.35$ damage mostly ceases to spread. (For one instead of ten themes, but still ten possible opinions, these two transitions happen at somewhat lower ε.)

How can damage spread if the simulation is made with parameters leading to a consensus? How can a stable society be influenced strongly by a single extreme event? For continuously varying opinions this is possible if even very tiny opinion differences are counted as damage. For discrete opinions, the concept of damage (agreement or disagreement) is better defined, and this is what was presented above. For this discrete case, the extreme event, simulated above by influencing initially only the network core, may change the whole population from one centrist opinion to the neighbouring centrist opinion, for example from 5 to 6 if the opinions range from 1 to 10. So, in both the damaged and the undamaged system, one has a widespread consensus, but the two consensus opinions are slightly different. At the time of this writing, this effect just seems to be happening in England as a result of terror attacks in July 2005.

Figure 6.16 illustrates this effect in the simpler case of only one theme, for which ten opinions are possible. The upper part shows that the final damage at

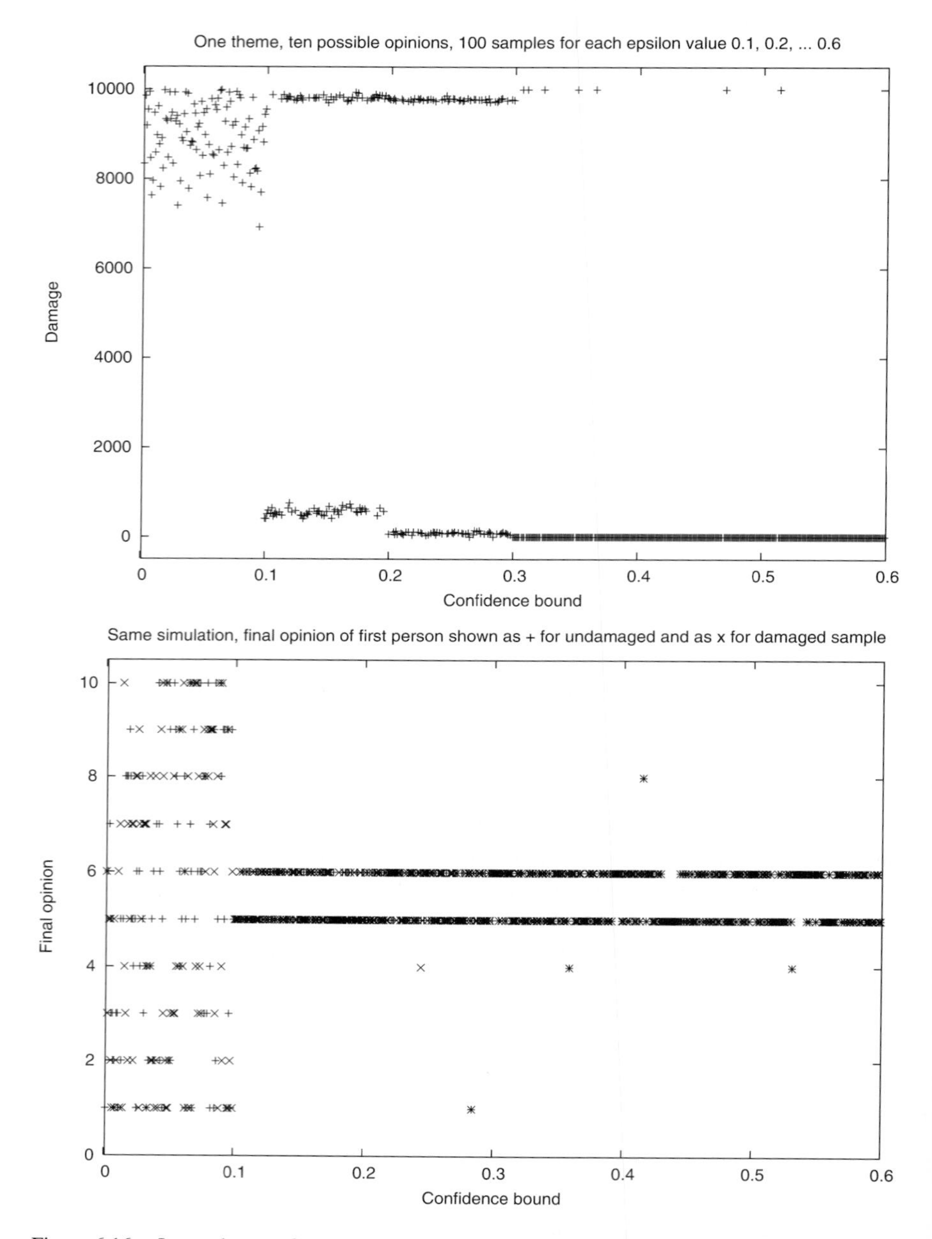

Figure 6.16. Large damage from extreme event may mean only a small opinion change. Damage (upper part) and final opinion (lower part) versus ε, for $10\,000 + 3$ people on directed BA network.

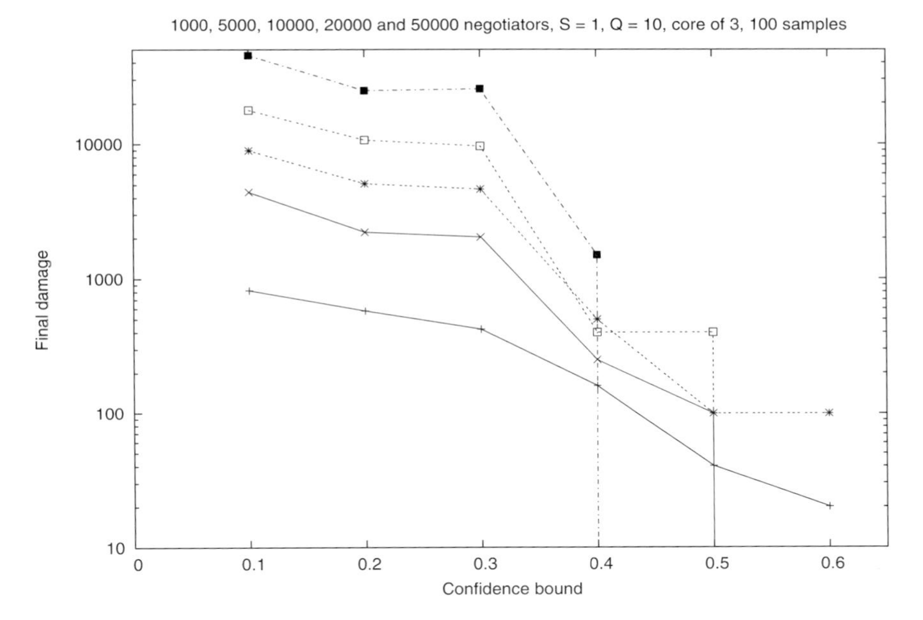

Figure 6.17. Size effects for final damage on directed scale-free network. On the left side the damage increases roughly proportionally to the system size, while on the right side it is much smaller and dies out for the larger systems. (One theme, ten possible opinions.)

low $\varepsilon \leqslant 0.2$ is scattered between zero and the population size 10^4, while at large $\varepsilon \geqslant 0.4$ it is mostly zero and in a few cases 10^4. The lower part shows for the same simulations the final opinions of the very first site, which are widespread for small and centrist for large ε. For 50 000 instead of 10 000 people, damage healed out in all 100 samples for the large ε values, Figure 6.17, for 100 000 also at $\varepsilon = 0.4$. (With nine instead of ten opinions, the final consensus was nearly always on opinion 5.) For our ten opinions, only confidence bounds of $\varepsilon = 0.1, 0.2, \ldots, 0.9$ make sense, corresponding to maximum opinion differences of $1, 2, \ldots, 9$. Thus a large damage is not necessarily a drastic shift of opinion, but can be a small change affecting everybody.

A very different simulation of human reaction to extreme events was given by Altmann, Hallerberg and Kantz (2005). When such an event occurs, like a flood, humans build protection against it. But then they slowly forget about it, the protection deteriorates, and the next catastrophe appears. In this way a nearly periodic recurrence of catastrophes becomes possible, instead of an exponential distribution of times between extreme events at constant protection.

In summary, the simulations of reactions to extreme events strongly depend on whether the opinions are continuous or discrete. For continuous opinions the damage increases smoothly from affecting only a few people to influencing nearly the whole population, if the confidence bound ε increases from zero to $1/2$, for both negotiators and opportunists: Figures 4 and 5 in Fortunato and Stauffer (2005). For discrete opinions, the influence of extreme events on the negotiators is more complicated and vanishes for large confidence bounds like 0.4 and higher, Figures 6.13 to 6.17 above. The damage spreading transition point for ε does not agree with the consensus transition point.

Parties

A thorough investigation of political elections in a southern part of Germany was made by Schneider and Hirtreiter (2005). They found that the membership in the two major political parties there increased (decreased) if that party gained (lost) votes in two consecutive elections for parliament. Then these authors modified the missionary model to include a possible party membership in addition to the usual opinion; party members are then more convincing than other people, and are less convinced to flip. The simulations showed that initially the election victories fluctuated between the two parties, but then one party (Hirtreiter's, not ours) always stayed in power, and the other stayed in opposition but did not die out, exactly as in reality of the last 60 years there.

These political applications do not tell us which party or candidate will win. They indicate general trends of parties, whatever the real political issues are. Similarly, a constitution and its election laws normally do not state which party should win the leading role; they leave that question to the electorate. Constitutions and election laws rule the general principles according to which political power is distributed, whatever the later policy issues will be. Or, as stated by George Orwell in "Animal Farm", all animals are equal.

But as we learned there, some animals are more equal than others, and in physics examinations we usually grade the student's answers as right or wrong, with intermediate possibilities. So, we assume that there is some truth in physics, but it is difficult to find and to learn. The physics truths of one century were often the half-truths of the next century in the sense that they were special cases of the later more general understanding. So, what about putting in some truth into these opinion dynamics?

Assmann (2004) and Hegselmann and Krause (2005b), see also Kuznetsov and Mandel (2005), did exactly that by assuming that some opinions are superior to others. Assmann studied discrete opinions of negotiators on directed Barabási–Albert scale-free networks. The opportunists of Hegselmann and Krause follow their standard rule (everybody can contact everybody) with one simple modification: at each time step, with some probability the agent does not follow the standard rule but adopts one opinion called Truth. This truth is the same for everybody while the probability may differ from person to person. Depending on parameters, a consensus to this truth may happen.

Political parties need advertising which reaches everybody, besides the personal contacts simulated in the standard models. Two independent simulations of missionaries (Schulze, 2003; Sznajd-Weron and Weron, 2003) showed that the larger the population is the smaller has the convincing power of the advertisement have to be in order to flip the general opinion. However, this advertising should not come too late when most have already fixed their opinion.

A non-political but similar problem was simulated by Brandau and Trimper (2006) who suggested: Let's have a party. Their parties are not political but weekend meetings of people for enjoyment. The participants at the various simultaneous parties tell each other on mobile phones how they like it, and people may switch from a boring to an interesting party. Will everyone at the end be at one single large party? Brandau is a mathematician; physicists of course work on weekends, for example by studying econophysics with minority games at the El Farol bar in Santa Fe (USA).

Simultaneous updating

Simulators of Ising models or cellular automata know the difference between simultaneous (= parallel, synchronous) and sequential updating. If element i is influenced by element k, then simultaneous updating means that i at time step t is influenced by the value of k at the preceding time step $t-1$, whereas for sequential updates i is influenced by the current value of k. That current value may be the one at time step $t-1$ if k has not yet been updated at the current time step t; but it is the new value $k(t)$ if k has been updated before i. Thus for sequential updates the order in which we go through the system may be relevant, while for parallel updates this does not matter. Programming, however, is simpler for sequential updates, and that choice may also be more realistic.

Simultaneous updating of Sznajd missionaries was studied, e.g., by Tu, Sousa, Kong and Liu (2005), who cite four earlier papers on this subject. The other simulations of missionaries in this Section 6.3 all refer to sequential updates. It is obvious that simultaneous updating may lead to frustration. What should an agent do if a pair from the left demands one choice, and a pair on the right demands the opposite choice? It does nothing, as in Galam conservatism of Section 3.3.1. Similarly, a student who is told to be at two different lectures at the same time feels frustrated. Thus for the missionaries, reaching a consensus under simultaneous updating is much more difficult, on square lattices and on various types of networks to be defined soon in Section 6.5. One may thus learn that committee meetings with simultaneous voting are better if one does not want a consensus, whereas continuous person-to-person contacts facilitate consensus.

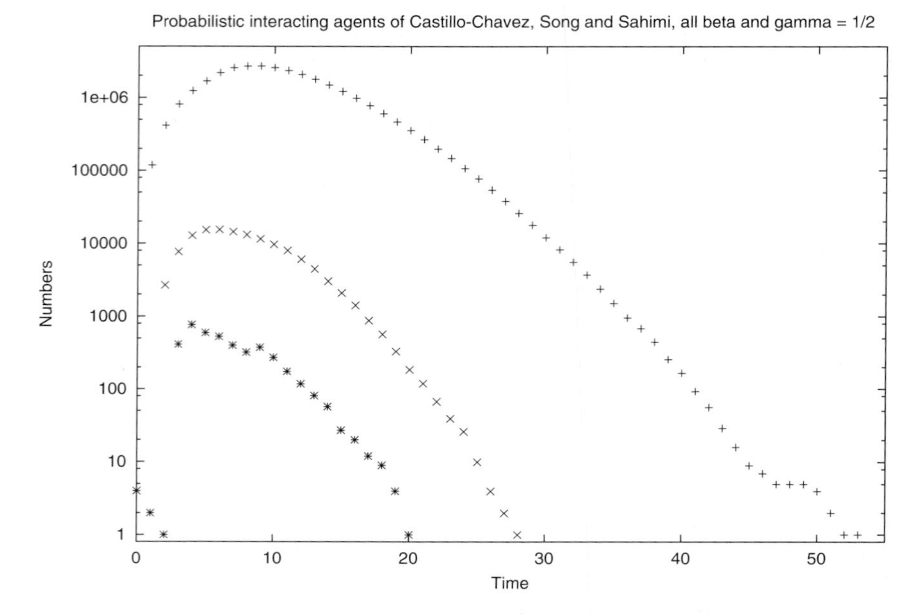

Figure 6.18. Support for terrorists: 40 million people are partially made susceptible (+), excited (×), or fanatic (*) by the network core of four initial fanatics in a directed Barabási–Albert network (Stauffer and Sahimi, 2005). The general population is not shown and varies between 40 and 37 million.

Terrorism

In connection with bio-terrorism, Castillo-Chavez and Song (2003) applied differential equations similar to epidemiology to opinion dynamics. The general population (G) rejects terrorism, some may become susceptible to it (S), these in turn can become excited (E), and the latter ones finally may become fanatics (F), with $G + S + E + F = 1$ for the sum of the four population fractions. All of them may also become directly part of G. People become more inclined towards terrorism only after being convinced by someone who is more inclined than they are at present; for example, S can be convinced to become E only by members of groups E and F. With the abbreviation $C = S + E + F = 1 - G$, their deterministic differential equations are:

$$\begin{aligned} \mathrm{d}S(t)/\mathrm{d}t &= \beta_1 CG - \beta_2 S(E+F)/C - \gamma_1 S, \\ \mathrm{d}E(t)/\mathrm{d}t &= \beta_2 S(E+F)/C - \beta_3 EF/C - \gamma_2 E, \\ \mathrm{d}F(t)/\mathrm{d}t &= \beta_3 EF/C - \gamma_3 E. \end{aligned}$$

Here, the various β and γ are suitable rate coefficients. The causes of terrorism are not part of these differential equations, which only deal with the opinion dynamics. Of course, instead of terror we may have in mind discussion of other questions for which different degrees of agreement and the directional convincing process make sense. For small ratios β/γ of the rate coefficients, finally everybody joins the normal population G. Probabilistic simulations of independent agents on lattices or networks gave results similar to these differential equations (Stauffer and Sahimi, 2005).

However, differential equations cannot check the influence of a single person, like Ghengis Khan on genetics, on the opinions of the whole population ("damage spreading", Fortunato and Stauffer (2005); see Section 6.3.3 above under "Extreme events"). Thus Stauffer and Sahimi (2005) asked: Can a few fanatics influence the opinion of a large segment of the society? They use a directed Barabási–Albert network with 4 neighbours selected by every person added to the network, and assume everybody except the initial core of 4 fanatics to belong to the general population. Figure 6.18 indicates that these 4 initial people can make 2.7 million others susceptible for some time, in a population of 40 million.

An empirical analysis of number of victims in single attacks from terrorism or guerilla warfare was complemented by Johnson, Spagat, Restrepo, Bohórquez, Suaárez, Restrepo and Zarama (2005) with a simple model: The number of victims is proportional to the size of the attack unit. At each step, an attacking person is selected randomly, and the corresponding unit is either split into single people (with probability of one percent), or joined to another unit (with probability of 99 percent) to which another randomly selected person belongs. This leads to a power law distribution (non-cumulative) with an exponent -2.5 for the number of victims. By assuming the above joining probability to decrease with a power of the unit size, more slowly decaying power laws were obtained in better agreement with reality.

6.3.4. Conclusions

The example of elections in Figure 6.8 makes clear what sociophysics can and cannot do: It can explain general (statistical) properties of elections, but cannot predict which candidate will win which election. Similarly we can explain the pressure of an ideal gas as a function of density and temperature, but not which air molecule will be where one minute from now, Section 2.2. Averaging over many people allows the application of statistical methods and computer simulation; each individual has its own unpredictable fate. Insurance companies have used such methods much longer than physicists have simulated social phenomena.

Moreover, many different models gave similar results; Figure 6.6 is only an example. Behera and Schweitzer (2003) and Galam (2005) showed that many properties of Sznajd missionaries can also be obtained from different models. This is nice and shows that details don't matter much, fully in line with universality experience for physical phase transitions, Section 2.5.

6.4. * Traffic jams

Traffic jams are a common sight in the big cities and on the expressways of rich countries, and China at present makes big steps to repeat the errors of the "West". The senile author, citizen of the country where the automobile was invented, is happy to have sold his last car 15 years ago. Others have to worry about the reasons why traffic jams occur, and this field is an important part of (socio-)physics since the publications in 1992 of Nagel, Schreckenberg, Biham, Middleton, Levine, ... We refer to the extensive reviews of Nagel, Esser and Rickert (2000), Chowdhury, Santen and Schadschneider (2000) or Helbing (2001), and to Mahnke, Kaupuzs and Lubashevsky (2005) for a recent paper. Here we mainly explain the Nagel–Schreckenberg model (to be abbreviated as NaSch, not as NS = National Socialism).

Just as for stock market crashes, people like to find specific reasons for a specific traffic jam. But just as no external reason was found for the 1987 crash on Wall Street, also some traffic jams appear without any specific reason. Of course, if an accident happens or if a road is narrowed by construction work, we have a clear reason for a jam. It is also obvious that jams occur more often at high than at low car densities. But why does it happen at the same road and the same traffic density today, when yesterday under the same conditions it was avoided? Randomness seems to play an important role here and was implemented in the NaSch model through a probability for drivers to needlessly slow down.

The rules of these NaSch "cellular automata" are simple: The street is a long chain of lattice sites, each accommodating at most one car. The speed of each car is allowed to be only an integer between zero and a maximum, say, five. Drivers are assumed to drive as fast as possible within safety constraints and other rules. At each time interval, all car positions and velocities are updated as follows: The velocity is increased by one unit if the distance to the car ahead is at least as large as the new velocity and if this new velocity is not larger than five. Then, however, with some probability the velocity is decreased by one unit (but not to a negative value), because the driver does not pay full attention. Finally, each car moves forward by the amount of sites given by its new velocity.

We get the "fundamental diagram" of Figure 6.19: Traffic flux versus traffic density. In the ideal case all cars would travel with maximum velocity five and thus never hit each other: Flux = five times density. However, since half of them at any time slow down to velocity four, the ratio flux/density should only be near 4.5. This is indeed the case at low densities in this figure, as indicated by the straight line there. But at higher densities we no longer can look only at an average driver; instead each driver may influence the cars behind. The figure shows

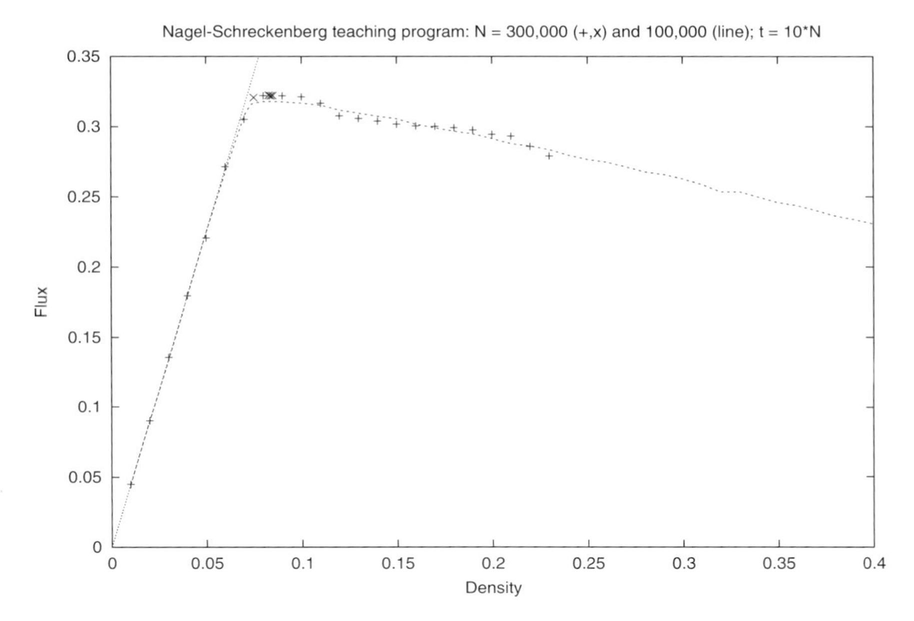

Figure 6.19. Car traffic on an expressway: The number of cars passing through per unit time, is plotted versus the car density, at a probability of $1/2$ to slow down needlessly. The program in the appendix, Section 9.6, simulated 100 000 and 300 000 cars on a circle of length 1 000 000.

that now the flux first deviates slightly from the straight line, then has a rather sharp maximum when about one in every 12 sites is occupied, and then for higher densities *decreases*. In this decreasing region, simulations of small systems differing only in their random numbers may give vastly different results, including metastable and jammed traffic, in agreement with the unpredictability of reality. More sophisticated simulations were applied, e.g., to Portland (Oregon, USA), as reviewed by Nagel, Esser and Rickert (2000).

Pedestrians also form traffic, not only cars. On a sidewalk, with people walking in opposite direction, separate "lanes" self-organise, where people walk in the same direction behind each other. Here also alternatives to cellular automata were used. Rules very different from NaSch and more similar to molecular dynamics simulations of fluids gave these and other aspects of reality (Helbing, 2001). A practical application is the rapid evacuation of a room filled with people: Jams at the door can be reduced if a pillar before the door prevents people from pushing and thus hindering each other.

We see in the traffic jams the same effect as in the hierarchies of the Bonabeau model of Section 6.2 of this chapter, the possible dominance of one language in Sections 5.3 and 5.5, the Eve effect in Sections 2.9 and 3.5, and many other complex systems: Randomness in one single element can influence the whole system, and not every victory or loss is based on justified reasons. "Life is unfair", as Nobel Laureate Jimmy Carter said in 1980.

6.5. * Networks

Solid state physics deals with regular lattices, or perhaps small lattice defects. Trees in an orchard may also be planted on a regular square lattice, and so sit students sleeping in our lectures. However, in general living beings do not form regular lattices, but more complex networks. One simple way to get such a network is to occupy randomly only a fraction of the lattice; this leads to the so-called percolation problem invented by Flory before any of the present authors existed (see Section 6.6 below), and improved by mathematicians and mathematical physicists at the time the junior authors of this book were born (random graphs are a limit of percolation for infinite interaction range). The brain is a complex network of neurons and was modelled first by Mc Culloch and Pitts when the senile author was born. More recently Watts and Strogatz (1998) invented a "small world" model to take into account that with a short chain of mutual acquaintances one can connect most people on Earth with each other. We emphasise here the scale-free networks of Barabási and Albert (1999) (pronounced approximately BOrobashi-OLbert according to Kertész who suggested the inclusion of this section). Also these networks have roots in the 1950s, as reviewed in a nice book (Barabási, 2002). Only at the end we go back to neural nets. In line with the intention of the whole book, we concentrate on simulations of networks, and not on the analysis of existing friendship networks in a Karate club or an Antarctic research station, of collaboration networks of film actors or scientists (Newman, 2001), or of connections between computers or web-sites on the Internet. As mentioned in Section 2.7, Albert and Barabási (2002) offered a much more thorough review.

6.5.1. Small world

The Watts–Strogatz networks start from a lattice with nearest-neighbour connections, often a one-dimensional chain. Then with some small probability p a site cuts the connection to one of its neighbours and replaces it by a connection to a randomly selected other site on the lattice. At the end one has a network with a mixture of nearest-neighbour bonds and infinite-range bonds. The two extreme cases $p = 0$ and $p = 1$ correspond to regular lattices and random graphs, respectively.

This clear separation between close and far-away neighbours makes these networks quite unrealistic; we talk with our office neighbours and family more often

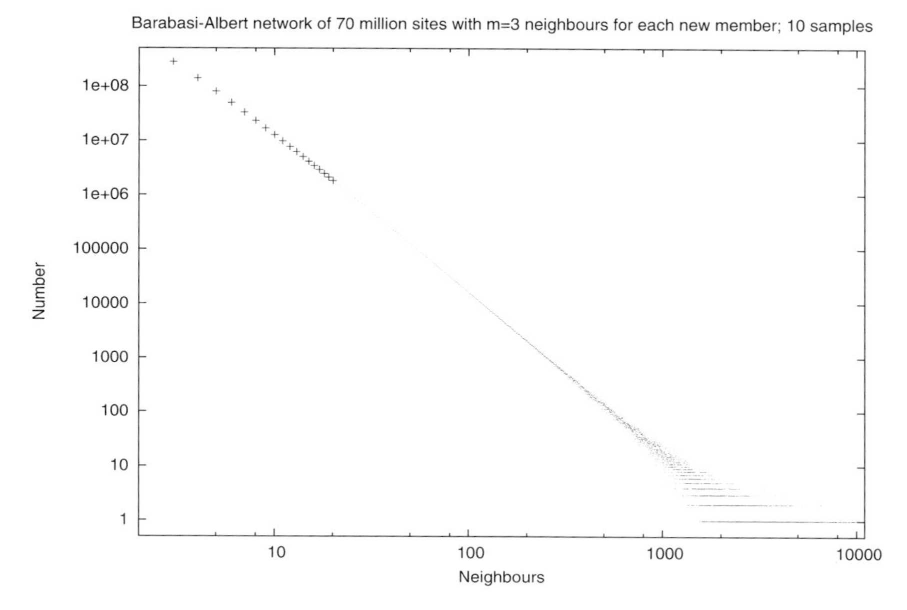

Figure 6.20. Distribution $N(k)$ of the number of sites which have k neighbours each on a scale-free BA network. The data in this log-log plot follow a straight line with slope -3. We summed over ten lattices of 70 million sites each, with $m = 3$ neighbours selected by every new member of the network. On average 10 sites in each network had more than 10 000 neighbours.

than with other people, but these other people are not all equally likely to be contacted by us, even in the age of e-mail. A more realistic way would be to have distance-dependent probabilities for these other connections, as assumed by Moukarzel and de Menezes (2002) and Huang, Zou, Shao, Tan and Jin (2004) for nodes and atoms, not for people. As an example relevant for the opinion dynamics summarised above we mention Elgazzar (2001) who put the missionaries of the Sznajd model onto a Watts–Strogatz network and found that even there they can reach a consensus. The main merit of these small-world networks seems that they were a stepping stone to the more realistic scale-free networks of Barabási and Albert (1999) described now.

6.5.2. *Scale free (BA)*

A Barabási–Albert (BA) network grows via preferential attachment, which takes into account that famous people attract new acquaintances easier than normal people. Thus one starts with a fully connected core of m nodes. Then, step by step, new nodes are added as follows: each new member of the network selects exactly m neighbours from the already existing network nodes. This selection is not made randomly; instead the m neighbour nodes are selected with a probability proportional to the number k of connections these nodes had already before: The well-connected get even better connected, the rich get richer. The simulation stops if the network has reached the predetermined size.

It would be inefficient and not accurate if each new network member would go m times through all previous nodes i, each having k_i neighbours, and select each with a probability equal to $k_i/\sum_j k_j$. Then only on average the number of neighbours of the new member would be m. It is better to select exactly m neighbours by a random selection from a list in which each node appears as often as it has neighbours. This algorithm is explained in the appendix, Section 9.7. No geometry is involved; the network is only topology, and the only distances one can define are the number of connections needed to link two nodes.

Such a simulation gives for each site i a number k_i of neighbours where $k_i \geqslant m$. The last-added sites have $k_i = m$ while most of the earlier sites have $k_i > m$ since they were selected later as neighbours. Typically, the m members of the initial core are among the most-connected sites, but usually also one of the later sites has more neighbours than one of the initial sites: The rich get richer on average but not in every single case, as you may have noticed in 2000 when you invested in information technology stocks during the 1990s.

The number $N(k)$ of sites having k neighbours each varies as $1/k^3$, except for small k; $N(k < m) = 0$. This power law, which is also explained theoretically (Barabási and Albert, 1999), gave the networks the name "scale free" since there is no characteristic scale in the distribution $N(k)$; if instead the distribution would have been a Gaussian centred at some K, or a decay $\propto \exp(-k/K)$, then this K would have been the scale of the distribution, and the width of the Gaussian could introduce a second scale. Figure 6.20 demonstrates this distribution $N(k)$, using the program of our Section 9.7. The parameter m does not influence qualitatively the results as long as $1 < m \ll \max$ where max is the total size of the network. This wide distribution of the number k of neighbours makes the scale-free networks much more realistic in the description of social and other networks than regular lattice, dilute lattices (percolation), random graphs, or Watts–Strogatz small worlds, which all have a rather narrow distribution of k.

Modifications have been published giving $N(k) \propto 1/k^\gamma$ with γ different from 3. In some sense $\gamma = 3$ is the most interesting case since it is the border between two different regimes. We leave these aspects to detailed reviews, e.g., of Albert and Barabási (2002) or Newman (2003). The simulations of reactions to extreme events and terrorism, presented in Section 6.3.3, used already these BA networks.

6.5.3. Selected properties of BA networks

Just as in small-world networks, also in the scale-free BA networks two randomly selected nodes have a rather short distance from each other, as measured by the number of network links needed to connect them. This number on average grows only logarithmically with the network size N. If we randomly cut a fraction p of links, the network as a whole still remains connected provided $1 - p$ is not smaller than a threshold vanishing as $1/\log(N)$. On the other hand, if we destroy the most connected fraction of nodes, then already a very small p suffices to cut the network into small parts.

Applied to computer network these effects explain why accidental failure of many computers barely affect the connectedness of the whole network, while targeted attacks by hackers on important computers can block the communications between the remaining ones (Cohen, Erez, Ben-Avraham and Havlin, 2000). Also airlines have noticed long ago the usefulness of networks: To bring passengers from A to B it may be inefficient to have a direct flight from A to B; instead it may save fuel to bring them from A to some hub C and with a different plane (and partly different passengers) from C to B, even if the distance A-C-B is much bigger than the distance A-B. You only have to look at the route map of airlines to see these hubs as centres of the network.

We mentioned already for opinion dynamics, Figure 6.8, that Sznajd missionaries were put onto BA networks to simulate successfully election results (Bernardes, Stauffer and Kertész, 2002; González, Sousa and Herrmann, 2004). And we mentioned in Section 6.3.3 damage spreading among networked negotiators. Galam voters, Kauffman genes, Ising spins, neural nets and other models also were combined with scale-free networks.

6.5.4. Modifications of BA networks

Pütsch (2003) looked at the 185 papers from 555 authors which up to October 2002 cited the original BA paper (Barabási and Albert, 1999); at the time of this writing there are about 1000 such papers. He defined clusters of authors (similarly to Newman (2001)) by defining two authors as connected if in this subset of 185 papers they had at least one publication together. One huge cluster from the human genome project coexisted with many smaller clusters. Thus this situation is very different from BA networks, where only one single cluster is grown. He then modified the BA algorithm by letting with a low probability p each newly added site form a new BA cluster, instead of joining the existing ones. In this way, an assembly of finite BA clusters was simulated, and Figure 6.21 shows a nice agreement of reality and simulation, quite rare in this sociophysics chapter. Note, however, that different collaboration networks obey different size statistics and were modelled differently (urn transfer) by Fenner, Levene and Loizou (2005)

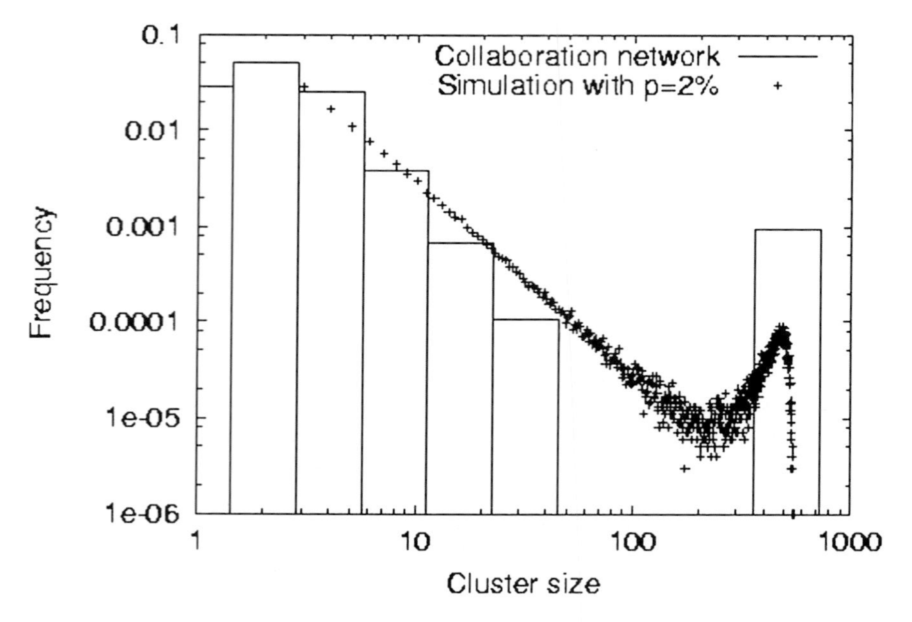

Figure 6.21. Cluster size distribution of real authors (bars) and averaged over 10 000 simulations (+) for scientific papers. Authors are defined as being part of one cluster if they published at least one paper together citing Barabási and Albert (1999). From Pütsch (2003), with permission.

and with focus on triangular author relations by Lambiotte and Ausloos (2005); see Simkin and Roychowdhury (2005) for modelling scientific citations.

Even though Section 3.3 explained everything you always wanted to know about sex, some people like González, Lind and Herrmann (2005) are still interested in this old-fashioned subject. They complain that BA networks do not describe well the monogamous people but only those with many sex partners. Thus they model the growth of sexual networks by particle aggregation: Male and female particles collide on a two-dimensional plane, and in contrast to granular materials they *gain* energy in a collision. The set of particles having had such collisions with each other forms the sexual network of past experiences. At the beginning one has only single particles of while one is marked, then the network containing this marked particle grows, and finally everybody belongs to this network. In contrast to scale-free networks, the distribution $N(k)$ of the number of k partners for each participant is a power law $\propto 1/k^3$ only for large k while for small k it varies much slower, as claimed in reality (Liljeros, Edling, Amaral and Stanley, 2001). A practical application of such studies of sexual networks is immunisation: Cohen, Havlin and Ben-Avraham, Chapter 4 in Bornholdt and Schuster (2003) suggested not to immunise the most-connected person (which is not known generally), but random people who then have a higher probability to be connected with the "hubs" of the network.

For computer networks it is nice that random break-downs still leave the network as a whole connected, as mentioned above in Section 6.5.3. However, the same effect is bad if applied to computer viruses or human epidemics: One sick person may infect directly, or indirectly with a few links, an appreciable fraction of the whole population, if the connections between people follow a BA network. We refer to Gallos, Cohen, Argyrakis, Bunde and Havlin (2005) and Xu, Wu and Wang (2006) for recent papers.

6.5.5. Neural networks

A little baby soon learns to distinguish the face of the mother from other faces. Did you try to log in on a computer recently by just sitting in front of it's screen? At least I had always to type in precisely my user name and password. Human and animal societies are based on recognising differing signals as always meaning the same thing: faces, words, smells, ... How can we teach computers to do that? That means, how can we for example present a picture to a computer and let the computer find out that this is just a modified version of one of the many pictures which we had stored in the computer before? Experts call this associative memory, and we explain here only the basic Hopfield model with Hebb rule, not the many modifications of the last half-century. We concentrate here on the efficient trick of Penna and de Oliveira (1989) for simulations and give a program in the appendix, Section 9.8.

Similarly to Ising magnets, or more precisely to spin glasses (Binder and Young, 1986), Hopfield neural networks assume an array of N variables $S_i = \pm 1$, coupled by synaptic strengths $J_{ik} = J_{ki}$ $(i, k = 1, 2, \ldots, N)$. In biological reality the S_i are neurons which either fire $(+1)$ or do not fire (-1) electrical impulses along their dendrites to the synapses connecting the neurons. Positive J means that the neurons strengthen each other in their present state (excitatory synapses), while negative J means they oppose each other (inhibitory synapses). Each neuron i feels the input

$$h_i = \sum_k J_{ik} S_k$$

from all other neurons and reacts accordingly at the next time step $t + 1$:

$$S_i(t+1) = \text{Sign}\, h_i(t)$$

where Sign = ± 1 is the sign function.

P different patterns $\xi_i^\mu = \pm 1$ $(\mu = 1, 2, \ldots, P)$ are stored in memory; these may correspond to the faces of P different people. To recognise them, the Hebb rule recommends the assumption $J_{ik} = \sum_\mu \xi_i^\mu \xi_k^\mu$ for all J, and this works: If now

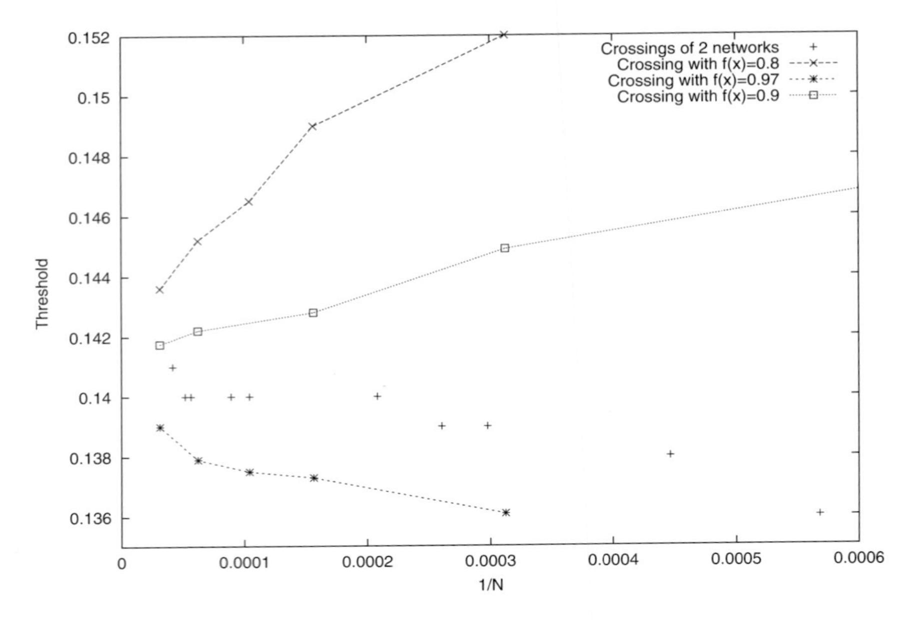

Figure 6.22. Critical ratio P/N of number of stored patterns to number of neurons, as a function of reciprocal network size, using different analysis techniques. If more patterns are stored, the neural network no longer can recognise them. From Ostfalk (2005).

we present to the computer a pattern $S(t = 0)$ which is similar but not identical to the stored pattern number ν, then after several of the above iterations $t = 1, 2, \ldots,$ the updated pattern $S(t)$ will become very similar to the desired ξ^ν. That means the computer has recognised pattern number ν from the somewhat incorrect initial pattern $S(t = 0)$, just as the baby recognised the mother's face. (We denote here the set of all S_i with $i = 1, 2, \ldots N$, by S without subscript; analogously for the ξ^μ.) More quantitatively, the similarity of two patterns is given by their overlap m, like

$$m_\mu(t) = \sum_i S_i(t)\xi_i^\mu$$

for the two patterns $S(t)$ and ξ^μ. (This overlap is related to the Hamming distance, Figure 2.19.) If a pattern is recognised completely, $S_i = \xi_i^\mu$ for all i, this overlap is N; if there is no correlation between two patterns, their overlap is $\ll N$.

Thus we want to start from a low overlap and want to get an overlap close to N. The Hopfield model with Hebb rule actually achieves this aim, provided the number P of stored patterns is not too high. If P/N is smaller than 0.14 for random patterns of ± 1, then most patterns are recognised to more than 90 percent; for larger P/N the overlaps jump down to about 20 percent. Figure 6.22 shows

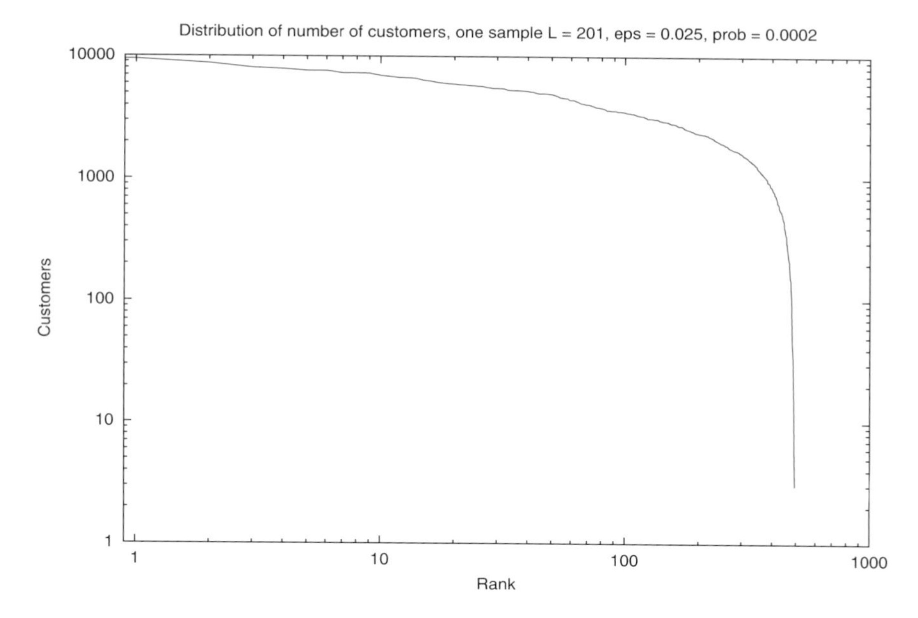

Figure 6.23. Rank plot of simulated movie successes; the film with the most visitors is placed on rank 1, that with the second-most number of visits on rank 2, etc. From Stauffer and Weisbuch (2003).

the latest estimates of the position of this first-order phase transition. Each student should get from these simulations that one should not learn too much before an exam; otherwise the brain capacity is overloaded and one cannot answer even simple questions.

For $N = 10^5$ neurons, we have $N^2 = 10^{10}$ matrix elements J. The Penna–Oliveira trick of 1989 avoids to store them and requires only to store the P overlaps, besides the stored P patterns of N bits each. Thus lots of memory is saved if $P \ll N$ as is the case for good recognition. It works for the above Hopfield model with Hebb rule:

$$h_i(t) = \sum_k J_{ik} S_k(t) = \sum_k \left[\sum_\mu \xi_i^\mu \xi_k^\mu \right] S_k(t)$$
$$= \sum_\mu \xi_i^\mu \left[\sum_k \xi_k^\mu S_k(t) \right] = \sum_\mu \xi_i^\mu m_\mu(t).$$

Thus we have to evaluate and store the P current overlaps m_μ, but they are of interest anyhow to observe the recognition progress. A simple program is given in the appendix, while the original paper describes the additional trick of single-bit handling, to save more memory and time (Penna and de Oliveira, 1989).

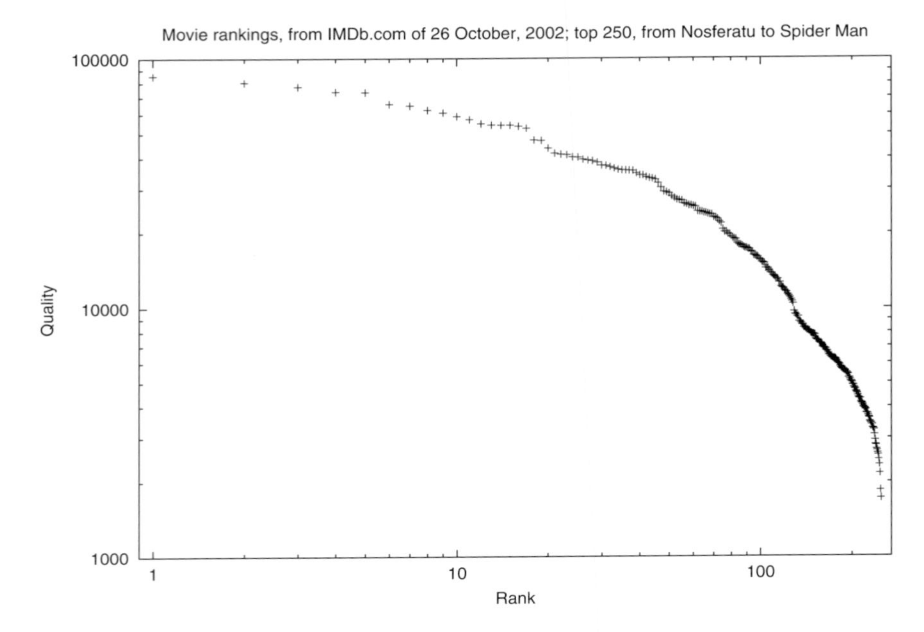

Figure 6.24. Rank plot of real movies, as evaluated on IMDb.com. From Stauffer and Weisbuch (2003); similar results were published earlier by de Vany and Walls (1996).

If Hopfield neurons are put onto a Barabási–Albert network of the preceding section, they work more efficiently if the size m of the fully connected network core is much larger than one though still much smaller than the total network size (Stauffer, Aharony, Costa and Adler, 2003).

6.6. * Social percolation

"Percolation" was already mentioned repeatedly, but never defined. This we do here and then apply it to social percolation, a field between sociophysics and econophysics. One paper has even made it to a marketing journal, and some work was partially supported by K-Mart International; but in contrast to Kai Nagel (public communication) we do not blame it for the 2002 "bankruptcy" of that company.

In standard percolation theory, each site of a large lattice is randomly occupied or empty. A cluster is a set of occupied neighbours, and an "infinite" cluster spans from one side of the lattice to the opposing side. For occupation probabilities p below some percolation threshold p_c, only finite clusters exist; for $p > p_c$ also an infinite cluster appears, usually one. At $p = p_c$ the largest cluster is

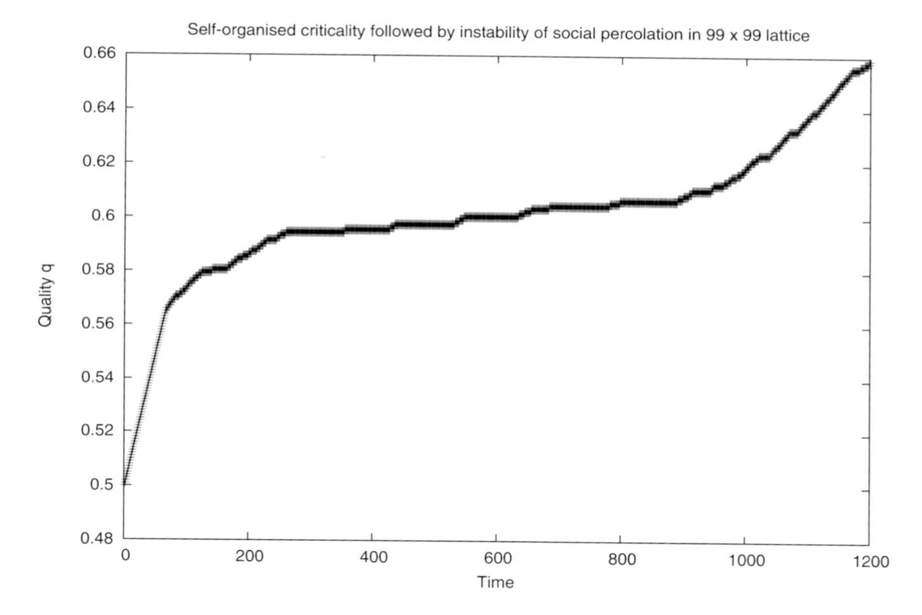

Figure 6.25. Increase of movie quality with time towards the percolation threshold $p_c = 0.592746$, followed by beginning of instability (increase towards infinity). The time is measured by the number of released movies.

fractal and sometimes spanning. Usually one looks only at nearest neighbours; if every site can be neighbour to any site independent of the distance, we get the so-called random graphs, which have the same critical exponents as the mean field theory. Percolation theory was invented in 1941, for sol–gel phase transitions which happen in rubber vulcanisation or when you boil an egg for breakfast.

Social percolation was suggested by Solomon and Weisbuch (1999) and simulated in many papers, as reviewed by Weisbuch and Solomon (2003). They modelled the success or failure of a Hollywood movie. At the start, a few people see it and tell their neighbours how they liked it. These neighbours trust this information and, if the reported quality q exceeds the personal quality standard p_i of neighbour i, this person goes to the movie and reports its quality q to other neighbours. In this way the information spreads through the population, and if the quality was high enough, and the quality standards low enough, the population of movie-goers percolates from Hollywood to Manhattan (infinite cluster on square lattice); otherwise the movie visitors remain a finite cluster, and the film becomes a flop instead of a hit. If the distribution of p_i is random between zero and one, then this social percolation is nothing new yet, and the border between hit and flop is $q = p_c$, the standard percolation threshold.

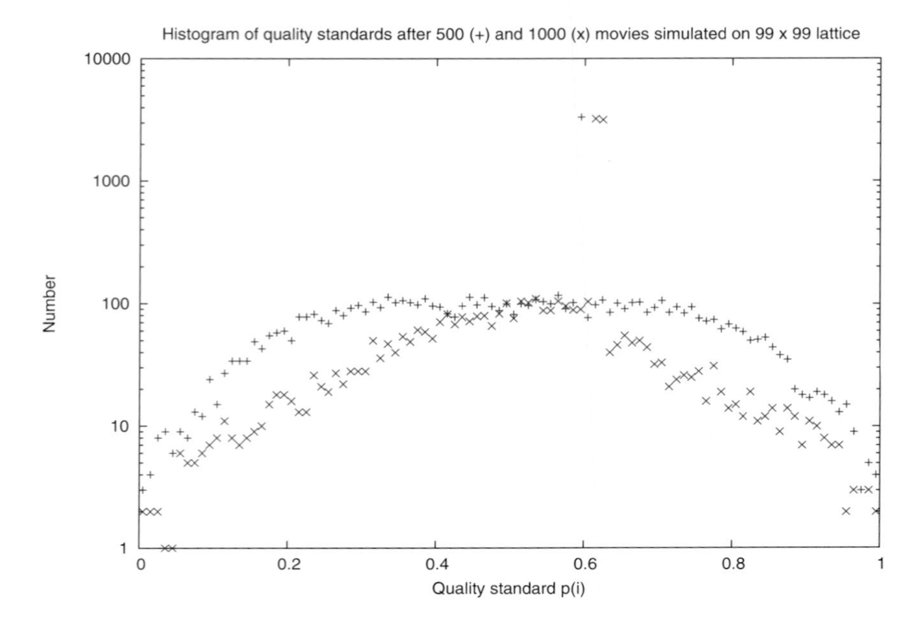

Figure 6.26. Histogram of the quality standards p_i for the simulation of the previous figure. At time 500 (+), the peak at p_c is accompanied by a broad distribution of p_i, while at time 1000 the peak has started to move to the right, and the distribution has narrowed.

The model becomes more interesting if q, and later also the p_i, become time dependent. Successful movies often have one or more sequels, which are sometimes of lesser quality. Thus social percolation assumes that a hit reduces the quality q of the next movie by a small amount 0.001, while a flop increases the quality of the next movie by the same amount. This effect is well known from some students who try to learn only as much as is needed to pass an examination. (Professors, of course, never write just as many pages for a book as the editor wants.) For students and movies alike, the q value now approaches the threshold p_c and then oscillates about it. Experts call this motion towards the critical p_c "self-organised criticality", see also Chapters 2 and 7, while in biology an analog is often described as evolution to the edge of chaos. Once q is near p_c, the clear distinction between hits and flops vanishes, and the number of clusters with s visitors each decays roughly as a power law, slightly faster than $1/s^2$.

Now also the p_i change, again by ± 0.001. If someone has just seen a movie, that person in general will not immediately see another one except if that second movie is better. On the other hand, if there were no good movies to see for some time, one may visit also some not so good movie. Thus for each lattice site (person) i which visited a movie, the p_i goes up by 0.001; for each person who

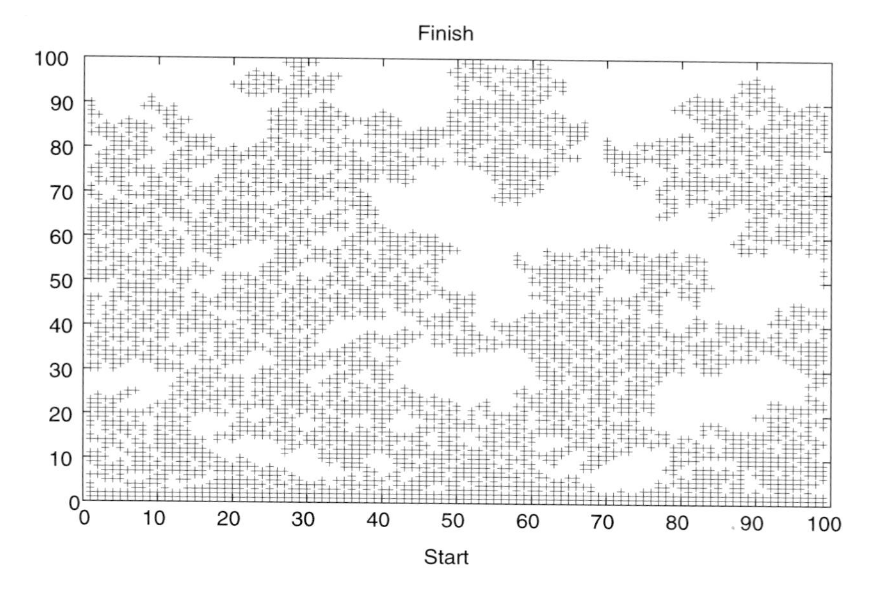

Figure 6.27. Example of a successful movie (number 501 in Figure 6.25) spanning from the bottom to the top. The simulation stopped when the top was reached.

was informed about the current movie but did not visit it, p_i goes down by 0.001. Now the dynamics gets coupled: First the quality q moves fast towards p_c, and then both q and the average p_i jointly change for each new movie by the same amount.

That latter effect is somewhat unrealistic, so it may be better to change only q and not the p_i. Also a little bit advertising, reaching everybody, has been included, leading to the simulations in Figure 6.23, in reasonable agreement (note the different vertical scales) with reality, Figure 6.24. For illustration only, the next three figures show social percolation on a small 99×99 lattice, with both q and p_i changing with time.

6.7. * Legal physics

Computer applications to legal questions were published by Yee (2001) and Hausken and Moxnes (2005). They were one of the new aspects at the Third Workshop on Dynamics of Social and Economic Systems, Mar del Plata (Argentina), June 2005, but not published in the proceedings, e.g., Rosenfeld and Martínez (2005). Perhaps here a new avalanche of papers started; interested readers should check for papers also by others: L. Devia, N.L. Olivera, N. Lipskier,

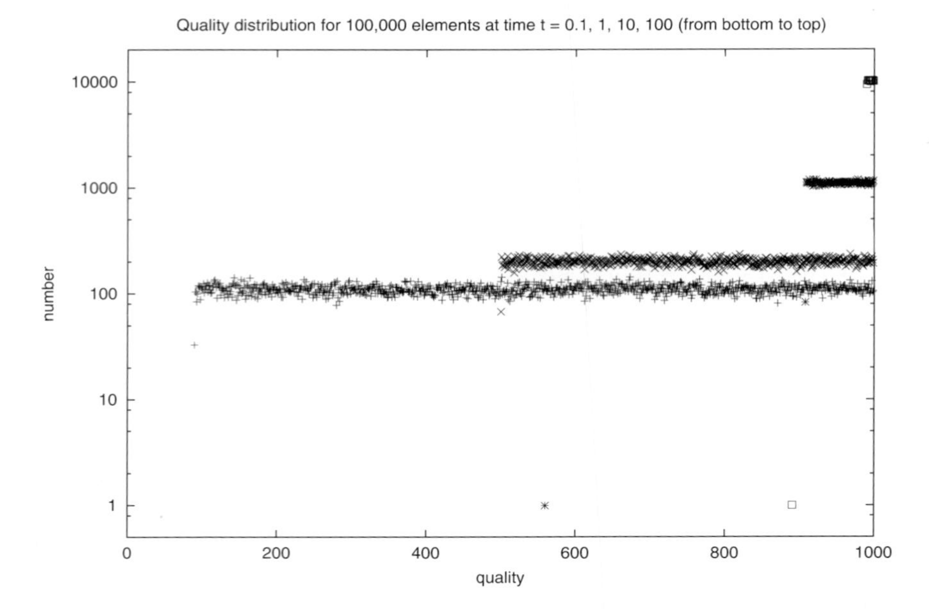

Figure 6.28. Increase of quality in judgments for the Yee model of 2001. 100 000 judgements are simulated and their quality histograms (1000 bins) are shown at times 0.1 (+), 1 (×), 10 (stars) and 100 (squares), from bottom to top. The longer the time (= Monte Carlo steps per judgement), the longer is the nearly empty interval of low quality.

R. Miro, all from Argentina, in the literature outside the natural and mathematical sciences. At present, no concentration on one problem of law is evident.

The Yee model of 2001 for the quality of court judgements is a simplification of the Bak–Sneppen model of 1993 for biological evolution, discussed in Section 4.3.1. The quality of a judgment, or the fitness of a biological species, is initially distributed randomly between 0 and 1. Then at each iteration the judgment with the lowest quality is removed and replaced by a new one with randomly selected quality. (In the Bak–Sneppen model, also the two neighbour species are removed, which makes the model lattice-dependent.) In spite of the random quality of the new judgment, the overall quality of the many judgments increases towards the maximum of 1, as seen in Figure 6.28. In contrast to the Bak–Sneppen model of Section 4.3.1 and Figure 4.25 there, in the Yee model at the end everybody is perfect, contradicting the conclusion of the movie "Some like it hot".

Chapter 7

Earthquakes

Earthquakes near densely populated urban centres are capable of causing enormous damages, both in human lives and in financial assets. In the last 400 years, the mean annual number of fatal earthquakes has increased roughly in proportion to population growth. The fatality rate in events involving fewer than 30 000 fatalities is approximately 6000 people/year by the late twentieth-century, and the historical data allows a fairly reliable estimate of a 30% increase in this rate in the next 30 years. Similar growth is noted for earthquakes involving larger number of fatalities, although future rates for these are less easy to forecast because the data become decreasingly regular (Bilham, 1996). Fatalities can also be caused by side effects of powerful earthquakes, and the memory of the tragic tsunami of December 26, 2004 in the eastern Indian Ocean, with its toll above 200 000 dead, still haunts us. Other recent examples are the January 26, 2001 7.6 earthquake that shook the Indian province of Gujarat, causing a death toll of more than 19 000 and more than 160 000 injured, with economic losses in excess of $ 1000 million; and the January 16, 1995 earthquake of magnitude only 6.9 that hit Kobe, Japan, and produced an estimated $ 200 000 million loss. That country has a very active earthquake prediction program, but it failed to predict the event. Since the Pacific plate boundary is a very active seismic region, a similar scenario is possible in a number of densely populated urban centres around it. Why is it so difficult to develop successful prediction or forecasting methods? The answer lies in the failure to identify so far any regular pattern in the occurrence of these great damaging earthquakes. These large events repeat at irregular intervals of hundreds to thousands of years, resulting in a limited historical record that has frustrated phenomenological studies. In recent years, coupled to the developments reported in early chapters of this book, emerged an alternative approach, based on a new understanding of earthquake physics arising from the construction and analysis of computational simulations. These computer simulations allow earthquake physics to be studied in numerical laboratories, where the simulation data they generate is used to develop theoretical understanding, that may be subsequently applied to observed data. Some refer to these developments as the birth of a new science of geocomplexity, that focuses on the temporal and spatial evolution of strongly correlated earth systems. Within this new science, the study of earthquakes stands out as central, since it is directly related to the modern ideas of self-organisation and criticality (Section 2.5), which were put together to label the class of so-called self-organised critical (SOC) models, and emergent structures. Many of the results and methods from within this field have shown impact in other areas of physics, beyond earth sciences. In particular, ideas about scaling regimes and friction are being used as paradigms in explaining the behaviour of threshold systems in general, that display sudden transitions and avalanche phenomena, such as neural networks, driven foams, charge density wave semiconductor devices, superconductors, and even the expansion of the early universe.

The purpose of this chapter is to show a few examples of how computer simulations are being used in this field to add to the understanding of earthquake source processes, and, in particular, to unravel a physical ingredient that may be the cause for the non-scaling regime observed in the seismic record of some faults. It is by no means exhaustive or complete, and reflects solely the (unfortunately little and marginal) experience that one of the authors has had as a post-doc and visiting scientist in one of the most active research groups in the field, led by Professor John B. Rundle, now at the University of California at Davis. It is mandatory to mention explicitly some other scientists that, while in Rundle's group, either as members or external collaborators, had an important role in the development of some of the results mentioned here. These were Professors William Klein and Kristy F. Tiampo, and Dr. Marian Anghel, to whom this chapter is dedicated.

7.1. Computational models for earthquakes

The power law frequency distribution of earthquake magnitudes, as synthesised by the Gutenberg–Richter phenomenological law, appears to be the most well documented evidence of critical behaviour in a natural system – see Figure 2.3. The first goal to be met by any model of earthquake source processes is to reproduce this scaling behaviour. In the early days of the field of earthquake modelling, Burridge and Knopoff (1967) succeeded in representing the basic physics involved in tectonic processes and were able to obtain a frequency distribution consistent with that law. In its first version, their model was a one-dimensional chain of a few massive blocks, representing the asperities or points of contact in the boundary between two moving tectonic plates, connected to each other by elastic forces (springs). The driving mechanism was modelled as a set of additional springs, connecting each block to a moving rod, and the threshold dynamics was ensured by the pinning caused by friction between blocks and the ground. The set of coupled second-order differential equations representing this physics was then solved numerically. Qualitatively, the main feature of the model can be described by focusing on the moment in which the resulting force on one of the blocks exceeds the threshold value of friction: the block slides, and the impact of its motion on the resulting forces on its neighbours may in turn cause their depinning. This chain reaction, or avalanche, is the model's proxy for an earthquake. Energy release during an avalanche can then be measured and its magnitude computed. The frequency distribution of the magnitude is then compared to a scaling law, which is the approach used in most models of this class. A power law statistics was obtained by Burridge and Knopoff when a friction force that decreases with velocity was used in the model.

Computational models are able to increase greatly, both in number of elements and dimensionality, the ability to investigate the above physics. The trade-off is to eliminate the massive terms of the equations, by considering that these are related to seismic waves, that carry typically less than 10% of the energy released in an earthquake. By so doing, the differential equations can be easily discretised, and the resulting model is a real-valued – or continuous – cellular automaton (CCA). Several models with these characteristics appeared in the literature in recent years – Nakanishi (1990), Rundle and Brown (1991), and Fisher, Dahmen, Ramanathan and Ben-Zion (1997) are a few of those – and the most successful of them, at least in the physics community, was that of Olami, Federsen and Christensen (1992), henceforth abbreviated by OFC, which were the first to claim the observation of SOC in a non-conservative model. A feature shared by all these models is a dynamical field, usually referred to as the stress, that is updated in each cycle of the automaton. Its value at each point in a discretised space increases by the action of some driving mechanism and relaxes as a threshold critical value is reached. As a result of this relaxation, stress is transfered to other points in space coupled by some interaction to the one where the threshold value first was reached. Because of this transfer, an avalanche of relaxations, or earthquake, may be triggered. One can divide these models into two classes, according to the range of the transfer interaction.

7.2. Short-range interactions

Using the language of the OFC model, the physics is contained in the dynamics of a single field, a real-valued dynamical variable E_i, the stress, defined on the $N = L^d$ sites of a d-dimensional cubic lattice. The dynamics of this field is completely deterministic, except for the initial configuration, which sets the initial value of the field at a site i by a random choice E_i between zero and a critical value E_c, usually set to one. The driving is uniform and homogeneous, and at each step of the driving time scale the field evolves with $E_i \to E_i + \nu$, where ν is an indirect measure of this time scale. Relaxation, which happens when a site i becomes critical $E_i \geqslant E_c$ and topples, follows the rule $E_i \to 0$ for this event initiator site and $E_{nn} \to E_{nn} + \alpha E_i$ for its q_i nearest neighbours. These sites may, in turn, become critical, and the process proceeds until $E_i < E_c$ for all i. This cascade, or avalanche, of topplings is an earthquake in the model, and data can then be collected on the statistical distribution of such events. The parameter α has an important role, since it measures the intrinsic dissipation ratio of these dynamics: the amount of stress that is lost by the system after site i topples is $E_{dis} = (q_i\alpha - 1)E_i$, and the system would be called conservative if $\alpha = \max_i(q_i)^{-1}$. One is usually interested in the zero-velocity limit $\nu \to 0$, since there is an intrinsically very large time scale separation between the driving and relaxation mechanism in the real earthquake process. This is easily implemented in the model by driving it to relaxation in a single step of the simulation: since the driving is homogeneous, the site with the largest stress at the completion of an avalanche will be the event initiator of the next. So, the new driving rule for the zero-velocity limit is to compute $E^* = \max_i(E_i)$ and then perform $E_i \to E_i + (E_c - E^*)$ for all i. Care must be taken when working with the model, for its approach to a statistically stationary state proceeds rather slowly, and transients are very long. One has to wait typically for $\sim 10^9$ avalanches in a $2D$, $L \sim 10^2$ model before collecting meaningful data.

The OFC model with nearest neighbours has been extensively studied lately, and a plethora of information gathered about the nature of its (quasi-) critical attractor state. The signatures of this state become first visible near the borders, and it spreads through the lattice from there on. The model has a strong tendency to synchronisation, which generates spatial correlations and is partially responsible for critical behaviour. This behaviour is lost when periodic boundary conditions are used, and synchronisation forces the system to periodic non-critical behaviour. The inhomogeneity induced by an open boundary is enough to destroy the periodic state while leaving intact correlations, thus allowing for the establishment of the (quasi-) critical attractor. The real nature of the attractor, and the issue of criticality for that matter, in the OFC model is still a matter of intense debate among experts. There is general agreement on the fact that the model is indeed critical in $2D$ if $\alpha > \alpha_c$, but there is no such agreement on the critical value itself. Estimates for α_c vary widely in the literature, ranging all the way from $\alpha_c = 0$ (Middleton and Tang, 1995), to $\alpha_c \simeq 0.18$ (Grassberger, 1994; Corral, Perez, Diaz-Guilera and Arenas, 1995), while strong arguments do exist in favour of $\alpha_c = 0.25$, which corresponds to the conservative limit mentioned above, and the notion of a quasi-critical state to describe the nature of the attractor when α is close to α_c (Prado and de Carvalho, 2000). On the other hand, there is no question about the fact that the OFC shares with a few other statistical models, such as the eight-vertices model, the unusual dependence of the exponent of the scale-free frequency distribution of earthquake sizes on the value of the conservation parameter α.

7.3. Long-range interactions

Other flavours of CCA models have been studied that involve long-range interactions, in particular by the geophysics community (Dahmen, Ertaş and Ben-Zion, 1998; Preston, Martins, Rundle, Anghel and Klein, 2000). These versions are supposedly more "realistic", since viscoelastic interactions in the earth's crust are known to be long-range, presumably decaying as the third power of distance. Although some of the excitement involved in the emergence of long-range order from short-range interactions is lost in these models, they still present interesting features and deserve the attention of the physics community. The infinite-range interaction model, where each site interacts with all other sites in the lattice, is one of those, and its mean-field character has served well the purpose of establishing a test ground for the exploration of new ideas.

As mentioned above, CCA models have usually very long transients, forcing the researcher in the field to be very demanding on the efficiency of the computer code used in the simulations. Useful comments on this subject for the beginner can be found in Grassberger (1994) and Preston, Martins, Rundle, Anghel and Klein (2000). To begin with, one should only deal with one-dimensional data structures, and draw the code aiming at an efficient use of these structures. As an example, the address of a site in a 2-D lattice, usually taken as an ordered pair (i, j), should be transformed into a single integer as in `index = i*L + j`. The integer operations of division and remainder are then used to get back the ordered pair, when needed, from this single integer. The dynamical field is then a 1-D real vector, and loops sweeping the lattice are controlled by a single integer variable. Another extremely useful strategy is to have moving failure and residual stresses: instead of sweeping the whole lattice each time the driving mechanism brings the site closest to failure – the initiator of the next avalanche – to its critical stress, the value of the critical stress itself is updated to the value of the stress of this initiator, and the residual stress is also moved to keep the difference between these two constant. The usage of "last in, first out" (LIFO) stacks is also good advice, in particular when running long-range models, for it allows a simple way of keeping track of only a few sites to examine for failure at each step of the relaxation avalanche. Here is a short description of an algorithm using these ideas for a 2-D model, expressed in a – hopefully – self-explanatory meta-language:

- site address structure: `(i,j)` $\rightarrow$ `index = i*L + j`
- initialisation: (e.g., E_i = random between 0 and E_c)
- main loop: *do*
  - selects site i with maximum E_i
  - moves failure and residual stresses: $E_r \rightarrow E_r + E_c - E_i,\ E_c \rightarrow E_i$
  - avalanche

- if time > transient, collects statistical data

until time is up.

The key routine is the relaxation algorithm, the avalanche procedure. In this example, we use two separate stacks (`stack1` and `stack2`), where `stack1` stores the addresses of the sites that will topple in the present relaxation step, while `stack2` keeps track of the sites that will topple in the next step.

– avalanche:
  - set `duration` to one, avalanche `size` to zero, and stacks the address of the site with maximum E_i: $i \to$ `stack1` (this is the initiator)
  - *while* `stack1` NOT empty *do*:
    · unstacks site address from `stack1`;
    · increments avalanche `size`;
    · computes stress drop at this site;
    · resets stress at this site to residual value;
    · dumps stress on neighbours (n): if $E_n > E_c$, $n \to$ `stack2`;
    · if `stack1` is empty, but `stack2` is not, exchanges `stack1` and `stack2` and increments `duration`.

The routine ends when `stack1` gets empty. Variable `duration` ends up with the number of cycles of stack changing, while `size` stores the number of sites that toppled during the avalanche.

7.4. The Rundle–Jackson–Brown model

A real geological fault is a highly disordered environment. Friction, for instance, is by no means uniform or homogeneous. Irregular gouge deposition and ageing act to create this disorder in space and time. In a computational model such as the ones we have described, disorder can be added to the dynamical behaviour by the introduction of some stochastic field into the equations of motion. We will call a CCA model with this feature a stochastic continuous cellular automaton (SCCA). We will briefly present here the Rundle–Jackson–Brown (RJB) SCCA model for an earthquake fault (Rundle and Jackson, 1977; Rundle and Brown, 1991), in its uniform long-range interaction, zero-velocity limit, mean-field version (Preston, Martins, Rundle, Anghel and Klein, 2000). Extensive work has recently focused on a near-mean-field version of this model, where the interaction range has a cutoff. The model can then be mapped into an Ising-like Langevin equation (Klein, Anghel, Ferguson, Rundle and Martins, 2000). In its original formulation, the RJB model has as dynamical variables two continuous real-valued fields, slip $s_i(t)$ and stress $\sigma_i(t)$, defined on the sites i of a lattice and coupled by a constitutive equation. It can be recast into a single-field model, which is how it is usually implemented in computer simulations. In this form, it closely resembles the OFC model, except for the interaction range and the noisy component. Its appeal to the geophysics community is enhanced by a careful choice of language. Thus, the interaction between sites is summed up by the stress Green's function matrix T_{ij}, $K_C = \sum_{j \neq i} T_{ij}$ is the model's equivalent of the combined effects of the coupling springs of the OFC formulation, K_L parameterises the loading interaction, $K = K_L + K_C$ is the elastic stiffness, related to the intrinsic stress dissipation measured by the factor $\delta = K_L/K$, and $T_{ii} = -K$. The inter-event, or loading, dynamics of the standard RJB model is thus very simple: in the same way as in the OFC model, the stress field undergoes linear growth. When the stress σ_i at a site i reaches a threshold for failure σ_F the site topples and a fast stochastic relaxation takes over. The stress drop $|\Delta\sigma_i|$ – which, in the OFC model, always resets σ_i discontinuously to its zero residual value – is now noisy: after toppling, the stress σ_i is

$$\sigma_i = \sigma_R + A\eta_i(t) \tag{7.1}$$

where σ_R is a uniform residual stress, A is the noise amplitude, and $\eta_i(t)$ is a random number from a uniform distribution in the interval (0, 1), tossed anew for each toppling site and each time it topples. The ensuing stress drop is then redistributed among all the sites of the system, according to the stress Green's function connecting each site j to the one that toppled and some internal dissipation:

$$\Delta\sigma_j = \frac{T_{ji}}{K}|\Delta\sigma_i| \tag{7.2}$$

where i is the site that failed. This increase may cause other sites to fail as well, and the process continues until all sites have stress below failure. This cascade of failures, or avalanche, is the model's equivalent for an earthquake. The RJB model may be used as the basis for a number of variations that serve the purpose of being a testing ground on which to simulate a number of ideas and conjectures about the physics of friction in earthquake source processes, the role of heterogeneity in loading and relaxation, to name but a few. One of such variations is described in the next section.

7.5. Precursory dynamics

The existence of an inter-event dynamical cycle is suggested by recent laboratory experiments addressing issues of solid-on-solid friction. A stable slip, with a slow velocity that increases with the stress level, is observed prior to failure, leading to a partial release of the accumulated stress (Tullis, 1996). This stress leakage mechanism is analogous to a temperature-dependent viscosity that has been observed in laboratory for the creeping of crystalline rocks, and can be modelled by the equation

$$\frac{ds(t)}{dt} = \alpha \frac{\sigma(t) - \sigma_R}{K} \tag{7.3}$$

where $s(t)$ and $\sigma(t)$ are the displacement and the stress at time t, σ_R is some residual stress value to which the system decays after $\sigma(t)$ reaches a failure threshold σ_F, K is the elastic stiffness and α measures the intensity of this stress leakage effect. This last parameter, with dimensions of inverse time, will in general be dependent on both the stress level and the temperature. An analogous leakage mechanism may also be present in other threshold systems, such as an integrate-and-fire neural network (Hopfield, 1994), and it is likely that the results here reviewed will also hold in that context.

One is thus led to study what effects the stress leakage mechanism would cause in threshold models in general. As an extra and important motivation, it was hinted that such an effect could be the physical ingredient behind the recurrent large-magnitude earthquakes observed in some faults in Southern California, such as those in the Parkfield region.

Building on the ideas put forth in the last section, the stress leakage dynamics was introduced into the framework of the infinite range uniform RJB model (Martins, Rundle, Anghel and Klein, 2002). In this mean-field version, the interaction matrix $T_{ij} = K_c/(N-1)$, where $N = 1/\Delta$ is the number of sites in the lattice, is uniform. The resulting equations of motion in the inter-event cycle can then be decoupled. The analytical solution for inter-event stress temporal evolution can then be added to the computational SCCA.

The important effect of the leakage stress dynamics in the pattern of failures of the system comes from the reduction it causes in the statistical spread of the stress field with time. A simple measure of this smoothing is obtained through the evolution of the variance of the dimensionless stress field η as a function of dimensionless time τ

$$\mathrm{var}\big[\eta(t)\big] = \mathrm{var}\big[\eta(0)\big] \exp\left\{-\left[\frac{1-\delta\Delta}{1-\Delta}\right]\phi\tau\right\}. \tag{7.4}$$

Because of this exponential smoothing of the stress field, the probability of a site to reach failure after receiving a transfer from a failing neighbour increases, and is an increasing function of the time-to-failure. As a consequence, the branching ratio, defined as the average number of failures caused by each failing site, also increases. The system is more likely to undergo larger avalanches, which may even be system-wide when the time-to-failure is large enough.

Figure 7.1 shows a log-log plot of the frequency distribution of events as a function of their size. The effect of the stress leakage dynamics shows up clearly in the excess over scaling obtained for large events as ϕ is increased from 0, together with a depletion of the distribution in the intermediate size range. The

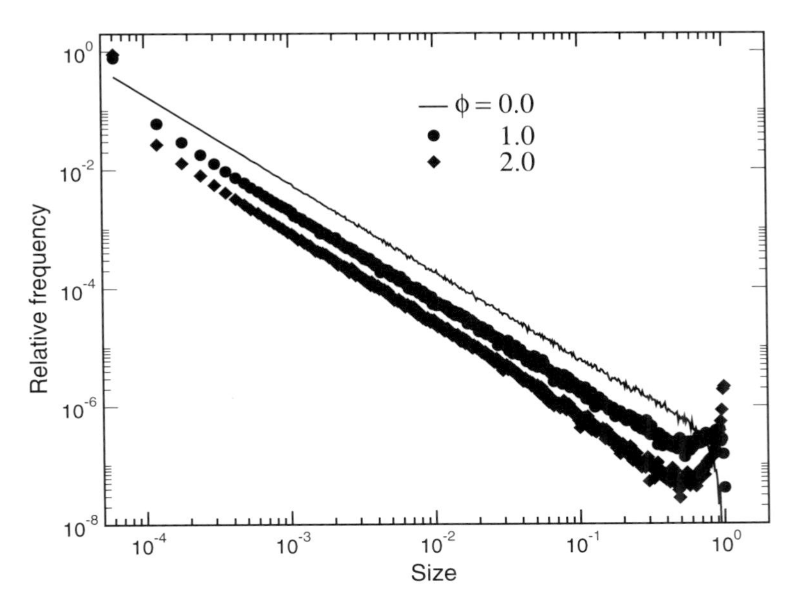

Figure 7.1. The frequency distribution of earthquakes synthesised by the RJB model is shown, as a function of their magnitude measured as the fraction of the lattice that suffered rupture. Parameter ϕ is a dimensionless measure of the intensity of the stress leakage effect, and its value in the plot runs from zero – no leakage – to 2. As it increases, the shape of the distribution changes, as commented in the text.

slope of the scaling part of the plot also gradually becomes steeper, starting from the mean field value 1.5. This means that a progressively larger depletion in the distribution occurs as the event size increases. The smoothing effect of the leakage dynamics, together with the resulting larger stress average that it causes in the field as a whole, results in a higher probability for large events to grow, eventually causing total rupture of the fault.

This increase in slope is reflected also in the distribution for the number of topplings. Here, a counter is updated each time a site fails, even if it had failed before in the same avalanche. This number reflects more closely the model's equivalent for the moment release in an event. The fraction of multiple failures vanishes in the exact mean-field limit, and the two distributions are equivalent. This is no longer true for the model with stress leakage. The model with leakage reflects an excess of longer events over scaling: events with some range of large durations are more frequent.

The SCCA models with no leakage dynamics show no signs of a characteristic-event regime. The power spectrum of the distribution of inter-event times for large size events is white. The inclusion of leakage dynamics however changes this

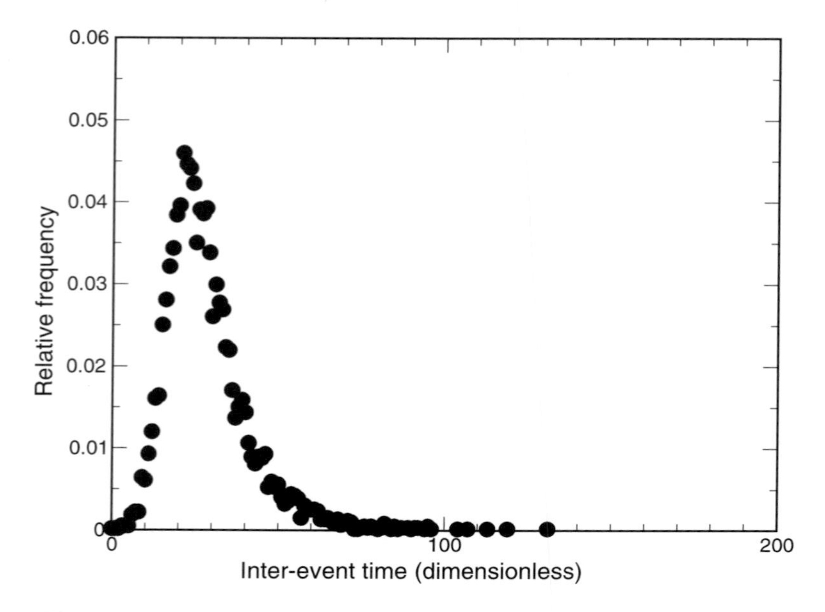

Figure 7.2. The distribution of (dimensionless) inter-event times for large earthquakes is shown for the synthetic seismic catalog generated by the RJB model with stress leakage, for $\phi = 2.0$. The appearance of a characteristic inter-event time is clear.

aspect radically, as shown in Figure 7.2. The distribution of inter-event times has a pronounced maximum, corresponding to a characteristic period, or a time lag more frequent than shorter or longer times, between large rupture events. This feature, whose presence was conjectured by Wesnousky (1994) after an analysis of the seismicity record of some individual faults in Souther California, is tuned by the leakage parameter, and will be more dominant as this parameter increases.

The introduction of a stress leakage process as an inter-event dynamics in a SCCA model for earthquake faults, such as the RJB, changes in a dramatic way the space-time patterns that it generates. In particular, the α value for a fault may determine its overall behaviour as of a scale-invariant or a nucleation type, with a mixed composition in between. The importance of this new parameter could in fact be recently evaluated. Its tuning to match the characteristics of each segment in a complex computer representation of the fault network of Southern California allowed the generation of space-time patterns of rupture of unprecedented realism (Rundle, Rundle, Tiampo, Martins, McGinnis and Klein, 2001). These patterns were also instrumental as test ground for a conjecture on the role of the eigenvalues and eigenvectors of a space-time correlation operator in the statistical forecasting of earthquakes. A method based on these ideas was put forth recently, and its results are so far encouraging (Rundle, Tiampo, Klein and Martins, 2002;

Tiampo, Rundle, Martins, Klein and McGinnis, 2004). Another conjecture, namely that the natural earthquake fault system may undergo ergodic dynamics, have benefited from the leaking SCCA model. The patterns of the time evolution of the Thirumalai–Mountain metric mined from real data have a close counterpart in those generated by the model (Tiampo, Rundle, Klein, Martins and Ferguson, 2003; Tiampo, Rundle, Klein and Martins, 2004).

7.6. Conclusion

Computer simulation models have changed dramatically the way research is done in various fields of Geosciences. The study of earthquake source processes, in particular, has progressed substantially since their introduction. Much has been learned on the basic physics underlying these processes and the trend towards their massive usage is clear. Several research groups around the globe share the expectation that the development of a General Earthquake Model, which would put together computer simulations and real-time analysis of seismicity, would bring us one step closer to earthquake forecasting. This expectation is supported by the dramatic increase in the quality of weather prediction caused by the General Circulation Models for the Earth's atmosphere, with which it shares both methods and scope. We presented in this chapter a brief introduction to some simple models of earthquakes, going all the way from the simple OFC nearest-neighbour automaton to a much more sophisticated one, the leaking RJB. We hope our presentation may encourage some of the readers to join this exciting effort.

Chapter 8

Summary

What have we learned from the research presented in this book? As usual, one can learn methods, and one can study results.

For *methods*, Chapter 2 warned against simple approximations, known in physics as mean-field theories, and represented, e.g., by differential equations for populations as a function of time. Instead, microscopic models should be used, based on individuals ("agents"), as done in most of our chapters (except Section 9.5 for demography). Each individual has its own properties: genome, opinion, ..., and simulations are easier if these properties are integers and not continuous variables. If they are binary ± 1 variables, they can even be stored in single bits, saving computer memory and time (Section 9.1). The latter techniques were used throughout Chapters 3 to 5.

Examples for *results* are: good agreement with the exponential increase of human adult mortality as a function of age, Figure 3.6; self-organisation of menopause and analogs without the need for any specifically human properties, Figure 3.9; establishment of sympatric speciation in many simulations of suitable environments, Figure 4.3; possible survival of many different languages without overwhelming dominance of any single one, Figure 5.6; explanation of typical vote distributions in political elections, Figure 6.8; a possible deviation from the earthquake power law, Figure 7.1.

Does it mean that biologists, linguists, sociologists and geologists take us serious? Some do: Biologists published papers with us, a famous linguist wanted to hear a talk about Chapter 5, the editor of a sociology journal repeatedly sent us papers to referee. But these are more the exception than the rule; the citations of these interdisciplinary papers of physicists by non-physicists are not many. Of course, why worry about not being cited by ageing biologists who don't cite each other?

There are good reasons for the less-than-overwhelming reception of interdisciplinary research of physicists by non-physicists, besides the natural rejection of anything alien known from immunology: Where is the really *big* effect? Physics itself is taken serious (not necessarily in a positive sense) since the explosion of the first atomic bombs six decades ago. Nothing like that has happened thus far for computational biology, linguistics, sociology or geology. If it would lead to anti-ageing medicine or to the reliable prediction of revolutions or earthquakes, the situation would be different, and perhaps this will come in the future. Mankind took centuries, from the description of planetary motion by Kepler to human landing on the Moon. The short-term task of Kepler was to write horoscopes for a famous general, and, in this task, he was not very successful: The general was assassinated. Similarly, the work summarised in this book may later lead to progress on unexpected questions, using the methods described here. These drawbacks and possibilities are shared by us together with computational biologists etc.

Thus let us follow Kepler and keep our minds open, without demanding immediate large-scale applications.

Chapter 9

Appendix: Programs

To encourage readers to start their own simulation we give here some selected programs. They may be simplifications compared to those which we really have used, for example by omitting the evaluation of less important quantities. Note that often the random numbers are produced not by a built-in random number generator like `rand` but by explicit multiplication of an odd integer with 16807, 13^{13} (for 64-bit integers only), or 65539 ($= 2^{16} + 3$, which may also be implemented by suitable shifts). Our programs are supposed to be understood, not to be merely used.

9.1. Single-bit handling

Good science says yes or no. Many variables in this book are binary, ± 1 (alternatively 0 or 1, or: true or false). It is a waste of computer memory to store such a variable in a computer word of 32 or 64 bits as is done if one uses the Boolean variable type `logical` in Fortran. By reducing the word length from `integer*8` (64 bits) or `integer*4` (32 bits) to 16 bits via `integer*2` or to 8 bits via (regrettably) `byte` in Fortran (or analogous declarations in C) for these binary variables, one can save some memory, but the most efficient way is to use the single bits. This was possible with C since the beginning, while only the Fortran 90 standard made these commands legal there, unfortunately not in the simple way in which some compilers handled these Fortran extensions decades earlier. For example, let us find the value (0 or 1) of the third bit in a word i. Then we define a bit mask which is zero everywhere except in its third position, The logical AND of i with this bit mask them gives only the third bit of i, at its original position; everything else is zero. Shifting this bit to the right end of the word, we get an integer which is zero or one, and equal to the original third bit of i.

These techniques were used for the genomes in Section 2.8 and Chapters 3, and 4, for the languages in Chapter 5, and for some papers on opinion dynamics and neural nets (Chapter 6). For Ising models or cellular automata like the "Game of Life" not discussed here, these techniques may not only save memory but also computer time since one single bit-string command may then deal with, say, 32 variables at once (de Oliveira, 1991). Physics simulations with these techniques started to our knowledge at the beginning of the 1970s.

We now list the relevant Fortran commands, giving their C analogs in parentheses, and explain them with true and false; "iff" means if and only if. Each command deals with all (32 ?) bits of the words at once:

`iand(i,k)` (in C: i & k): True iff both are true.
`ior (i,k)` (in C: i | k): False iff both are false.
`ieor(i,k)` (in C: i $\wedge$ k): True iff the two are different (exclusive or).
`not(i)` (in C: $\sim$ i): True iff bit is false (reverses all bits).
`ishft(i,k)` (in C: i << k): Shift to left by k positions if $k > 0$.
`ishft(i,-k)` (in C: i >> k): Shift to right by k positions if $k > 0$.

Usually the bits shifted over the ends of the word by `ishft` are lost, and zeroes are inserted at the other end of the word. But it is better to check this, if it is important. Circular shifts can be constructed by combining shifts to the left and right with an `ior`.

Counting all the bits in a word i can be done in a simple inefficient way by shifting the word 32 times to the right after reading its rightmost bit and adding it to a sum through `+iand(i,1)`. More efficient is a table of length 256 for each 8-bit section of the word (or of length 65536 for each 16-bit section) which contains for the binary number in that section the number of bits set to one, like `itable(7) = 3`. This small table is filled at the beginning of the program once, using the primitive method. With suitable shifts and ANDs with bit masks, the word i is divided into its few sections, for each section the number of one-bits is read off, and the results are added. Perhaps your compiler has a bit-count built in saving you the trouble.

9.2. Ageing in Penna model

This program and its description are taken from the appendix of our book (de Oliveira, de Oliveira and Stauffer, 1999) except that the genetic death age is calculated right at birth. Only the asexual version is given here, where the last letter f of a variable name means "female"; the male variables are missing.

```
c      Asexual; genetic death age calculated at birth,
c      stored in DATA
       implicit none
       integer popdim
       parameter(popdim=80000)
       integer popmax,inipop,maxstep,medstep,minage,fage,
     1 lim,fmut,birth,n6,t,verhu,i,p,seed,ibm,gene1,fa,
     2 n,imut,age,nmut,fpop, fymed(0:32),fnumber(0:32),
     3 bit(0:32),gen1f(popdim),dataf(popdim),dage,
```

```
     4 gd(0:33),gdeath(0:33)
      parameter(popmax=200000   , inipop=popmax/10,
     1 maxstep=10000,medstep=maxstep/2,minage= 8,fage=32,
     2 lim=3, fmut=1, birth=3, seed=1, n6=63)
      data gdeath/34*0/
c
      ibm=2*seed-1
      print *, popmax,inipop,maxstep,medstep,minage,
     1         fage,lim,fmut,birth,seed
      fpop=inipop
      bit(0)=1
      do 2 i=0,32
        if(i.gt.0) bit(i)=ishft(bit(i-1),1)
        fymed(i)=0
        fnumber(i)=0
 2      ibm=ibm*16807
      fnumber(0)=fpop
      do 6 i=1,fpop
        dataf(i)=ishft(33,6)
 6      gen1f(i)=0
c     dataf: age at bits 0 to 5, genetic death
c     age at 6 to 11
      do 7 t=1,maxstep
      verhu=(fpop*2.0/popmax-1.0)*2147483647
      print *, t,fpop
      do 3 age=0,33
 3      gd(age)=0
      i=1
      fa=fpop
 9    age =iand(n6,dataf(i))
      dage=iand(n6,ishft(dataf(i),-6))
      fnumber(age)=fnumber(age)-1
      age=age+1
      if(age.ge.dage) gd(age)=gd(age)+1
      ibm=ibm*16807
      if(ibm.lt.verhu.or.age.ge.dage) then
c     death
        if(fpop.le.1) goto 1
        gen1f(i)=gen1f(fpop)
        dataf(i)=dataf(fpop)
        fpop=fpop-1
        if(fpop.ge.fa) then
          i=i+1
        else
          fa=fa-1
        endif
```

```
      else
c     survival
        fnumber(age)=fnumber(age)+1
        dataf(i)=ior(age,ishft(dage,6))
        if(age.ge.minage.and.age.le.fage) then
          do 12 n=1,birth
c         birth
            gene1=gen1f(i)
            do 13 imut=1,fmut
              ibm=ibm*16807
              p=bit(ishft(ibm,-27))
c             mutations in mother
 13           gene1=ior(gene1,p)
              fnumber(0)=fnumber(0)+1
              fpop=fpop+1
              if(fpop.gt.popdim) goto 1
              gen1f(fpop)=gene1
              nmut=0
              do 11 age=1,32
                nmut=nmut+iand(gene1,1)
                if(nmut.ge.lim) goto 21
 11             gene1=ishft(gene1,-1)
 21          dataf(fpop)=ishft(age,6)
 12       continue
c         if(female suitable) then
        endif
        i=i+1
      endif
c     if(death) .. else (survival, birth) ..
      if(i.le.fa) goto 9
c     end of selection and birth, now start averages
      if(t.lt.maxstep-medstep) goto 7
      do 10 i=0,32
        gdeath(i)=gdeath(i)+gd(i)
 10     fymed(i)=fymed(i)+fnumber(i)
 7    continue
      print 100, (i,fymed(i)*1.0/medstep,
     1 gdeath(i)*1.0/medstep,i=0,32)
 100  format(1x,i3,2f11.2)
      stop
 1    print *, 'error'
      stop
      end
```

The crucial array is called `dataf` and is basically the passport for the living females (no males yet for `datam`). In the rightmost (least significant) 6 bits of the

32-bit word dataf is stored the age, and in the next 6 bits to the left the genetic death age.

Loop 7 goes over the time steps, from 1 to maxstep, with averages taken only over the last medstep iterations. The Verhulst factor is newly calculated for each new time step, but within one time step stays constant for all survivors. After label 9 we read off from dataf (by shifts to the right and logical AND with the bit-string 111111) the current age and the genetic death age. The number fnumber of females at this age is decreased by one, and the age increased by one. A new random number ibm is drawn, and compared with the Verhulst factor. If the random number is smaller than the Verhulst factor (normalised to the interval from -2^{31} to $+2^{31}$ to speed up the program), or if the new age is no longer below the genetic death age dage, then this individual dies; otherwise it survives and can give birth. Already before this if-condition, a genetic death is counted in the array gd(age).

Death means that the information (gen1f and dataf) for the last individual simulated is placed into the array positions for the dead individual and the population is decreased by one.

Survival means that the number fnumber of females of the new age is increased by one. The genetic age and the current age are stored in the passport dataf, via shifts to the left and logical OR operations. Now, if the age is between the minimum and maximum ages for reproduction (minage and fage), births can happen.

At birth, each of the birth daughters produced in loop 12 gets the genome gen1f of the mother, except that at fmut randomly selected bit positions between 0 and 31 (obtained by looking at the five most significant bits of the random number ibm) the bit in the daughter genome gene1 is set equal to one through a logical OR with a suitable array element p. The number of babies, fnumber(0), then is increased by one, and so is the population fpop. The genetic death age of the new baby fpop is evaluated in loop 11 by going through the bits of the baby genome gene1 and checking at which age the number nmut of set bits up to the position age reaches the lethal limit lim. The new baby gets a passport dataf(fpop) containing the genetic death age and also the current age of zero.

The 48 lines following i=1, up to goto 9, are the loop over all individuals $i = 1, 2, \ldots,$ fpop. Since the population fpop varies due to the death and birth processes, we did not deal with them in a fixed loop i=1,fpop, and instead used the backward jump goto 9 and the number fa of adult individuals; at the beginning of the iteration all individuals are adult: fa = fpop.

After all individuals have been dealt with at this time, we collect the age distribution fnumber into the array fymed during the latter part of the simulation. After the time loop 7 is finished the average population and the number of genetic deaths are printed out as a function of age. For quality simulations, 64-bit integers should be used for ibm, verhu, fymed, gdeath, as in the language program of Section 9.4.

9.3. Bak–Sneppen evolution

```
      parameter(n=1000   ,np1=n+1)
      dimension ir(0:np1), ihist(0:1000)
      data iseed/123456789/,max/100000   /,maxn/2147483647/
     1 ,ihist/1001*0/
      print *, n,iseed,max
      ibm=2*iseed-1
      factor=0.5d0/2147483648.0d0
      do 1 i=0,np1
      ibm=ibm*16807
1       ir(i)=ibm
      do 2 istep=1,max
        min=maxn
        do 3 i=1,n
          if(ir(i).gt.min) goto 3
          imin=i
          min=ir(i)
3       continue
        ibm=ibm*65549
        ir(imin+1)=ibm
        ibm=ibm*16807
        ir(imin   )=ibm
        ibm=ibm*65539
        ir(imin-1)=ibm
2     continue
      do 4 i=1,n
        k=1000*(0.5+factor*ir(i))
4       ihist(k)=ihist(k)+1
      do 6 k=0,1000
6       print *, k, ihist(k)
      stop
      end
```

The program fragments in Section 4.3.1 would be quite useful for learning how to keep track of the smallest element in a large evolving set, had they been written in an understandable language. However, since they were written in C, this section presents a simple program for teaching Bak–Sneppen simulation in Fortran, without binary trees. Brazilians may call this primitive search for the minimum the Portuguese method, while Germans say one should not use a big cannon to shoot at a sparrow. The program took 35 minutes on a fast workstation for $N = 100\,000$ species up to time $= 100$, as in Figure 4.25. A ten times bigger simulation for the same time would last hundred times longer and thus is not recommended.

The species fitness is taken here as an integer ir between -2^{31} and $+2^{31}$ and initially is random. The outer loop 2 runs over all steps, of which each time unit needs n for n species. The innermost loop 3 determines ir(imin) as the smallest of them, with the value min. When loop 3 is finished, loop 2 continues with replacing ir(min) and its two neighbours by three new random numbers. After loop 2 is finished, we bin all the fitness ir into thousand intervals $k = 1, 2, \ldots, 1000$ and print out the histogram ihist.

9.4. Language competition

```
      implicit none
      integer popdim,nbyte,nbit,nshift,len,irun,nrun,
     1 nhist(1000000)
      real fmut,rand,select
      parameter(nbyte=1,popdim=20000     ,nbit=8*nbyte,
     1          len=2**(nbit-1))
c               integer*2 gen1f(popdim), gene1, p, bit(0:nbit)
                byte      gen1f(popdim), gene1, p, bit(0:nbit)
      integer popmax,inipop,maxstep,fage,k,nlog(0:30),j,
     1 birth,t,i,seed,fa,n,fpop,nlabel(-len:len),number,kmut
      parameter(popmax=popdim,inipop=1,maxstep=1000,nrun=1,
     1 fage=nbit,birth=1,seed=1)
      integer*8 ibm,verhu,mult,imut
      nshift=0
      if(nbyte.eq.2) nshift=60
      if(nbyte.eq.1) nshift=61
      if(nshift.eq.0) stop 9
      ibm=2*seed-1
      mult=13**7
      mult=mult*13**6
      fmut=rand(seed)
      print *, popmax,inipop,maxstep,
     1          fage,birth,seed,nbit,nrun
      do 18 kmut= 600,600,100
      fmut=kmut*nbit*0.0001
      print *, fmut,kmut
      if(fmut.ge.1.0) stop 9
      imut=2147483648.0d0*(fmut*4.0d0-2.0d0)*2147483648.0d0
      bit(0)=1
      do 2 i=0,nbit
        if(i.gt.0) bit(i)=ishft(bit(i-1),1)
```

```
 2        ibm=ibm*16807
       do 15 k=1,1000000
 15       nhist(k)=0
       do 11 irun=1,nrun
       fpop=inipop
       do 6 i=1,fpop
 6        gen1f(i)=0
       select=2.0/popmax
c
       do 7 t=1,maxstep
       verhu=2147483648.0d0*(fpop*4.0/popmax-2.0)
     1 *2147483648.0d0
       do 3 i=-len,len
 3       nlabel(i)=0
       do 4 i=1,fpop
 4        nlabel(gen1f(i))=nlabel(gen1f(i))+1
       if(t.eq.(t/100)*100) then
         number=0
         do 5 i=-len,len
 5          if(nlabel(i).ge.10) number=number+1
         print 8,irun,t,fpop,number,nlabel(0),
     1    (nlabel(2**i),i=0,4)
 8       format(2i5,3i10,5i8)
       end if
       i=1
       fa=fpop
 9     if(rand(0).lt.fpop*(1.0-(nlabel(gen1f(i))*1.0/fpop)
     1 **2)*select) then
 14      k=1+rand(0)*fpop
         if(k.le.0.or.k.gt.fpop) goto 14
         gen1f(i)=gen1f(k)
       end if
       ibm=ibm*16807
       if(ibm.lt.verhu) then
c      death
         if(fpop.le.1) goto 1
         gen1f(i)=gen1f(fpop)
         fpop=fpop-1
         if(fpop.ge.fa) then
           i=i+1
         else
           fa=fa-1
         endif
       else
c      survival
         do 12 n=1,birth
```

```
            gene1=gen1f(i)
            fpop=fpop+1
            if(fpop.gt.popdim) goto 1
            ibm=ibm*mult
            if(ibm.gt.imut) goto 13
c           Exactly one mutation is made with probability fmut
            ibm=ibm*16807
            p=bit(ishft(ibm,-nshift))
            gene1=ieor(gene1,p)
 13         continue
            gen1f(fpop)=gene1
 12       continue
          i=i+1
        endif
c       if(death) .. else (survival, birth) ..
        if(i.le.fa) goto 9
 7      continue
        do 10 i=-len,len
          if(nlabel(i).eq.0) goto 10
          k=min0(1000000,nlabel(i))
          nhist(k)=nhist(k)+1
c         if(irun.eq.nrun) print *,i,nlabel(i)
 10     continue
 11     continue
        do 19 k=0,30
 19       nlog(k)=0
        do 16 j=1,1000000
        k=1.0+alog(float(j))/0.69315
 16     nlog(k)=nlog(k)+nhist(j)
        do 17 k=0,30
 17       if(nlog(k).gt.0) print *, 0.707*2**k, nlog(k)
 18     continue
        stop
 1      print *, 'error',fpop
        end
```

The program allows for bit-strings of length $\ell = 8$ stored in words of type `byte`, or of length $\ell = 16$ using type `integer*2`. This choice has to be made in the `parameter` line and the line following it. We have 2^ℓ possible languages, each of which can be stored easily. For $\ell = 32$ and 64 we used a different, more time consuming program available from us as `language20.f`.

Our random numbers are 64-bit integers `ibm` with $-2^{63} <$ `ibm` $< 2^{63}$ produced by multiplication with 16807 or with `mult` $= 13^{13}$; in addition we use a built-in random number generator `rand(0)` to give real numbers between 0 and 1. If only 32-bit integers are available, readers have to adjust the lines where $2147483648 = 2^{31}$ appears.

Loop 7 is the main time loop, and we now describe in the order of the program what happens at each iteration. Loops 3 and 4 count in `nlabel(.)` how many individuals speak a given language. Every 100 time steps the total `number` of languages spoken by at least ten people is determined and printed out together with some of the language sizes `nlabel`.

The 37 lines following `i=1`, up to `goto 9`, are the loop over all individuals $i = 1, 2, \ldots,$ `fpop`. Since the population `fpop` varies due to the death and birth processes, we did not deal with them in a fixed loop `i=1,fpop`, and instead used the backward jump `goto 9` and the number `fa` of adult individuals; at the beginning of the iteration all individuals are adult: `fa = fpop`. (We explained this already for the ageing program, from which the present program was developed.)

The six lines starting with label 9 simulate the switching from a rare language to that of a randomly selected individual `k`. Then comes an `if then else endif` choice between Verhulst death and survival. In case of death, the last individual `fpop` is put into the place of the now dead individual `i`, and if `fpop` was a child born during the same iteration, then the counter `i` for the individual is increased by one since this child is not subject now to Verhulst deaths. Otherwise the number `fa` of adults to be treated decreases by one.

In the case of survival instead of death, the counter `i` always increases by one, and loop 12 allows for the birth of several children. Each child increases `fpop` by one, gets a random bit position `ishft(ibm,-nshift)` between 1 and ℓ, and has the bit at that position changed with an exclusive-or command `ieor`.

After the `if then else endif` choice between death and survival, we jump back to label 9 if the counter `i` is not larger than the number `fa` of adults to be treated. In this way all the adults, including the ones which replace the dead ones, are treated once, while the children born during this iteration neither die nor give birth.

9.5. Retirement demography

If the mortality function $\mu(a) = -d[\ln S(a)]/da$ obeys equation (6.1b), then the probability $S(a)$ to survive from birth up to age a is (Wachter and Finch, 1997)

$$S(a) = \exp\left[-A \cdot e^{-bX} \cdot \left(e^{ba} - 1\right)\right].$$

In the computer program, μ is denoted by `q` and a by `iage`. We assume the characteristic age X to be 102 years before 1971 and to increase by 0.15 years every year afterwards, while the Gompertz slope b from the year 1821 up to 1971 increases linearly from $b_0 = 0.07$ to $b_{\max} = 0.093$ and stays constant thereafter. These assumptions make our extrapolations different from traditional ones like Bomsdorf (2004).

The number of births per woman is assumed to increase linearly from 2.17 in 1821 to 2.2 in 1971 and to follow equation (6.1c) thereafter; more precisely it takes the smaller of these two values. Births happen with equal probabilities for ages 21 to 40 of the mother. These birth rates are not the fecundities of real demography but the number of children reaching age 21, when they can get children themselves. For rich countries today about 99 percent of all babies reach this age; in earlier times much more babies were born and died during childhood, making them negligible for population growth. Immigrants and emigrants are assumed to be between 6 and 40 years old, with equal probability for all these ages.

```
      real*8 S(0:130),pop(0:130),q(130),A,b,X,X0,babies
      data X0/103/,d/0.15/,A/7.0/,b0/0.07/,bmax/0.093/,
     1 birth0/2.17/,menop/40/,c/.0000/
c     Yashin X=X0+d*t; Gavrilov-Azbel-Gompertz
c     q/b = A*exp(b(a-X))
      print *, X0, a, birth0, menop, b0, c, d
      b=b0
      X=X0
      const=(2.2-birth0)/150.
      do 3 iyear=1321,2100
      birth=birth0+min0(150,iyear-1821)*const
     1     -0.4*(1+tanh((iyear-1971)*0.33))
      if(iyear.ge.1821.and.iyear.le.1971)
     1  b=b0 + (iyear-1821)*(bmax-b0)/150
      if(iyear.gt.1971) X=X0+d*(iyear-1971)
      do 1 iage=0,130
        S(iage)=dexp(-A*dexp(-b*X)*(dexp(b*iage)-1.0d0))
c       S = survival probability for Gompertz law
c       pop = actual survivors, can be larger than one
c       q = mortality function calculated from s
        if(iyear.eq.1321) pop(iage)=S(iage)
 1      if(iage.gt.0) q(iage)=dlog(S(iage-1)/S(iage))
      babies=0.0d0
      do 4 iage=21,menop
 4      babies=babies+pop(iage)*(0.5d0/(menop-20))*birth
      pop(0)=babies
      do 6 iage=130,1,-1
 6      pop(iage)=pop(iage-1)*(S(iage)/S(iage-1))
      if(iyear.gt.2005) then
        do 2 iage=6,40
 2        pop(iage)=pop(iage)+tot*c/35.0
      endif
      worker=0.0
      pensio=0.0
      tot =0.0
```

```
      expect=0.0
c     numbers of: workers, pensioneers, population,
c     life exp. at 65
      do 5 iage=1,130
        if(iage.gt.65) expect=expect+S(iage)/S(65)
        ss=pop(iage)
        tot=tot+ss
        if(iage.gt.20.and.iage.le.60) worker=worker+ss
c5      if(iage.gt.60. or.iage.le.20) pensio=pensio+ss
 5      if(iage.gt.60) pensio=pensio+ss
 3    if(iyear.gt.1820)
     1 print 100,iyear,tot,birth,pensio/tot,pensio/worker,
     2 expect,b,X
 100  format(i4,7f8.3)
      stop
      end
```

The first lines set the various parameters. The main loop 3 starts in the year 1321 to have a reasonably stable age distribution when the real simulation starts in the year 1821. Then for each year, the births per woman `birth`, the Gompertz slope `b`, and the characteristic age `X` are calculated according to the above assumptions (Azbel', 1996).

Loop 1 first calculates the survival probability `S(iage)` from the formula given at the beginning of this section, and then the mortality function (which is not used later but could be printed out if desired); it also initialises the age distribution `pop(iage)` such that the population of babies less than one year old is normalised to unity.

Loop 4 simulates the birth of daughters; sons are negligible for this program. Their numbers are put into `pop` at zero age. (Sons would have to be treated separately if we assume that the fraction of women employed for work changes with time, while that for men does not.) Loop 6 calculates the new age distribution from the survival probabilities `S(iage)`. Loop 2 adds immigrants to the population, a fraction `c` of the total population each year. Now the population dynamics is finished.

The remaining lines of loop 3 calculate in loop 5 the life expectancy $\int S(a)\,da$ at age 65, the total population (arbitrary units), the number of people in the working ages from 21 to 60 (taking into account that many retire before the official retirement age if that exists), and the number of retired people. (The omitted line c5 would include the children among the pensioneers.)

9.6. Car traffic

This car-following program of the NaSch model is a simplification of what Andreas Schadschneider gave to the senile author years ago, and was improved by students. It stores the velocity of each car but not its position, and thus simplifies the periodic boundary conditions and does not require a memory element for each lattice site. Instead of the position, the distance between each car and the one before it is stored.

```
c      Nagel-Schreckenberg car following program
       implicit none
       integer vmax,L,N,step,j,vsum,dist,iter0,iter1,ibm,
      1 iseed,ip,distjm1
       real p,vav
       parameter (N=2000   ,L=10000   )
       dimension vel(0:N),dist(0:N)
       byte vel
       data p/0.5/, vmax/5/, iter0/5000 /, iter1/20000/,
      1 iseed/4711/
       print *, p,vmax,N,L,iter0,iter1,iseed
       if(N.ge.L .or. iter1.le.iter0) stop 9
       ibm=2*iseed-1
       ip=(2*p-1)*2147483648.0d0

c      Initialisation from jam; dis(j-1) is gap between
c      j-1 and j. 1 2 3 4 5 ... N moving to right.
c      At the start only N moves
       do 1 j=1,N
         vel(j)=0
 1       dist(j)=0
       dist(N)=L-N
       dist(0)=L-N
       vav=0.
c      End of initialisation; now iter1 steps

       do 2 step=1,iter1
c      Begin to update velocities and distances
         vsum=0
         vel(0)=vel(N)
         do 3 j=1,N
           vel(j)=min(vel(j)+1,vmax,dist(j))
           ibm=ibm*16807
           if (ibm.lt.ip) vel(j)=max(0,vel(j)-1)
           dist(j-1)=dist(j-1)+vel(j)-vel(j-1)
 3         vsum=vsum+vel(j)
           dist(N)=dist(N)+vel(1)-vel(N)
c      End updating velocities and distances;
c      calculate average
c      Ignore first iter0 iterations for averages
         if (step.gt.iter0) vav=vav+vsum
```

```
      if(mod(step,100).eq.0) print *, step, vsum*1.0/N
 2    continue

      vav=vav*1.0/(iter1-iter0)
      print *,  'density, current,
     1 velocity = ', N*1.0/L,vav/L,vav/N
      stop
      end
```

Initially, all N cars stand bumper-by-bumper in one huge jam at the left end of the single-lane one-way street, loop 1. (Other initial conditions may give different results because of metastability.) Loop 2 goes over all time steps, after which three averages are printed out. The crucial part is the innermost loop 3. This loop 3 starts with the deterministic possible increase of the velocity, if the distance `dist` ahead and the maximum velocity allow it. Then a random number `ibm` determines whether or not the velocity `vel` needlessly decreases by one unit. Finally, the velocity is summed up in `vsum` for later averaging.

9.7. Scale-free networks

This program only builds a network of `max` nodes; for applications one may wish to have for each site the list of its neighbours, not only their number as is counted here. For small networks this is easily done by a $m_{nb} \times$ `max` neighbour matrix where m_{nb} must at least as large as the largest number of neighbours for any node. Large networks then need a more efficient but complicated way of storage. If one wants to avoid these complications for large networks and nevertheless needs a list of neighbours for each site one can work with directed networks (Newman, Strogatz and Watts, 2001; Dorogovtsev, Mendes and Sanukhin, 2001), where a new site selects m bosses from the existing networks, but these bosses do not have the new site as a boss. Then every node has exactly m neighbours which are easily stored in a $m \times$ `max` matrix. Such a program was published by Sumour and Shabat (2005). Now we deal with the simple case without neighbour list.

```
      parameter(nrun=1 ,maxtime=7000000 ,m=3,iseed=1,
     1 max=maxtime+m,length=1+2*m*maxtime+m*(m-1))
      integer*8 ibm
      dimension k(max), nk(10000), list(length)
      data nk/10000*0/
      print *, nrun, maxtime, m, iseed
      ibm=2*iseed-1
      factor=(0.25/2147483648.0d0)/2147483648.0d0
c     factor=0.5/2147483648.0d0
      do 5 irun=1,nrun
        do 3 i=1,m
```

```
          do 7 j=(i-1)*(m-1)+1,(i-1)*(m-1)+m-1
 7          list(j)=i
 3        k(i)=m-1
        L=m*(m-1)
        if(m.eq.1) then
          L=1
          list(1)=1
        endif
c       All m initial sites are connected with each other
        do 1 n=m+1,max
          do 2 new=1,m
 4          ibm=ibm*16807
            j=1+(ibm*factor+0.5)*L
            if(j.le.0.or.j.gt.L) goto 4
            j=list(j)
            list(L+new)=j
 2          k(j)=k(j)+1
          do 8 j=1,m
 8          list(j+L+m)=n
          L=L+2*m
 1        k(n)=m
        print *, irun
        do 5 i=1,max
          k(i)=min0(k(i),10000)
 5        nk(k(i))=nk(k(i))+1
      do 6 i=1,10000
 6    if(nk(i).gt.0) print *, i,nk(i)
      stop
      end
```

The initial core of m nodes needs to be treated separately in loop 3 since there every node has only $m-1$ neighbours. (One may simplify the code by making each node also neighbour to itself.) Therefore the case $m=1$ needs to be treated separately after loop 3. The important part of the program is the loop 1 over all the nodes which are added to the network after the core has been built.

In that loop 1, we go for each new node through m neighbour selections. This is done in loop 2, where `n` is the new node. First, `j` is a random index between 1 and `L` where `L` is the current length of the Kertész `list`. (In the rare case that rounding errors put `j` outside this interval, a new random integer `ibm` is selected.) Then `j` changes its meaning to become that node which stands at position `j` of the Kertész list. This node has now been selected as a neighbour, it is added to the list, and the number `k(j)` of neighbours of this node is increased by one since the new site is also a neighbour of `j`. After loop 2 is finished, loop 8 adds the new site m times to the list, and the number `k(n)` of neighbours for this new node `n` is set to m.

In this way, the neighbour relations are symmetric (undirected), that means if n is neighbour to j then also j is neighbour to n. And the Kertész `list` contains, in a rather disordered way, each site exactly as often as it is a neighbour. Thus selecting randomly an element of that list gives us a node with a probability proportional to the number of neighbours i that node has.

In the final lines of the program, we calculate the number `nk` of sites having k neighbours and print out its nonzero elements up to a maximum neighbourhood of 10 000 nodes. The program allows to average (sum) over `nrun` networks through loop 5, since a single network is not self-averaging (Stauffer and Aharony, 2004).

9.8. Neural Hopfield–Hebb networks

In order to simplify the programming and to allow understandable print-outs of small patterns, we do not store many random patterns but only two nonrandom ones: the first one is a cross $\times$ and the second one a plus $+$. Loops 1 and 2 initialise these two patterns ξ^{μ}. They, and the later time-dependent pattern S are printed out only if the patterns of size $L \times L$ have the linear dimension $L = 38$. The "fuzzy" pattern S, which later should agree with one of the stored patterns, is initialised in loop 3: With probability of 55 percent it agrees with pattern $\mu = 1$, and in 45 percent of the L^2 sites it disagrees. Now the iterations of loop 4 start. Loop 5 calculates the overlaps $m_{\mu}(t)$ between S and the stored patterns $\mu = 1, 2$. If nothing changed in the last iteration, nothing will change in the future since a fixed point has been reached (`ifixed` = 0), and the simulation stops. Otherwise the inner loop 4 calculates the new `ifixed` and the fields $h_i = \sum_{\mu} \xi_i^{\mu} m_{\mu}(t)$, and from the sign of the fields the new $S_i(t)$.

```
      program neuron
      parameter(L=16000,n=L*L)
      dimension ixi(n,2),is(n),m(2)
      byte ixi,is
      prob=0.450
      ibm=2*1-1
      iprob=(2.0*prob-1.0)*2147483648.0d0
c     probability for initial pixels to differ from
c     desired pattern
      print *, prob,ibm
      ifixed=0
      do 1 mu=1,2
      do 1 i=1,n
 1      ixi(i,mu)=1
      do 2 line=1,L
```

```
          Lm1=line-1
          ibm=ibm*65539
          ixi(line+Lm1*L ,1) = -1
          ixi(L-Lm1+Lm1*L,1) = -1
          ixi(N/2+line   ,2) = -1
 2        ixi(L/2 + LM1*L,2) = -1
        do 3 i=1,n
          ibm=ibm*16807
          is(i)=ixi(i,1)
 3        if(ibm.lt.iprob) is(i)=-is(i)
        if(L.eq.38) print 100, (ixi(i,1),i=1,n)
c       initialisation of 2 patterns + 1 noisy version
c       of 1st pattern
        do 4 itime=1,100
        if(L.eq.38) print 100, (is(i),i=1,n)
        do 5 mu=1,2
          m(mu)=0
          do 5 k=1,n
 5          m(mu)=m(mu)+is(k)*ixi(k,mu)
        print *, itime,m(1),m(2),ifixed
        if(ifixed.eq.n) stop
        ifixed=0
        do 4 i=1,n
          isold=is(i)
          ifield=ixi(i,1)*m(1)+ixi(i,2)*m(2)
          is(i)=1
          if(ifield.lt.0) is(i)=-1
 4        ifixed=ifixed+isold*is(i)
 100    format(1x,38i2)
        end
```

In three iterations and less than ten seconds on a fast workstation with a Gigabyte of memory, pattern $\mu = 1$ is recovered without any error. The above patterns each have 256 million bits ± 1, and so the synaptic matrix J_{ik} has nearly 10^{17} elements, 10^8 times more than what is stored here. We leave it to the critical reader to see that only in this teaching example and not in general the Penna–Oliveira trick saves memory by a factor 10^8.

References*

Abrams, D.M., Strogatz, S.H., 2003. Nature 424, 900.
Ackermann, M., Stearns, S.C., Jenal, U., 2003. Science 300, 1920.
Albert, R., Barabási, A.-L., 2002. Rev. Mod. Phys. 74, 47.
Altevolmer, A.K., 1999. Int. J. Mod. Phys. C10, 717.
Altmann, E.G., Hallerberg, S., Kantz, K., 2005. Preprint.
Anderson, S.R., 2004. Doctor Dolittle's Delusion: Animals and the Uniqueness of Human Language. Yale University Press, New Haven.
Anderson, P.W., Arrow, K.J., Pines, D., 1988. The Economy as an Evolving Complex System. Addison-Wesley Publishing, New York.
Angle, J., 1986. Social Forces 64, 293.
Arnopoulos, P., 2005. Sociophysics: Cosmos and Chaos in Nature and Culture. Nova Science Publishers, New York.
Arthur, B., 1990. Scientific American 262 (February), 80.
Assmann, P., 2004. Int. J. Mod. Phys. C 14, 1439.
Austad, S.A., 1993. J. Zool. 229, 695.
Austad, S.N., 2001. The Comparative Biology of Aging. Cristafalo, V.J., Adelman, R. (Eds.), Annual Review of Gerontology and Geriatrics, vol. 21. Springer, New York.
Aviv, A., Levy, D., Mangel, M., 2003. Mech. Ageing. Dev. 124, 829.
Axelrod, R., 1997. J. Conflict Resolut. 41, 203.
Azbel', M.Ya., 1996. Proc. Roy. Soc. B263, 1449.
Azbel', M.Ya., 2005. Physica A353, 625.
Bagnoli, F., Bezzi, M., 2000. An Evolutionary Model for Simple Ecosystems. Stauffer, D. (Ed.), Annual Reviews of Computational Physics, vol. VII. World Scientific, Singapore. P. 265.
Bagnoli, F., Guardini, C., 2005. Physica A347, 489 and 534.
Bak, P., 1997. How Nature Works: the Science of Self-Organized Criticality. Oxford University Press.
Bak, P., 2004. Physica A340, entire volume.
Bak, P., Sneppen, K., 1993. Phys. Rev. Lett. 71, 4083.
Barabási, A.L., 2002. Linked: The New Science of Networks. Perseus Books Group, Cambridge, MA.
Barabási, A.L., Albert, R., 1999. Science 286, 509.

* Citations like physics/0501097 refer to the e-print server at www.arXiv.org which is freely accessible. Listing is alphabetical according to the first author's family name, then first initial, then year. For each first author, single-authored papers are listed before collaborations.

Baxter, R.J., 1982. Exactly Solved Models in Statistical Mechanics. Academic Press, New York.

Beggs, J.M., Plenz, D., 2003. J. Neurosci. 23, 1167.

Behera, L., Schweitzer, F., 2003. Int. J. Mod. Phys. C14, 1331.

Ben-Naim, E., Krapivsky, P., Redner, S., 2003. Physica D183, 190.

Ben-Naim, E., Redner, S., 2005. J. Stat. Mech. L11002.

Bernardes, A.T., Stauffer, D., Kertész, J., 2002. Eur. Phys. J. B 25, 123.

Berquó, E., Cavenaghi, S., 2005. Ciência Hoje 37 (9), 28.

Bilham, R., 1996. Reduction and Predictability of Natural Disasters. Addison-Wesley, Reading. P. 19.

Binder, K., Young, A.P., 1986. Rev. Mod. Phys. 58, 801.

Boag, P.T., Grant, P.R., 1978. Nature 274, 793.

Boag, P.T., Grant, P.R., 1981. Science 214, 82.

Bomsdorf, E., 2004. Exp. Gerontology 39, 159.

Bonabeau, E., Theraulaz, G.G., Deneubourg, J.-L., 1995. Physica A217, 373.

Bordogna, C.M., Albano, E.V., 2006. In: DYSES05 Proceedings. Int. J. Mod. Phys. C17 (1).

Bornholdt, S., Schuster, H.G. (Eds.), 2003. Handbook of Graphs and Networks. Wiley–VCH, Weinheim.

Branco, M.A., Sherman, P.W., 2005. Trends Ecol. Evol. 20, 271.

Brandau, M., Trimper, S., 2006. Int. J. Mod. Phys. C17 (2), physics/0507179.

Brigatti, E., Martins, J.S. Sá, Roditi, I., 2004. Eur. Phys. J. B42, 431.

Briscoe, E.J., 2000. Language 76, 245.

Brown, G.P., Shine, R., Madsen, T., 2002. J. Tropical Ecology 18, 549.

Burridge, R., Knopoff, L., 1967. Bull. Seism. Soc. Am. 57, 341.

Caiafa, C.F., Proto, A.N., 2006. In: DYSES05 Proceedings. Int. J. Mod. Phys. C17 (1).

Callen, E., Shapero, D., 1974. Physics Today 27 (2), 23.

Cangelosi, A., Parisi, D. (Eds.), 2002. Simulating the Evolution of Language. Springer, New York;
See also Culicover, P., Nowak, A., Dynamical Grammar. Oxford University Press, Oxford, 2003, and current literature in http://www.isrl.uiuc.edu/amag/langev.

Cann, R.L., Stoneking, M., Wilson, A.C., 1987. Nature 325, 31.

Carey, J.R., 2002. Exp. Gerontology 37, 567.

Carey, J.R., Liedo, P., Orozco, D., Vaupel, J.W., 1992. Science 258, 457.

Carey, J.R., Judge, D.S., 2000. Longevity Records: Life Spans of Mammals, Birds, Amphiphiles, Reptiles, and Fish. Odense University Press, Odense, Denmark.

Castellano, C., Loreto, V., Barratt, A., Cecconi, F., Parisi, D., 2005. Phys. Rev. E71, 066107.

Castillo-Chavez, C., Song, B., 2003. Models for the transmission dynamics of fanatic behaviors. In: Banks, H.T., Castillo-Chavez, C. (Eds.), Bioterrorism — Mathematical modeling applications in homeland security. SIAM, Philadelphia, ISBN 0-89871-549-0, p. 155.
Cavalli-Sforza, L.L., 1996. Genes, Peuples et Langues. Odile Jacob, Paris.
Cavalli-Sforza, L.L., 1997. Proc. Natl. Acad. Sci. USA 94, 7719.
Cebrat, S., Łaszkiewicz, A., 2005. J. Insurance Medicine 37, 3.
Cell, 2005. Cell 120, 435–567. (Collection of reviews on ageing by several authors.)
Charlesworth, B., 2001. J. Theor. Biol. 210, 47.
Christensen, K., Moloney, N.R., 2005. Complexity and Criticality. Imperial College Press, London.
Chowdhury, D., Santen, L., Schadschneider, A., 2000. Physics Reports 329, 199.
Chowdhury, D., Stauffer, D., 2005. J. Biosci. (India) 30, 277.
Chowdhury, D., Stauffer, D., Kunwar, A., 2003. Phys. Rev. Lett. 90, 068101.
Coe, J.B., Mao, Y., Cates, M.E., 2002. Phys. Rev. Lett. 89, 288103. Phys. Rev. E 72 (2005) 051925.
Cohen, R., Erez, K., Ben-Avraham, D., Havlin, S., 2000. Phys. Rev. Lett. 85, 4626. Phys. Rev. Lett. 86 (2001) 3682.
Coniglio, A., Klein, W., 1980. J. Phys. A13, 2775.
Corral, A., Perez, C.J., Diaz-Guilera, A., Arenas, A., 1995. Phys. Rev. Lett. 74, 118.
Couzin, I.D., Krause, J., Franks, N.R., Levin, S.R., 2005. Nature 433, 513.
Coyne, J.A., Orr, H.A., 2004. Speciation. Sinauer Associates, Sunderland.
Curtsinger, J.W., Fukui, H.H., Townsend, D.R., Vaupel, J.W., 1992. Science 258, 457.
Dahmen, K., Ertaş, D., Ben-Zion, Y., 1998. Phys. Rev. E58, 1494.
Darwin, C., 1859. On the Origin of Species by Means of Natural Selection. John Murray, London.
Dasgupta, S., 1994. J. Physique I4, 1563;
Phys. Scr. T106 (2003) 19.
Davenport, R.J., 2004. Sci. Aging Knowl. Environ. 48. Paper nf106.
de Almeida, R.M.C., de Oliveira, S. Moss, Penna, T.J.P., 1998. Physica A253, 366.
Deffuant, G., Amblard, F., Weisbuch, G., Faure, T., 2002. Journal of Artificial Societies and Social Simulation 5 (4). Paper 1 (jasss.soc.surrey.ac.uk).
de La Lama, M.S., López, J.M., Wio, H.S., 2005. Europhysics Letters 72, 851.
de Oliveira, P.M.C., 1991. Computing Boolean Statistical Models. World Scientific, Singapore, London, New York.
de Oliveira, P.M.C., 2002. Physica A306, 351.
de Oliveira, P.M.C., 2005. Ciência Hoje 37 (9), 20.
de Oliveira, P.M.C., de Oliveira, S. Moss, Bernardes, A.T., Stauffer, D., 1998. Lancet 352, 911.

de Oliveira, P.M.C., Martins, J.S. Sá, Stauffer, D., de Oliveira, S. Moss, 2004. Phys. Rev. E70, 051910.
de Oliveira, S. Moss, Bernardes, A.T., Martins, J.S. Sá, 1999. Eur. Phys. J. B7, 501.
de Oliveira, S. Moss, de Oliveira, P.M.C., Stauffer, D., 1995. Physica A221, 453.
de Oliveira, S. Moss, de Oliveira, P.M.C., Stauffer, D., 1999. Evolution, Money, War and Computers. Teubner, Leipzig.
de Oliveira, S. Moss, de Oliveira, P.M.C., Stauffer, D., 2003. Physica A322, 521.
de Oliveira, S. Moss, Penna, T.J.P., Stauffer, D., 1995. Physica A215, 298.
de Oliveira, S. Moss, Stauffer, D., de Oliveira, P.M.C., Martins, J.S. Sá, 2004. Physica A332, 380.
de Oliveira, V.M., Gomes, M.A.F., Tsang, I.R., 2006. Physica A, in press, physics/0505197.
de Vany, A., Walls, W.D., 1996. Economic J. 106, 1493.
Demetrius, L., 2003. Physica A322, 477.
Desai, R., James, E., Lui, E., 1999. Theory in Biosciences 118, 98.
Dieckman, U., Doebeli, M., 1999. Nature 400, 354.
Doebeli, M., Dieckmann, U., 2003. Nature 421, 259.
Doebeli, M., Ruxton, G.D., 1997. Evolution 51, 1730.
Dornic, I., Chaté, H., Chavé, J., Hinrichsen, H., 2001. Phys. Rev. Lett. 87, 045701.
Dorogovtsev, S.N., Mendes, J.F.F., Sanukhin, A.N., 2001. Phys Rev. E64, 025101.
Droz, M., Pękalski, A., 2004. Physica A336, 86.
Efros, A.L., Désesquelles, P., 2005. Int. J. Mod. Phys. C16, 1561.
Eigen, M., 1971. Naturwissenschaften 58, 465.
Eigen, M., McCaskill, J., Schuster, P., 1989. Adv. Chem. Physics 75, 149.
Einstein, A., 1998. In: Stachel, J.I. (Ed.), Einstein's Miraculous Year: Five Papers that Changed the Face of Physics. Princeton University Press, New Jersey.
Elgazzar, A.S., 2001. Int. J. Mod. Phys. C12, 1537.
Endler, J.A., 1973. Science 179, 243.
Enquist, B.J., Brown, J.H., West, G.B., 1998. Nature 395, 163.
Excoffier, L., 1997. La Recherche 302, 82.
Fenner, T., Levene, M., Loizou, G., 2005. Physica A355, 641.
Ferrenberg, A.M., Landau, D.P., 1991. Phys. Rev. B44, 5081.
Files, J., 2005. New York Times. June 10, page A 11.
Finch, C.E., 1998. J. Gerontology: Biol. Sci. 53A, B235.
Fisher, D.S., Dahmen, K., Ramanathan, S., Ben-Zion, Y., 1997. Phys. Rev. Lett. 78, 4885.
Fortunato, S., 2004a. Int. J. Mod. Phys. C15, 1021.
Fortunato, S., 2004b. Int. J. Mod. Phys. C15, 1301.
Fortunato, S., Stauffer, D., 2005. In: Albeverio, S., Jentsch, V., Kantz, H. (Eds.), Extreme Events in Nature and Society. Springer, Berlin–Heidelberg, p. 231.
Fortunato, S., Latora, V., Pluchino, A., Rapisarda, A., 2005. Int. J. Mod. Phys. C16, 1535.

Galam, S., 1990. J. Stat. Phys. 61, 943.
Galam, S., 1997. Physica A238, 66.
Galam, S., 2004. Physica A336, 49.
Galam, S., 2005. Europhys. Lett. 70, 705.
Galam, S., Chopard, B., Masselot, A., Droz, M., 1998. Eur. Phys. J. B4, 529.
Galam, S., Gefen, Y., Shapir, Y., 1982. J. Mathematical Sociology 9, 1.
Galam, S., Moscovici, S., 1991. Eur. J. Social Psychology 21, 49.
Galam, S., Wonczak, S., 2000. Eur. Phys. J. B18, 183.
Galam, S., Zucker, J.-D., 2000. Physica A287, 644.
Gallos, L.K., 2005. Int. J. Mod. Phys. C16, 1329.
Gallos, L.K., Cohen, R., Argyrakis, P., Bunde, A., Havlin, S., 2005. Phys. Rev. Lett. 94, 188701.
Geritz, S.A.H., Kisdi, E., Meszéna, G., Mertz, J.A., 1998. Evo. Ecol. 12, 35.
Gavrilets, S., 1997. Trends Ecol. Evol. 12, 307.
Gavrilets, S., 2004. Fitness Landscapes and the Origin of Species. Princeton University Press, Princeton.
Gavrilov, L.A., Gavrilova, N.S., 1991. The Biology of Life Span. Harwood Academic, New York. J. Theor. Biology 213 (2001) 527. Biologia prodolzhitel'nosti zhizni. Moscow, Nauka, 1986 (first published in Russian).
Gavrilov, L.A., Gavrilova, N.S., 2005. Newsletter of American Aging Association, April 2005, page 6, http://www.americanaging.org/news/apr05.html.
Gavrilova, N.S., Gavrilov, L.A., 2005. In: The Living to 100 and Beyond Symposium, Society of Actuaries, Orlando, Florida, Jan. 2005, page 1, http://library.soa.org/library-pdf/m-li05-1_V.pdf.
Goddard, M.R., Godfray, H.C.J., Burt, A., 2005. Nature 434, 571.
Gomes, M.A.F., Vasconcelos, G.L., Tsang, I.S., Tsang, I.R., 1999. Physica A271, 489.
Gong, T., Wang, W.S.Y., 2005. Language and Linguistics 6, 1.
González, M.C., Lind, P.G., Herrmann, H.J., 2005, physics/0508145.
González, M.C., Sousa, A.O., Herrmann, H.J., 2004. Int. J. Mod. Phys. C15, 45.
Grant, P.R., 1986. In: Ecology and Evolution of Darwin's Finches. Princeton University Press, Princeton, USA.
Grant, P.R., Grant, B.R., 1989. In: Otte, D., Endler, J.A. (Eds.), Speciation and Its Consequences. Sinauer Associates, Sunderland, MA.
Grant, B.R., Grant, P.R., 1996. Evolution 50, 2471.
Grassberger, P., 1994. Phys. Rev. E49, 2436.
Hausken, K., Moxnes, J.F., 2005. Int. J. Mod. Phys. C16, 1701.
Hayflick, L., 2003. Exp. Geront. 38, 1231.
He, M.F., Pan, Q.H., Wang, S., 2005. Int. J. Mod. Phys. C16, 177.
He, M.F., Ruan, H.B., Yu, C.L., Yao, L., 2004. Int. J. Mod. Phys. C15, 289.
He, M.F., Yu, G., 2006. Int. J. Mod. Phys. C17 (2).
Hegselmann, R., Krause, U., 2002. Journal of Artificial Societies and Social Simulation 5 (3). Paper 2 (jasss.soc.surrey.ac.uk).

Hegselmann, R., Krause, U., 2005a. Computational Economics 25, 381.
Hegselmann, R., Krause, U., 2005b. Truth and cognitive division of labor. Preprint.
Helbing, D., 2001. Rev. Mod. Phys. 73, 1067.
Heumann, M., Hötzel, M., 1995. J. Stat. Phys. 79, 483.
Hołyst, J.A., Kacperski, K., Schweitzer, F., 2001. In: Stauffer, D. (Ed.), Annual Reviews of Computational Physics, vol. IX. World Scientific, Singapore, pp. 253–274.
Hopfield, J.J., 1994. Phys. Today 47 (2), 40.
Howard, R.S., Lively, C.M., 1994. Nature 367, 554. Nature 368 (1994) 358 (Erratum).
Huang, S.Y., Zou, X.W., Shao, Z.G., Tan, Z.J., Jin, Z.Z., 2004. Phys. Rev. E69, 067104.
Ising, E., 1925. Z. Phys. 31, 253.
Ito, N., 1996. Int. J. Mod. Phys. C7, 107.
Jacobmeier, D., 2005. Int. J. Mod. Phys. C16, 633.
Jacquard, A., 1978. Éloge de la Difference: La Génétique et les Hommes. Seuil, Paris.
Jan, N., 1994. J. Stat. Phys. 77, 915.
Johnson, N., Spagat, M., Restrepo, J., Bohórquez, J., Suaárez, N., Restrepo, E., Zarama, E., 2005, physics/0506213.
Kauffman, S.A., 1993. Origins of Order: Self-Organization and Selection in Evolution. Oxford University Press, Oxford.
Kinouchi, O., Martinez, A.S., Lima, G.F., Laurenço, G.M., Risau-Gusman, S., 2001. Physica A315, 665.
Klein, W., Anghel, M., Ferguson, C.D., Rundle, J.B., Martins, J.S. Sá, 2000. In: Geocomplexity and the Physics of Earthquakes. American Geophysical Union, Washington, p. 43.
Klement, P., Doubal, S., 1997. Mech. Ageing Dev. 98, 167.
Klemm, K., Eguíluz, V.M., Toral, R., San Miguel, M., 2003. Physica A327, 1.
Kirkpatrick, M., Barton, N.H., 1997. Am. Nat. 150, 1.
Kohring, G.A., 1996. J. Physique I6, 301.
Komarova, N.L., 2004. J. Theor. Biology 230, 227.
Kondrashov, A.S., Kondrashov, F.A., 1999. Nature 400, 351.
Kosmidis, K., Halley, J.M., Argyrakis, P., 2005. Physica A353, 595.
Kunwar, A., 2004. Int. J. Mod. Phys. C15, 1449.
Kuznetsov, D.V., Mandel, I., 2005, physics/0506217. Preprint for Proceedings of Joint Statistical Meeting.
Lack, D., 1983. In: Darwin's Finches. Cambridge University Press, Cambridge, England.
Lahdenperä, M., Lummaa, V., Helle, S., Tremblay, M., Russell, A.F., 2004. Nature 428, 178.

Lamarck, J., 1802. Recherche sur l'organisation des corps vivants. Maillard, Paris. Hydrogéologie. Agasse et Maillard, Paris, 1802. La Phylosophie Zoologique. Dentu, Paris, 1809.
Lambiotte, R., Ausloos, M., 2005, physics/0507154. Preprint for Phys. Rev. E.
Lande, R., 1982. Evolution 36, 213.
Langaney, A., 1999. La Phylosophie . . . Biologique. Belin, Paris.
Łaszkiewicz, A., Cebrat, S., Stauffer, D., 2005. Adv. Complex Syst. 8, 7.
Łaszkiewicz, A., Szymczak, Sz., Kurdziel, A., Cebrat, S., 2002. Int. J. Mod. Phys. C13, 97.
Latané, B., 1981. Am. Psychologist 36, 343.
Lee, T.D., Yang, C.N., 1952. Phys. Rev. 87, 410.
Liggett, T.M., 1985. Interacting Particle Systems. Springer, New York.
Liljeros, F., Edling, C.R., Amaral, L.A.N., Stanley, H.E., 2001. Nature 411, 907.
Luz-Burgoa, K., Dell, T., de Oliveira, S. Moss, 2005. Phys. Rev. E72, 011914.
Luz-Burgoa, K., de Oliveira, S. Moss, Martins, J.S. Sá, Stauffer, D., Sousa, A.O., 2003. Braz. J. Phys. 33, 623.
Luz-Burgoa, K., Schwämmle, V., Martins, J.S. Sá, de Oliveira, S. Moss, 2005. Doctoral thesis, q-bio.PE/0504006.
Lynch, M., Gabriel, W., 1990. Evolution 44, 1725.
Magdon-Maksymowicz, M.S., Sitarz, M., Bubak, M., Maksymowicz, A.Z., Szewczyk, J., 2002. Comp. Phys. Comm. 147, 621.
Mahnke, R., Kaupuzs, J., Lubashevsky, I., 2005. Phys. Rep. 408, 1.
Majorana, E., 1942. Sciencia 36, 58.
Makowiec, D., 2001. Physica A289, 208.
Makowiec, D., Stauffer, D., Zieliński, M., 2001. Int. J. Mod. Phys. C12, 1067.
Malarz, K., 2000. Int. J. Mod. Phys. C11, 309.
Malarz, K., Stauffer, D., Kułakowski, K., 2005, physics/0502118.
Martins, J.S. Sá, 2000. Phys. Rev. E61, 2212.
Martins, J.S. Sá, Cebrat, S., 2000. Theory in Biosciences 119, 156.
Martins, J.S. Sá, de Oliveira, P.M.C., 2004. Braz. J. Phys. 34, 1077.
Martins, J.S. Sá, de Oliveira, S. Moss, 1998. Int. J. Mod. Phys. C9, 421.
Martins, J.S. Sá, de Oliveira, S. Moss, de Medeiros, G.A., 2001. Phys. Rev. E64, 021906.
Martins, J.S. Sá, Rundle, J.B., Anghel, M., Klein, W., 2002. Phys. Rev. E65, 056117.
Martins, J.S. Sá, Stauffer, D., 2001. Physica A294, 191.
Martins, J.S. Sá, Stauffer, D., 2004. Ingenierias Univ. Nuevo Leon, Mexico 7 (Jan–Mar), 35.
Masa, M., Cebrat, S., Stauffer, D., 2005. Physica A, in press.
Maynard Smith, J., 1998. Shaping Life: Genes, Embryos and Evolution. Weidenfeld & Nicolson, London.
Mayr, E., 1963. Animal Species and Evolution. Belknap Press.

McBrearty, S., Jablonski, N.G., 2005. Nature 437, 105.
Medeiros, N.G.F., Onody, R.N., 2001. Phys. Rev. E64, 041915.
Mendel, G., 1866. Verhandlungen des Naturforschenden Vereins. Also in www.netspace.org/MendelWeb/MWNotes.html.
Metropolis, N., Rosenbluth, A.W., Rosenbluth, M.N., Teller, A.H., Teller, E., 1953. J. Chem. Phys. 21, 1087.
Meyer-Ortmanns, H., 2001. Int. J. Mod. Phys. C12, 319.
Meyer-Ortmanns, H., 2003. Int. J. Mod. Phys. C14, 311.
Michard, Q., Bouchaud, J.-P., 2005. Eur. Phys. J. B47, 151.
Middleton, A.A., Tang, C., 1995. Phys. Rev. Lett. 74, 742.
Mira, J., Paredes, Á., 2005. Europhys. Lett. 69, 1031.
Monod, J., 1973. Le Hasard et la Nécessité. Seuil, Paris.
Moukarzel, C., de Menezes, M.A., 2002. Phys. Rev. E65, 056701.
Mueller, L.D., Rose, M.R., 1996. Proc. Natl. Acad. Sci. USA 93, 15249.
Nagel, K., Esser, J., Rickert, M., 2000. In: Stauffer, D. (Ed.), Annual Reviews of Computational Physics, vol. VII. World Scientific, Singapore, p. 151.
Nakanishi, H., 1990. Phys. Rev. A 43, 6613.
Naumis, G.G., del Castellino-Mussot, M., Pérez, L.A., Vázquez, G.J., 2006. In: DYSES05 Proceedings. Int. J. Mod. Phys. C17 (1) and preprint.
Nettle, D., 1999a. Proc. Natl. Acad. Sci. USA 96, 3325.
Nettle, D., 1999b. Lingua 108, 95.
Newman, M.E.J., 2001. Phys. Rev. E64, 01631 and 01632.
Newman, M.E.J., 2003. SIAM Review 45, 167.
Newman, M.E.J., Roberts, B.W., 1995. Proc. Roy. Soc. B260, 31.
Newman, M.E.J., Strogatz, S.H., Watts, D.J., 2001. Phys. Rev. E64, 026118.
Novoseltsev, V.N., Novoseltseva, J.A., Yashin, A.I., 2003. Mech. Ageing Dev. 124, 605.
Novotny, V., Drożdż, P., 2000. Proc. Roy. Soc. London B267, 947.
Nowak, M.A., Komarova, N.L., Niyogi, P., 2002. Nature 417, 611.
Odeen, A., Florin, A.-B., 2000. Proc. R. Soc. London B Biol. Sci. 267, 601.
Olami, Z., Feder, H.J.S., Christensen, K., 1992. Phys. Rev. Lett. 68, 1244.
Olsen, E.M., Heino, M., Lilly, G.R., Morgan, M.J., Brather, J., Ernando, B., Dieckmann, U., 2004. Nature 428, 932.
Onsager, L., 1944. Phys. Rev. 65, 117.
Ostfalk, B., 2005. Staatsexamensarbeit (bachelor thesis). Cologne Univ.
Otto, S.P., Nuismer, S.L., 2004. Science 304, 1018.
Paczuski, M., Maslov, S., Bak, P., 1995. Phys. Rev. E53, 414.
Paevskii, V.A., 1985. Demography of Birds. Nauka, Moscow. In Russian.
Panhuis, T.M., Butlin, R., Zuk, M., Tregenza, T., 2001. Trends Ecol. Evol. 16, 364.
Parisi, G., 1999. Physica A263, 557.
Patriarca, M., Leppänen, T., 2004. Physica A338, 296.

Penna, T.J.P., 1995. J. Stat. Phys. 78, 1629.
Penna, T.J.P., 2005, in preparation.
Penna, T.J.P., de Oliveira, P.M.C., 1989. J. Phys. A 22, L719.
Penna, T.J.P., de Oliveira, S. Moss, Stauffer, D., 1995. Phys. Rev. E52, 3309.
Penna, T.J.P., Racco, A., Sousa, A.O., 2001. Physica A295, 31.
Pletcher, S.D., Neuhauser, C., 2000. Int. J. Mod. Phys. C11, 525.
Podos, J., 2001. Nature 409, 185.
Podos, J., Nowicki, S., 2004. Bioscience 54, 501.
Prado, C.P.C., de Carvalho, J.X., 2000. Phys. Rev. Lett. 84, 4006.
Preston, E.F., Martins, J.S. Sá, Rundle, J.B., Anghel, M., Klein, W., 2000. Comp. Sci. Eng. 2, 34.
Proctor, C.J., Kirkwood, T.B.L., 2002. Exp. Gerontology 36, 351.
Porter, A.H., Johnson, N.A., 2002. Evolution 56, 2103.
Pütsch, F., 2003. Adv. Complex Syst. 6, 477.
Raup, D.M., 1986. Science 231, 1528.
Raup, D.M., 1991. Extinction: Bad Genes or Bad Luck. Norton, New York.
Redfield, R.J., 1994. Nature 369, 145.
Reznick, D.N., Bryant, M.J., Roff, D., Ghalambor, C.K., Ghalambor, D.E., 2004. Nature 431, 1095.
Rice, W.R., Hoster, E.E., 1993. Evolution 47, 1637.
Ridley, M., 2003. Evolution. Blackwell, Oxford.
Rikvold, P.A., 2005, q-bio.PE/0508025. J. Theor. Biology, submitted for publication.
Robine, J.-M., Vaupel, J.W., 2001. Exp. Gerontology 36, 915.
Rodrigues, F.A., Costa, L. da F., 2005. Int. J. Mod. Phys. C16, 1785.
Rohde, K., Stauffer, D., 2005, q-bio.PE/0505016, q-bio.PE/0507033.
Rose, M.E., 1991. The Evolutionary Biology of Aging. Oxford University Press, New York.
Rosenfeld, A., Martínez, A.N., 2005. Data processing and environmental protection – e-waste. Eprint from proiap@ciudad.com.ar.
Rundle, J.B., Brown, S.R., 1991. J. Stat. Phys. 65, 403.
Rundle, J.B., Jackson, D.D., 1977. Bull. Seismol. Soc. Am. 67, 1363.
Rundle, P.B., Rundle, J.B., Tiampo, K.F., Martins, J.S. Sá, McGinnis, S., Klein, W., 2001. Phys. Rev. Lett. 87, 148501.
Rundle, J.B., Tiampo, K.F., Klein, W., Martins, J.S. Sá, 2002. Proc. Natl. Acad. Sci. USA 99, 2514.
Sanderson, N., 1989. Evolution 43, 1223.
Sanderson, W.C., Scherbov, S., 2005. Nature 435, 811.
San Miguel, M., Eguíluz, V.M., Toral, R., Klemm, K., 2005. Comp. Sci. Engin. 7.
Sato, A., O'h Uigin, C., Figueroa, F., Grant, P.R., Grant, B.R., Tichy, H., Klein, J., 1999. Proc. Natl. Acad. Sci. U.S.A. 96, 5101.
Savage, V.M., Gillooly, J.F., Woodruff, W.H., West, G.B., Allen, A.P., Enquist, B.J., Brown, J.H., 2004. Functional Ecology 18, 257.

Scharf, F., 2004. Computer simulations of exons and introns in the Penna ageing model. Masters Thesis, Cologne University.
Schelling, T.C., 1971. J. Mathematical Sociology 1, 143.
Schluter, D., 2001. Trends Ecol. Evol. 16, 372.
Schneider, J.J., 2004. Int. J. Mod. Phys. C15, 659.
Schneider, J., Cebrat, S., Stauffer, D., 1998. Int. J. Mod. Phys. C9, 721.
Schneider, J.J., Hirtreiter, C., 2005. Int. J. Mod. Phys. C16, 1165.
Schulze, C., 2003. Int. J. Mod. Phys. C14, 95.
Schulze, C., 2004. Int. J. Mod. Phys. C15, 569.
Schulze, C., 2005. Int. J. Mod. Phys. C16, 351.
Schulze, C., Stauffer, D., 2004. Adv. Complex Syst. 7, 289.
Schulze, C., Stauffer, D., 2005. Int. J. Mod. Phys. C16, 781. In: AIP Conf. Proc. (8th Granada Seminar), Vol. 779, p. 49.
Schwämmle, V., 2005. Int. J. Mod. Phys. C16, 1519.
Schwämmle, V., 2006. Int. J. Mod. Phys. C17, 3.
Schwämmle, V., Brigatti, E., 2005, q-bio.PE/0509032.
Schwämmle, V., de Oliveira, S. Moss, 2005. Phys. Rev. E 72, 031911.
Schwämmle, V., Luz-Burgoa, K., Martins, J.S. Sá, de Oliveira, S. Moss, 2005. Physica A, in press, q-bio.PE/0508016.
Schwämmle, V., Sousa, A.O., de Oliveira, S.M., 2005. Submitted to Phys. Rev. E, q-bio.PE/0508017.
Schweitzer, F., 2003. Browning Agents and Active Particles. On the emergence of complex behavior in the natural and social sciences. Springer, Berlin.
Science, 2003. Science 303, 1315–1335. (Collection of reviews on languages by several authors.)
Sepkoski, J.J. Jr., 1993. Paleobiology 19, 43.
Sharpless, N.E., de Pinho, R.A., 2005. Nature 436, 636.
Shklovskii, B.I., 2005. Theory in Biosciences 123, 431.
Siller, S., 2001. Nature 411, 689.
Simkin, M.V., Roychowdhury, V.M., 2005, physics/0504094.
Sitarz, M., Maksymowicz, A., 2005. Int. J. Mod. Phys. C16 (12).
Sneppen, K., 1992. Phys. Rev. Lett. 69, 3539.
Solé, R., 2005. Nature 434, 289.
Solé, R.V., Manrubia, S.C., 1996. Phys. Rev. E54, R42.
Solomon, S., Weisbuch, G., 1999, adap-org/9909001.
Sousa, A.O., 2003a. Physica A326, 233.
Sousa, A.O., 2003b. Theory in Bioscience 122, 303.
Sousa, A.O., 2004. Eur. Phys. J. B39, 521.
Sousa, A.O., 2005. Priv. comm.
Sousa, A.O., de Oliveira, S. Moss, 1999a. Eur. Phys. J. B9, 365.
Sousa, A.O., de Oliveira, S. Moss, 1999b. Eur. Phys. J. B10, 781.
Sousa, A.O., de Oliveira, S. Moss, 2001. Physica A294, 431.

Sousa, A.O., de Oliveira, S. Moss, Martins, J.S. Sá, 2003. Phys. Rev. E67, 032903.
Sousa, A.O., de Oliveira, S. Moss, Stauffer, D., 2001. Int. J. Mod. Phys. C12, 1477.
Sousa, A.O., Malarz, K., Galam, S., 2005. Int. J. Mod. Phys. C16, 1507.
Sousa, A.O., Stauffer, D., 2000. Int. J. Mod. Phys. C11, 1063.
Stanley, H.E., 1971. Introduction to Phase Transitions and Critical Phenomena. Oxford University Press, Oxford.
Stauffer, D., 2002a. Exp. Gerontology 37, 1131.
Stauffer, D., 2002b. Int. J. Mod. Phys. C13, 975.
Stauffer, D., 2002c. Journal of Artificial Societies and Social Simulation 5, 1. Paper 4 (jasss.soc.surrey.ac.uk);
Stauffer, D. AIP Conf. Proc. 690 (2003) 147.
Stauffer, D., 2002d. Review of Biological Ageing on the Computer. In: Lässig, M., Valleriani, A. (Eds.), Biological Evolution and Statistical Physics. Springer, Berlin–Heidelberg, ISBN 0-89871-549-0, p. 258.
Stauffer, D., 2003. Int. J. Mod. Phys. C14, 237.
Stauffer, D., 2005. In: AIP Conf. Proc. (8th Granada Seminar), Vol. 779, pp. 56 and 75.
Stauffer, D., Aharony, A., 2004, cond-mat/0412612.
Stauffer, D., Aharony, A., da Fontoura Costa, L., Adler, J., 2003. Eur. Phys. J. B32, 395.
Stauffer, D., Kunwar, A., Chowdhury, D., 2005. Physica A352, 202.
Stauffer, D., Martins, J.S. Sá, 2003. Adv. Complex System. 6, 558.
Stauffer, D., Martins, J.S. Sá, 2004. Physica A334, 558.
Stauffer, D., Radomski, J.P., 2001. Exp. Gerontol. 37, 175.
Stauffer, D., Sahimi, M., 2005. Physica A, in press, physics/0506154.
Stauffer, D., Schulze, C., 2005. Phys. of Life Rev. 2, 89.
Stauffer, D., Sousa, A.O., Schulze, C., 2004. Journal of Artificial Societies and Social Simulation 7, 3. Paper 7 (jasss.soc.surrey.ac.uk).
Stauffer, D., Weisbuch, G., 2003. Int. J. Mod. Phys. B17, 5495.
Strehler, B.L., Mildvan, A.S., 1960. Science 132, 14.
Suchecki, K., Eguíluz, V.M., San Miguel, M., 2005. Europhys. Lett. 69, 228.
Suematsu, K., Kohno, M., 1999. J. Theor. Biol. 201, 231.
Sumour, M.E., Shabat, M.M., 2005. Int. J. Mod. Phys. C16, 585.
Sun, Q., Luo, L.D., Mao, Z.W., He, M.F., 2005. Int. J. Mod. Phys. C16, 1745.
Sutherland, W.J., 2003. Nature 423, 276.
Sznajd-Weron, K., 2005. In: First Polish Symposium on Econo- and Sociophysics. Acta Physica Polonica B36, 2537.
Sznajd-Weron, K., Sznajd, J., 2000. Int. J. Mod. Phys. C11, 1157.
Sznajd-Weron, K., Sznajd, J., 2005. Physica A351, 593.
Sznajd-Weron, K., Weron, R., 2003. Physica A324, 437.
Tan, Z., 2005. Exp. Gerontology 36, 89.

Tauber, C.A., Tauber, M.J., 1989. In: Otte, D., Endler, J.A. (Eds.), Speciation and Its Consequences. Sinauer Associates, Sunderland, MA.

Teşileanu, T., Meyer-Ortmanns, H., 2006. Int. J. Mod. Phys. C17 (3), physics/0508229.

Tessone, C.J., Toral, R., Amengual, P., Wio, H.S., San Miguel, M., 2004. Eur. Phys. J. B39, 535.

Thatcher, A.R., Kannisto, V., Vaupel, J.W., 1998. The Force of Mortality at Ages 80 to 120. Odense University Press, Odense;
See also Thatcher, A.R. J. Roy. Statist. Soc. A162 (1999) 5, and private communication.

Thoms, J., Donahue, P., Jan, N., 1995. J. Physique I5, 935.

Tiampo, K.F., Rundle, J.B., Klein, W., Martins, J.S. Sá, 2004. Pure Appl. Geophys. 161, 1957.

Tiampo, K.F., Rundle, J.B., Klein, W., Martins, J.S. Sá, Ferguson, C.D., 2003. Phys. Rev. Lett. 91, 238501.

Tiampo, K.F., Rundle, J.B., Martins, J.S. Sá, Klein, W., McGinnis, S., 2004. Pure Appl. Geophys. 161, 1489.

Ticona, A., de Oliveira, P.M.C., 2004. Phys. Rev. E69, 021903.

Tregenza, T., Butlin, R.K., 1999. Nature 400, 311.

Tu, Y.S., Sousa, A.O., Kong, L.J., Liu, M.R., 2005. Int. J. Mod. Phys. C16, 1149.

Tuljapurkar, S., 2005. Sci. Aging Knowl. Environ. (14). Paper pe9.

Tullis, T.E., 1996. Proc. Nat. Acad. Sci. USA 93, 3803.

Turelli, M., Barton, N.H., Coyne, J.A., 2001. Trends Ecol. Evol. 16, 330.

van Doorn, G.S., Weissing, F.J., 2001. Selection 2, 17.

Vaupel, J.W., Carey, J.R., Christensen, K., Johnson, T.E., Yashin, A.I., Holm, N.V., Iachine, I.A., Kannisto, V., Khazaeli, A.A., Liedo, P., Longo, V.D., Zeng, Y., Manton, K.G., Curtsinger, J.W., 1998. Science 280, 855.

Via, S., 2001. Trends in Ecol. Evol. 16, 381.

Voland, E., Chasiotis, A., Schiefenhövel, W. (Eds.), 2005. Grandmotherhood: The Evolutionary Significance of the Second Half of Female Life. Rutgers University Press, New Brunswick NJ, USA.

Wachter, K.W., Finch, C.E., 1997. Between Zeus and the Salmon. The Biodemography of Longevity. National Academy Press, Washington DC.

Wang, W.S.Y., Ke, J., Minett, J.W., 2004. In: Huang, C.R., Lenders, W. (Eds.), Computational linguistics and beyond. Academica Sinica Institute of Linguistics, Taipei, www.ee.cuhk.edu.hk/~wsywang.

Wang, W.S.Y., Minett, J.W., 2005a. Trends Ecol. Evol. 20, 263.

Wang, W.S.Y., Minett, J.W., 2005b. Trans. Philological Soc. 103, 121.

Watts, D.J., Strogatz, S.H., 1998. Nature 393, 440.

Weidlich, W., 2000. Sociodynamics: A Systematic Approach to Mathematical Modelling in the Social Sciences. Harwood Academic, Chur.

Weisbuch, G., Deffuant, G., Amblard, F., 2005. Physica A353, 555.

Weisbuch, G., Solomon, S., 2003. Chapter 15 in Bornholdt and Schuster (2003).
Wesnousky, S.G., 1994. Bull. Seism. Soc. Am. 84, 1940.
West, G.B., 1999. Physica A263, 104.
West, G.B., Brown, J.H., Enquist, B.J., 1997. Science 276, 122.
Wilmoth, J.R., Deegan, L.J., Lundström, H., Horiuchi, S., 2000. Science 289, 2366.
Wilson, K.G., 1971. Phys. Rev. B4, 3174 and 3184.
Wilson, K.G., 1979. Scientific American 241 (August), 140.
Xu, X.J., Wu, Z.X., Wang, Y.H., 2006. Int. J. Mod. Phys. C17 (2).
Yashin, A.I., Begun, A.S., Boiko, S.I., Ukraintseva, S.V., Oeppen, J., 2001. Exp. Gerontology 37, 157.
Yee, K.K., 2001, nlin.AO/0106028.
Zekri, L., Stauffer, D., 2005. In: AIP Conf. Proc., Vol. 779, p. 69, q-bio.PE/0503015.
Zhang, J., 2004. J. Mathematical Sociology 28, 147.

Subject Index